AF663701

Ethnobotanical Studies on Trees, Shrubs and Climbers of Himalaya

Ethnobotanical Studies on Trees, Shrubs and Climbers of Himalaya

By

S. K. SOOD, SANJAY K. SHARMA
NEELAM KUMAR, HARISH KUMAR

Department of Biosciences
Himachal Pradesh University, Shimla-171005 (India)

2009

SATISH SERIAL PUBLISHING HOUSE

403, Express Tower, Commercial Complex, Azadpur, Delhi-110033 (India)
Phone : 011-27672852, Fax : 91-11-27672046
E-mail :info@satishserial.com, hkjain1975@yahoo.com
Website : www.satishserial.com

Published by :

SATISH SERIAL PUBLISHING HOUSE
403, Express Tower, Commercial Complex, Azadpur, Delhi-110033 (INDIA)
Phone : 011-27672852 Fax : 91-11-27672046
E-mail : info@satishserial.com, hkjain1975@yahoo.com

ISBN 81-89304-67-4

Typeset at: Laxmi Art Creations # 9811482328
Printed at: Salasar Imaging Systems, Delhi-35

Preface

Human beings have been utilizing plant genetic resources for their sustenance since times immemorial. It is undeniable fact that Indian region with over 67.7 million tribal people belonging to 573 tribal groups living in different geographic locations within various subsistence patterns, is very rich in ethnobotanical heritage. Unfortunately, this traditional wisdom developed over years of quest is vanishing consequent to the assault of modern civilization into tribal areas, causing much damage both to biodiversity and well being of man. Therefore, for the effective conservation, management and sustainable strategies, the need of the hour is to develop a sound database on their ethnobotanical prudence. With this sole objective in mind, the preparation of this volume has been necessitated, and it is based on intensive ethnobotanical explorations undertaken by the authors in Jogindernagar tehsil (Mandi, H.P.) with more than 384 villages covering an appropriate area of 770.2 sq km -a region of Himalaya significantly rich in flora embodying an ancient heritage, endemism and cultural diversity. Layout of textural matter encompasses information on General Introduction (Ethnobotany-A General Account, Review of the Previous Work, The Study Area); Ethnobotany of Trees, Shrubs and Climbers of Jogindernagar Region of Sino-Himalaya; Epilogue; Literature Cited; and Appendices providing useful information on Index to Local Names (Appendix-I), Index to Uses (Appendix-II) and Index to Families (Appendix-III). The information is supplemented with about 75 photographic plates, 6 figures, 11 tables and total importance values (TIV) of ethnobotanically important plants of the area presently investigated, which enhance the value of the compilation significantly.

Our grateful thanks are due to Prof. S.K. Gupta, Vice-Chancellor, Himachal Pradesh University, Shimla for his incessant encouragement and kind interest in this study. Grateful thanks are due to Prof. Dalip Malhotra, Prof. T.N. Lakhanpal, Prof. D.C. Kalia and Prof. V.K. Mattu for according their full co-operation and well wishes during the compilation of this work. A special word of thanks is due to Mr Suresh, Mr Munish, Mr. Dhiraj, Dr Kuldip and Mr Bal Krishan for their extensive help in the completion of this task. Last but not the least, we express our indebtedness to the members of our families for bearing the opportunity cost.

Shimla

S.K. Sood, Sanjay K. Sharma
Neelam Kumar & Harish Kumar

List of Figures and Plates

Abbreviations

AD	*anno domini*
Alc.	Alcoholic
Alk.	Alkaline
Anon.	Anonymous
Aq.	Aqueous
Ass.	Assam
B.	Bengali
BC	before Christ
B.P.	Blood Pressure
Bo.	Bombay
Bu-	Butyl
Ca	Calcium
C	Central
cm	Centimeter
CNS	Central Nervous System
Cu	Copper
E.	East
EBH	Ethnobotanical Herbarium
Eng.	English
ENT	Eye, nose and throat
et al.	(*et alia*) and other authors
etc.	(*et cetera*) and so on
Et	Ethyl
Extr.(s)	Extract (s)
Fig. (s)	Figure (s)
g	Gram
GIT	Gastrointestinal tract
G.	Gujarat
H.	Hindi
H.P.	Himachal Pradesh
"	Inches
i.e.	(L. *id est*) that is
K	Potassium
Kan.	Kannada
Kash.	Kashmir
kg	Kilogram
m	metre
Mal.	Malyalam
Mar.	Marathi
Me-	Methyl
mg	milligram
Mg	Magnesium
M.P.	Madhya Pradesh
ml	millilitre
N.	North

(-)ve	Negative
N	Nitrogen
N.W.	North West
O	Oxygen
-OH	Hydroxy
P	Phosphorous
(+)ve	Positive
P.	Punjab
%	Per cent
Ph-	Phenyl
Pl. (s)	Plate (s)
S.	South
S.E.	South-East
Sans.	Sanskrit
Sp	Specie
Spp	Species
Sq km	Square Kilometre
Syn.	Synonym
Tam.	Tamil
Tel.	Telgu
U.P.	Uttar Pradesh
Var.	Variety
Vern.	Vernacular
viz.	*Vide licet* (namely)
Vit.	Vitamin
Vol. (s)	Volume (s)
W.	West
WHO	World Health Organisation
&	And

Contents

Chapter 1

Introduction

Ethnobotany- A General Account

Ethnobotany is the holistic approach which involves the reciprocal and dynamic aspect of the interaction of indigenous people with plants (Schultes, 1962; Ford, 1980), and includes studies such as on food, fibre, dyes, tannins, medicinal and other useful plants, harmful plants, taboos and magico-religious beliefs about plants, material use, and the acts of domestication, conservation, improvement or destruction of plants (Ford, 1978; Jain, 1987b, 1995). In fact, this study of the plants in relation to people includes both the wild and domesticated plants (Heiser, 1995). The scope, concepts and implications of ethnobotany have been expanding at a very fast rate (Schultes, 1960; Jain, 1967, 1986, 1987a). According to Wambebe (1990), approximate 80 per cent of the world's population depends wholly or partially on traditional medicine for its primary health care needs. At present, more emphasis has been laid on the study of the impact of local knowledge and the social and cultural aspects (Cohen *et al.*, 1991; Boef, 1992), to further promote the utilization and conservation of plant genetic resources held by native people.

There could be no two opinions on utility of ethnobotanical studies as these bring out more useful information on unexplored or unexploited natural resources and new uses of existing resources, conservation of natural resources, and the possible factors responsible for the threatened extinction of many species of plants and animals. Besides, ethnobiological knowledge is equally important for policy makers for devising solutions to local and regional problems (Alcorn, 1995). In overall perspective, ethnobotany has a very direct role in supplementary, emerging or inexpensive food, wild relatives of crop plants, new sources of medicinal herbs, inexpensive preparations and more importantly in the foods and medicines easily acceptable to the primitive people (Jain, 1987b).

The term 'indigenous peoples' referred to those who have close, personal relationships with the lands where they are resident (or regularly migrate across)- including the waters, forests, sky, hills and mountains associated with those lands (Alcorn, 1996). Their number has been estimated to be around 250 million all over the globe. The harmony with nature is inherent in their culture and way of life of these societies. In fact, these societies are human conservatories which can't be duplicated by application of science and technology. Considerable amount of knowledge on the traditional uses of plants is available with the indigenous people who are considered as the repository of accumulated experience and knowledge of indigenous flora and fauna. Both indigenous culture and biodiversity are inseparable. All over the world, the indigenous societies have protected the biodiversity but development programmes have threatened the very survival of traditional cultures. A stark observation with regard to this is that the biodiversity tends to disappear with the disappearance of the indigenous culture. As such, there exists no record of this unique knowledge and this loss seems irreversible. Therefore, the chief priority is to collect and inventorise the traditional knowledge and their resource management practices, before these are lost forever. It has been an undeniable fact that the knowledge of indigenous people is invaluable in the present day context of biodiversity conservation and its sustainable utilization (Ravishankar, 1996), and the field of cultural and biodiversity remains a relatively new and uncharted territory (Arizpe, 1996).

The Indian subcontinent has a rich biological and ethnic diversity. Although its total land area is only 2.4% of the whole world, it amounts for 8% of the total global biodiversity (Khoshoo, 1995). Of the estimated wealth of 15,000 species of flowering plants in the country, economic utility of only 9000 plants are hitherto known, and about 5000 species are endemic to this land (Anon., 1994; Jain & Sastry, 1980). There are over 550 natives or tribal communities under 227 ethnic groups covering 5000 villages in India (Anon., 1994). Our age-old culture and traditions with obligation to natural entities have further contributed to the richness of heritage (Kosambi, 1962).

The picturesque state of Himachal Pradesh spreads over an area of 55,673 sq kms (10.54 per cent of the Himalayan landmass) in the lap of north-western

Himalaya between latitude of 30° 22′40′′N and longitude 75° 45′55′′E to 79° 04′22′′E, is ethnobotanically and phytogeographically a very diverse terrain and inhabited by various ethnic groups. A large number of indigenous and endemic plant species are found in this region. The land locked state is bounded by Jammu and Kashmir in the north, Punjab and Haryana in the south-west and nothern part of Uttar Pradesh in the south-east and Tibetian part of China in the north-east. It is rather unfortunate that very few ethnobotanical works have been undertaken in Himachal Pradesh so far, and the region needs to be systematically explored to consolidate our rich traditions of plants and eco-system management.

Review of the Previous Work

The pioneering work on Indian ethnobotany has been that of Janaki Ammal (1955), who studied subsistence on food plants from certain tribals of south India. In the last five decades, there has been an upsurge of activities in search of ethnobotanical information all over the world (Gunther, 1945; Faulks, 1958; Towle, 1961; Schultes, 1962, 1967, 1979, 1983, 1986; Hartwell, 1967-71, Martin, 1972; Renfrew, 1973; Craford, 1983; Felger & Moser, 1985; Duke, 1968, 1986; Kindscher, 1987; Schultes & Raffauf, 1990; Taylor, 1990; Turner *et al.*, 1990, Bellany, 1995; Balick & Cox, 1996; Barford & Kvist, 1996; Neuwinger, 1996, 2000; Newall *et al.*, 1996, Redford & Mansaur, 1997; Schultes & Reis, 1997; Brussell, 1998; Schoen & Wynu, 1998; Pieroni, 1999; Gefu *et al.*, 2000; Minnis, 2000; Payee, 2000; Bisset & Vichtl, 2001; Martin *et al.*, 2001; Seitel, 2001; Grotenhermen & Russo, 2002) with Mudgal (1987) and Chandra (1990) compiling the diverse ethnobotanical literature existing in various languages of the world. From the Indian context, the first bibliography providing information on the use of plants by different communities was published in the early eighties (Jain *et al.*, 1984) followed by a graphic review of Indian ethnobotanical literature with analysis highlighting possible directions for future work (Jain & Srivastava, 2001) and a recent detailed one listing 1600 references of which over 1250 relate strictly to ethnobotanical work or very closely allied themes on India and the others on concept, global evolution of the discipline, and methodolgy of new areas from work of some outstanding foreign ethnobotanists (Jain, 2002).

In India, valuable works documenting the various utilitarian aspects of plants by the tribals and rural inhabitants have been those of Jain (1981)- 'Glimpses of Indian Ethnobotany'; Maheswari (1992)- 'Ethnobotany in India'; Joshi (1995)- 'Ethnobotany of the Primitive Tribes in Rajasthan'; Arora & Pandey (1996)- 'Wild Edible Plants of India-Diversity, Conservation and Use'; Rama Rao & Henry (1996)- 'Ethnobotany of Eastern Ghats in Andhra Pradesh,

India'; Kaushik (1988)- 'Indigenous Medicinal Plants (Including Microbes and Fungi)'; Pandey (1998)- 'Ethno-forestery (Local Knowledge for Sustainable Forestery and Livelihood Security)'; Singh & Pandey (1998)- 'Ethnobotany of Rajasthan, India'; Dagar & Dagar (1999)- 'Ethnobotany of Aboragines of Andaman-Nicobar, Islands'; Mudgal *et al.* (1999)- 'Ethnobotany of Totopora'; Singh & Kumar (2000a)- 'Ethnobotanical Wisdom of Gaddi Tribe of Western Himalayas'; Sood *et al.* (2001)- 'Ethnobotany of Cold Desert Tribes of Lahaul-Spiti (N.W. Himalaya)'; Singh *et al.* (2003a)- 'Ethnomedicine of North-East India'; Singh *et al.* (2003b)- 'Herbal Medicine of Manipur: A Colour Encyclopedia'; and Sood & Thakur (2004)- 'Ethnobotany of Rewalsar Himalaya'.

Work on North-West Himalaya

In Himachal Pradesh, information pertaining to folk and ethnobotanical practices is scanty; mention can be made of the publications of Kapur (1993)- 'Ethnomedico plants of Kangra valley (Himachal Pradesh)' and Singh & Kumar (2000b)- 'Observations in ethnoveterinary medicine among the Gaddi tribe of Kangra valley (Himachal Pradesh)' in Kangra region; Negi & Subramani (2002)-'Ethnobotanical study in the village Chhitkul of Sangla valley, District Kinnaur, Himachal Pradesh' on plants of Kinnaur district; Uniyal & Chauhan (1973)- 'Traditionally important medicinal plants of Kullu forest division, H.P.', Singh (1990)- 'Ethnobotanical studies of useful plants of the Kullu district in the north-western Himalaya, India', Jain & Puri (1994)- 'An ethno-medico-botanical survey of Parvati valley in Himachal Pradesh (India)', and Singh *et al.*, (1998)- 'Ethno-cultural analysis of sacred species and ecosystems in Chhakinal watershed, Himachal Pradesh' in District Kullu; Koelz (1979)- 'Notes on ethnobotany of Lahaul, a province of the Punjab', Duhoon *et al.*, (1996)- Seabuckthorn (*Hippophae* spp)- A less known wonder plant of ethno-medico-botanical importance in cold desert of India', Sood *et al.*, (2001)- 'Ethnobotany of cold desert tribes of Lahaul-Spiti (N. W. Himalaya)', and Chandrasekhar & Srivastava (2003)- 'Ethnobotanical studies in Pin valley National park, Lahaul-Spiti, Himachal Pradesh' in twin District of Lahaul-Spiti; Gaur & Singh (1993)- 'Ethno-medicinal plants Mandi district', Singh, (1996b)- 'Wild edible plants of Mandi district in north-west Himalaya', and Sood & Thakur (2004)- 'Ethnobotany of Rewalsar Himalaya' on plants of Mandi district; Chauhan (1996)- 'Ethno-medico-botany of Pabbar valley in District Shimla, Himachal Pradesh (India)' in Shimla region; Chauhan (1988)- 'Ethnobotanical study of medicinal plants of Himachal Pradesh', Lal *et al.*, (1996)- 'Plants used as ethnomedicine and supplementary food by Gaddies of Himachal Pradesh', Sharma & Rana (1999)- 'Traditional medicinal uses of plants in Himachal hills', and Singh & Kumar (2000a)- 'Ethnobotanical wisdom of Gaddi tribe of Western Himalayas' in the Himalayas.

From the floristic viewpoint, the state of Himachal Pradesh has been exhaustively explored and the various treatises, like 'The Himalayan District of North West Provinces of India' (Atkinson, 1882), 'Flora of British India' (Hooker, 1872-1897), 'Flora Simlensis' (Collett, 1902), 'Flora of Bushahar Himalayas' (Nair, 1964), Flora of Himachal Pradesh' (Chowdhery & Wadhawa, 1984), 'Concise Flowers of the Himalaya' (Polunin & Stainton, 1984), 'Flowers of the Himalaya- A Supplement' (Stainton, 1988), 'Flora of Lahaul-Spiti- A Cold Desert in North West Himalaya' (Aswal & Mehrotra, 1994), and 'Flora of Kullu District' (Dhaliwal & Sharma, 1999) have added to the wealth of information, and are noteworthy. Studies on diversity of medicinal and aromatic plants in different regions of the state have been carried out by various workers, viz. Kangra valley (Ahluwalia, 1952; Uniyal & Chauhan, 1971), Kullu (Rastogi, 1960; Uniyal & Chauhan, 1972; Dobriyal *et al.*, 1997), Lahaul-Spiti (Rau, 1961; Uniyal *et al.*, 1973, Arora *et al.*, 1980, Chauhan, 1989), Chamba (Gupta, 1961, 1971; Shabnam, 1964), Una (Chauhan, 1974), Mandi (Singh, 1993), and Himachal Pradesh (Sarin & Chopra, 1985; Chauhan, 1989b, 1990, 1999, 2003; Singh & Aswal, 1992; Sharma & Sood, 1997; Singh, 1999; Badola, 2001). Besides, other publications dealt with aspects involving present status and future prospects for cultivation and collection of medicinal plants (Gupta, 1988), role of forestery extension in exploitation of medicinal and aromatic plants (Sagwal & Gupta, 1988), Conservation of threatened medicinal plants (Badola & Pal, 2003) and collection and marketing strategies of important medicinal and aromatic plants in H.P. (Negi & Bhalla, 2002). Moreover, reports on the use of wild edible plants are those of Aitchson (1868), Gupta (1966), Sarin (1967), Kapoor (1978) and Parmar & Kaushal (1982).

Scrutiny of the above literature reveals that no attempt has been made so far to conduct ethnobotanical studies on trees, shrubs and climbers of Jogindernagar (Mandi, H.P.)- a region of north-western Himalaya significantly rich in flora embodying an ancient heritage, endemism and cultural diversity. Hence, an attempt has been made to document the precious indigenous wisdom on the multifact usage of plants from the region with a view not only to conserve it from being lost irreversibly to growing pressures of population, urbanisation and development but also for using them as valuable clues for social forestry endeavours, therapeutic agents, supplementary foods and sustainable management of species as well as their habitats.

The Study Area

Etymology

Prior to 1925 the whole area was called 'Sukrahatti' which was then renamed as Jogindernagar after Raja Joginder Sen of Mandi for funding an elaborated hydel power scheme involving tunnelling and piping the water over several kms from the river Uhl to the area.

Geographical Extent of Study Area

Jogindernagar tehsil with more than 384 inhabited villages covers an approximate area of 770.2 sq km (inclusive Lad Bharol), lies between 31° 58′ 60N and 76° 46′ 0E in Distt. Mandi of North-West Himalaya (Figs. 1-3). The place is situated at an altitude of 1220m on Mandi-Pathankot National Highway and is 56 km from Mandi, 21 km from Baijnath, 38 km from Palampur and 59 km from Dharamshala, and also connected by a narrow gauge rail track to Pathankot.

Climate

Unlike winters when temperature can hover just above freezing point, the climate is mild during summers. Monsoon occurs during July-September due to South-West currents.

Land and People

The population of Jogindernagar tehsil is predominantly rural (70,627: 17,203 SC, 1,578 ST) and semiurban (5,048: 1,203 SC, 43 ST). The ratio of women/1000 males in the area is 1,070 females (Anon., 2002-2003). Topographically, the area falls under mid-hills of Himalaya and is dominated by nature loving rural based communities with a predominance of Hindus mainly comprising of Brahmins, Rajputs, Khatris, Julahas, Chamars, Lohars, Tarkhans, Kumhars, etc. Cultivable land is very marginal and the people practise primitive agriculture and cattle rearing as important means of livelihood. A major part of their life sustains on a large number of plant species which play an important role in their day-to-day needs and healthcare. They gather food, fuel, fodder and medicinal plants traditionally for meeting their requirements and hold the forests around them in great esteem.

Some of the important fairs of the region are Jogindernagar (March), Machhayal (April), Gumma (April), Urla (June), Malara (October), Baba Balak Rupi (November-December), etc. The rural populace participate in all the fairs with a lot of zeal and vigour.

Collection of Ethnobotanical Data

Intensive and extensive ethnobotanical field surveys were undertaken to various rural pockets of Jogindernagar, Distt. Mandi (H.P.) in different seasons during the years 1998-2003 (Table I). First hand information on ethnobotanical aspects was collected through interviews from knowledgeable people (traditional practitioners, old experienced farmers, family heads, housewives, eminent elderly persons of the community, etc.), as suggested by Jain (1987a). Some of them were also taken to the study sites in the forest for this purpose. While collecting the plant specimens, their uses and local names were also ascertained and recorded carefully in the field notebook with the assistance of local informants. The data were verified in different villages among the interviewers and showing the same plant sample, and even with the same informants on different occasions. Ethnobotanical lore was considered valid if at least three informants made similar comments. An attempt was also made to document the information on the mode of preparation of the drugs and methods of their administration along with the doses .

Table 1: Flowering and fruiting season of the ethnobotanically important spp collected from the study area

Name of the Spp	Place	Altitude (m)	Flowering & Fruiting Season
Abies pindrow Royle⁺	Winch Camp	2220	April-May
Abrus precatorius L.⁺	Langna	850	May-August
Acacia catechu Willd.	Langna	850	May-July
Acacia concinna DC⁺	Jogindernagar	1220	January-April
Achyranthes aspera L⁺.	Macchayal	870	May-October
Achyranthes bidentata Blume⁺	Harabagh	1470	June-August
Actinidia chinense var. *deliocosa* (A.Chev.) A.Chev.©	Jogindernagar	1220	February-November
Adhatoda vasica Nees⁺	Gumma	1500	April-November
Aegle marmelos (L.) Correa	Langna	850	April-July
Aesculus indica Colebr.⁺	Winch Camp	2220	May-September
Agave americana L.⁺	Galu	1620	March-May
Albizzia julibrissin Durazz.⁺	Dak Bangla	1080	May-August

Contd...

Table 1. Contd...

Albizzia lebbeck Benth.+	Makridi	1290	April-May
Albizzia stipulata Boiv.	Kunkar	1260	May-June
Alnus nepalensis Don+	Sui	2040	May-December
Alnus nitida Endl.	Digli	2040	August-October
Aloe vera (L.) Webb. & Berth	Harabagh	1470	March-April
Andrachne cordifolia (Decne.) Muell.-Arg.+	Sui	2040	April-October
Araucaria columnaris (Forst.) Hook.©	Hargunain	1280	September-October
Arundinaria falcata Nees	Winch Camp	2220	May-November
Asclepias currasavica L.+	Bassi	930	Flowers throughout the year
Asclepias physocarpa Schlecht©	Harabagh	1470	April-July
Atyloshia platycarpa Benth.+	Basahi	1350	May-August
Bambusa nutans Wall.+	Seinthal	700	Flowers sporadically
Bauhinia vahlii Wight & Arn.+	Makridi	1290	April-June
Bauhinia variegata L.+	Ropadi	1600	March-August
Berberis aristata DC+	Winch Camp	2220	May-September
Berberis lycium Royle	Galu	1620	April-July
Boehmeria platyphylla Don	Hargunain	1280	July-September
Bombax ceiba L.+	Shanan	1100	February-May
Bougainvillea spectabilis Willd.	Bassi	930	January-June
Broussonetia papyrifera Vent.+	Winch Camp	2220	April-May
Buddleia crispa Benth.*	Dhelu	1220	March-May
Buddleia paniculata Wall.*	Winch Camp	2220	March-April
Butea monosperma Kuntze	Ropari	1600	March-May
Buxus wallichiana Baill.+	Winch Camp	2220	April-November

Contd...

Table 1. Contd...

Caesalpinia decapetala (Roth.) Alston	Machhayal	870	April-July
Callicarpa macrophylla Vahl.+	Drahal	1070	July-September
Callistemon lanceolatus DC©	Bassi	930	March-April
Calotropis procera R. Br.+	Langna	850	February-May
Camellia sinensis (L.) Kuntze©	Chauntra	1040	June-September
Canna indica L.+ ©	Aehju	1240	January-February
Carica papaya L.+ ©	Macchayal	870	Greater part of the year
Carissa congesta Wight+	Basahi Dhar	1350	April-July
Carpinus viminea Wall. *ex* Lindl.	Winch Camp	2220	April-August
Carya illenoesis (Wang.) Koch.+ ©	Makridi	1290	May-October
Caryopteris wallichiana Schauer+	Digli	2040	April-August
Cassia fistula L.+	Basahi	1350	March-July
Cassia laevigata Willd.	Bihun	900	June-December
Cedrella serrata Royle+	Sui	2040	June-October
Cedrella toona Roxb. *ex* Rottl.+	Seinthal	700	May-June
Cedrus deodara (Roxb.) Louden+	Winch Camp	2220	June-July
Celtis australis L.	Galu	1620	April-October
Cestrum nocturnum L.+©	Ladhruhin	1280	August-October
Cinnamonum camphora (L.) Nees & Eberm+©	Jogindernagar	1220	April-August
Cinnamomum tamala Nees & Eberm+	Harabagh	1470	June-August
Cissampelos pareira L.+	Matru	800	April-September
Citrus aurantium L.©	Makridi	1290	October-March

Contd...

Table 1. Contd...

Citrus grandis (L.) Osbeck+©	Darkoti	1300	November-March
Citrus jambhiri Lush.*	Ladhruhin	1280	October-March
Citrus limon (L.) Burm. f.+©	Thorat	1400	November-February
Citrus medica L.+©	Dhelu	1220	April-September
Clematis buchananiana DC+	Winch Camp	2220	July-October
Clematis grata Wall.+	Gumma	1500	July-September
Clerodendron philippinum Schauer+©	Shanan	1100	August-October
Colebrookia oppositifolia Sm.+	Galu	1620	April-November
Cordia dichotoma Forst.+	Langna		870 April-July
Coriaria nepalensis Wall.*	Harabagh	1470	March-June
Cornus macrophylla Wall.+	Harabagh	1470	May-October
Costus speciosus Smith©	Langna	870	June-August
Cotoneaster bacillaris Wall. *ex* Lindl.+	Winch Camp	2220	May-October
Cotoneaster microphylla Wall. *ex* Lindl.+	Brij Mandi	1300	May-June
Cryptolepis buchanani Roem. & Sch.+	Shanan	1100	April-November
Cucumis sativus L.©	Sukabagh	800	May-September
Cucurbita hispida Thunb.+©	Dhelu	1220	June-September
Cucurbita melo L.*©	Makridi	1290	July-September
Cucurbita pepo L.+©	Jalphehar	1100	June-September
Cuscuta reflexa Roxb.+	Bassi	930	July-October
Cyathula tomentosa (Roth.) Moq.*	Digli	2040	July-December
Cycas circinalis L.+©	Shanan	1100	February-May
Cyphomandra betacea (Cav.) Sendt.	Khuddar	880	May-September

Contd...

Table 1. Contd...

Dalbergia sissoo Roxb.⁺	Basahi	1350	May-August
Daphne cannabina Wall.⁺	Winch Camp	2220	April-October
Datura metel L.⁺	Groru	940	June-January
Datura stramonium L.⁺	Galu	1620	June-October
Datura suaveolens Humb. & Bonpl. *ex* Willd.*	Gumma	1500	May-January
Debregeasia hypoleuca Wedd.⁺	Harabagh	1470	May-August
Dendrocalamus hamiltonii Nees & Arn. *ex* Munro⁺	Groru	940	Flowering sporadically
Desmodium multiflorum DC*	Winch Camp	2220	June-October
Deutzia corymbosa R. Brown*	Shanan	1100	May-October
Dioscorea bellophylla Voigt⁺	Seinthal	700	September-November
Dioscorea bulbifera L.⁺	Seri	1080	June-August
Dioscorea deltoidea Wall.	Digli	2040	May-August
Diospyros kaki L.⁺ ©	Winch Camp	2220	May-June
Dodonea viscosa Jacq.⁺	Shanan	1100	February-April
Duranta repens L.⁺	Chauntra	1040	April-August
Ehretia acuminata R. Br.	Jogindernagar	1220	May-July
Ehretia laevis Roxb.	Galu	1620	January-June
Elaeagnus umbellata Thunb.	Sui	2040	May-October
Embellia tsjeriam-cottam A. DC⁺	Harabagh	1470	July-September
Emblica officinalis Gaertn.⁺	Bassi	930	April-September
Eriobotyrica japonica Lindl.⁺ ©	Makridi	1290	July-September
Erythrina suberosa Roxb.⁺	Basahi	1350	March-June
Eucalyptus umbellata Domin⁺	Aehju	1240	March-April
Eugenia jambolana Lam.⁺	Gharwasra	1100	April-August
Eyonymus echinatus Wall. *ex* Roxb.	Winch Camp	2220	May-November

Contd...

Table 1. Contd...

Euphorbia pulcherrima Willd.⁺	Makridi	1290	November-January
Euphorbia royleana Boiss.⁺	Langna	850	June-July
Ficus bengalensis L.⁺	Basahi	1350	February-June
Ficus clavata Wall.	Seinthal	700	June-September
Ficus foveolata Wall.	Harabagh	1470	March-July
Ficus glomerata Roxb.⁺	Macchayal	870	April-July
Ficus hispida L.⁺	Seri	1080	June-August
Ficus palmata Forssk.⁺	Shanan	1100	April-September
Ficus religiosa L.⁺	Langna	850	March-October
Ficus roxburghii Wall. *ex* Miq.⁺	Dhelu	1220	April-August
Ficus rumphii Bl.⁺	Langna	850	April-July
Ficus semicordata Buch.-Ham. *ex* Smith⁺	Khuddar	880	May-September
Flacourtia ramontchi L' Herit⁺	Langna	850	March-April
Flemingia macrophylla Blume *ex* Miq.⁺	Jogindernagar	1220	June-September
Flemingia strobilifera R. Br.⁺	Shanan	1100	August-September
Fuchsia splendens (Plum.) L.©	Jogindernagar	1220	May-September
Gardenia angusta (L.) Merr.	Groru	940	July-October
Glochiodion velutinum Wight⁺	Dhelu	1220	May-August
Gmelina arborea Roxb.⁺	Jogindernagar	1220	March-June
Gossypium arboreum L.⁺©	Chauntra	1040	July-September
Grevillea robusta A. Cunn *ex* R. Br.⁺©	Jogindernagar	1220	March-July
Grewia oppositifolia Roxb.	Matru	820	April-June
Hedera helix auct (non L.) Clarke⁺	Jogindernagar	1220	June-November
Hibiscus rosa-sinensis L.⁺	Bhadyada	1240	February-October
Holarrhena antidysentrica Wall.	Langna	850	June-February

Contd...

Table 1. Contd...

Hydrangea grandiflora Sieb.©	Jogindernagar	1220	April-September
Hypericum cernuum Roxb.+	Gumma	1500	April-October
Hypericum patulum Thunb.+	Winch Camp	2220	August-September
Ilex dipyrena Wall.+	Digli	2040	April-October
Ilex excelsa (Wall.) Hook. f.+	Harabagh	1470	May-October
Indigofera gerardiana Wall.+	Winch camp	2220	May-July
Indigofera pulchella Roxb.+	Gumma	1500	July-November
Inula cappa DC+	Basahi Dhar	1350	May-October
Ipomoea cairica (L.) Sweet+	Jogindernagar	1220	June-November
Ipomoea carnea Jacq.+	Jhalwan	1085	July-September
Ipomoea eriocarpa R. Br.	Macchayal	870	August-October
Ipomoea muricata Jacq.+	Gumma	1500	June-August
Ipomoea quamoclit L.+	Langna	850	August-October
Jacaranda acutifolia Humb. & Bonpl.+	Jogindernagar	1220	April-June
Jasminum dispersum Wall.+	Galu	1620	April-November
Jasminum grandiflorum L.+	Harabagh	1470	April-September
Jasminum humile L.+	Dhelu	1220	May-September
Jasminum multiflorum (Burm. f.) Andrews+	Sui	2040	November-April
Juglans regia L.	Jimjima	1500	April-October
Kydia calycina Roxb.+	Jogindernagar	1220	July-September
Lagerstroemia indica L.+©	Jogindernagar	1220	June-October
Lantana camara L.+	Chauntra	1040	April-June
Leea aspera Wall. *ex* Roxb.+	Seinthal	700	July-August
Leptodermis lanceolata Wall.+	Dhelu	1220	June-October
Lettsomia thomsonii Clarke*	Basahi	1350	January-February
Leucaena glauca Benth.+	Jogindernagar	1220	September-November

Contd...

Table 1. Contd...

Leycesteria formosa Wall.⁺	Winch Camp	2220	May-November
Litchi chinensis Sonner.⁺©	Makriri	1290	April-June
Litsea polyantha Juss.	Dhelu	1220	March-July
Litsea sebifera Pers.⁺	Bhararu	780	May-September
Litsea umberosa Nees⁺	Winch Camp	2220	March-May
Lonicera japonica Wall.⁺©	Jogindernagar	1220	May-June
Lonicera quinquelocularis Hardw.⁺	Shanan	1100	April-September
Loranthus ligustrinus Wall.*	Bassi	930	April-September
Luffa cylindrica (L.) Roem.©	Chauntra	1040	May-September
Lyonia ovalifolia (Wall.) Drude⁺	Harabagh	1470	April-May
Machilus odoratissima Nees⁺	Shanan	1100	March-April
Mallotus phillipinensis (Lam.) Muell.-Arg.	Macchayal	870	April-October
Malvaviscus conzatii Greenm.⁺	Seri	1080	Major part of the year
Mangifera indica L.⁺	Langna	850	March-July
Melia azedarach L.⁺	Machayal	870	May-October
Melothria heterophylla (Lour.) Cogn.⁺	Sui	2040	May-October
Michelia champaca L.⁺	Shanan	1100	April-October
Milletia extensa Benth. *ex* Baker⁺	Basahi	1350	July-September
Momordica charantia L.©	Jogindernagar	1220	July-October
Moringa pterygosperma Gaertn.	Langna	850	February-April
Morus alba L.⁺	Darkoti	1300	March-May
Mucuna pruriens (L.) DC⁺	Bhararu	780	July-September
Murraya koenigii Spreng.⁺	Chauntra	1040	March-June
Myrica esculenta Buch.-Ham. *ex* D. Don⁺	Shanan	1100	November-April
Myrsine africana L.⁺	Harabagh	1470	April-October

Contd...

Table 1. Contd...

Nerium indicum Mill.⁺	Jogindernagar	1220	April-November
Opuntia dillenii Haw.⁺	Langna	850	April-June
Oroxylum indicum Vent.⁺	Langna	850	May-August
Osbeckia stellata Wall.	Harabagh	1470	July-October
Osyris arborea Wall.⁺	Dhelu	1220	August-April
Pavetta indica L.⁺	Drahal	1070	July-October
Phoenix sylvestris Roxb.⁺	Groru	940	May-June
Phyllanthus parvifolius Buch.-Ham. *ex* D. Don*	Jimjima	1500	June-October
Picea smithiana (Wall.) Boiss.⁺	Winch Camp	2220	April-May
Pinus roxburghii Sarg.⁺	Makriri	1290	September-October
Pinus wallichiana A. B. Jackson⁺	Sui	2040	April-June
Pistacia integrimma J. L. Stewart *ex* Brand.⁺	Seri	1080	March-May
Plumbago zeylanica L.⁺	Makriri	1290	April-September
Plumeria rubra L.©	Jogindernagar	1220	June-September
Pogostemon benghalensis Burm. f.⁺	Jogindernagar	1220	December-April
Polygonum chinense L.⁺	Digli	2040	September-October
Populus ciliata Wall.	Winch Camp	2220	April-July
Populus nigra L.⁺	Jogindernagar	1220	April-July
Premna latifolia Roxb.⁺	Shanan	1100	July-September
Prinsepia utilis Royle⁺	Dhelu	1220	June-October
Prunus amygdalus Baill.⁺©	Winch camp	2220	May-September
Prunus armeniaca L.⁺	Winch camp	2220	May-August
Prunus cerasoides D. Don⁺	Harabagh	1470	November-May
Prunus domestica L.⁺©	Harabagh	1470	April-August
Prunus persica Benth. & Hook. f.⁺	Hargunaian	1280	February-May

Contd...

Table 1. Contd...

Psidium guajava L.+	Shanan	1100	July-September
Punica granatum L.+	Basahi	1350	May-September
Pyrus communis L.+	Chauntra	1040	June-August
Pyrus malus L.+©	Winch Camp	2220	May-October
Pyrus pashia Ham.+	Bhararu	780	May-August
Quercus glauca Thunb.+	Jimjima	1500	March-July
Quercus leucotrichophora A. Camus+	Jimjima	1500	April-October
Quercus semicarpifolia Sm.	Winch camp	2220	May-August
Randia dumetorum (Retz.) Lam+	Langna	850	April-September
Reinwardtia indica Dumort.+	Galu	1620	November-May
Rhamnus persicus Laws.+	Basahi	1350	April-October
Rhamnus triquetra Wall.+	Jimjima	1500	May-August
Rhododendron arboreum Sm.+	Winch Camp	2220	March-May
Rhus cotinus L.+	Winch Camp	2220	March-August
Rhus semialata Murr.+	Shanan	1100	April-August
Rhus succedanea L.	Digli	2040	June-September
Ricinus communis L.+	Aehju	1240	March-May
Robinia pseudo-acacia L.+	Jogindernagar	1220	March-May
Rosa alba L.+©	Jogindernagar	1220	June-August
Rosa moschata Mill.+	Winch Camp	2220	April-October
Roylea cinerea (D. Don) Baill.+	Seri	1080	March-November
Rubia cordifolia L.+	Jogindernagar	1220	July-October
Rubus ellipticus Sm.+	Machhayal	870	January-May
Rubus lasiocarpus Sm.+	Winch Camp	2220	March-May
Rubus niveus Wall.+	Digli	2040	April-June
Sachharum officinarum L.+©	Makriri	1290	January-April

Contd...

Table 1. Contd...

Sageretia oppositifolia (Roth. *ex* Schult.) G. Don	Jhalwan	1085	June-September
Salix babylonica L.+©	ogindernagar	1220	February-May
Salix karelinii Turcz.*	Winch Camp	2220	May-July
Salix tetrasperma Roxb.+	Bassi	1350	April-July
Sapindus mukorosii Gaertn.+	Dhelu	1220	April-September
Sapium insigne (Royle) Benth.+	Machayal	870	May-August
Sapium sebiferum (L.) Roxb.+	Winch Camp	2220	June-August
Saraca asoca (Roxb.) De Wilde+©	Jogindernagar	1220	March-September
Sarcococca saligna (D. Don) Muell.-Arg.+	Winch Camp	2220	May-October
Saurauia nepaulensis DC	Winch Camp	2220	March-October
Sechium edule Sw.+©	Jogindernagar	1220	June-September
Senecio rufinervis DC*	Winch Camp	2220	August-October
Sida cordifolia L.	Approach Road	1300	August-September
Skimmia laureola Sieb. & Zucc. *ex* Walp.+	Chauntra	1040	April-May
Smilax aspera L.+	Shanan	1100	September-November
Solanum erianthum Don+	Chauntra	1040	July-October
Solanum melongena L.©	Jogindernagar	1220	June-September
Solanum pseudocapsicum L.*©	Harabagh	1470	June-October
Solanum torvum Swartz+	Harabagh	1470	April-August
Solidago canadensis L.	Shanan	1100	October-December
Sorbaria tomentosa (Lindl.) Rehder*	Winch Camp	2220	June-August
Spirarea canescens D. Don+	Winch Camp	2220	May-October
Spondias mangifera Willd.	Basahi Dhar	1350	April-June
Stephania elegans Hook. f. & Thoms.+	Jogindernagar	1220	August-October

Contd...

Table 1. Contd...

Stephania glabra (Roxb.) Miers.	Ladhruhin	1280	June-August
Sterculia villosa Roxb.	Basahi Dhar	1350	February-April
Strobilanthes glutinosis Nees*	Harabagh	1470	October-November
Symplocos racemosa Roxb.⁺	Harabagh	1470	October-May
Symplocos spicata Roxb.⁺	Galu	1620	May-July
Tabernaemontana divaricata (L.) R. Br. *ex* Roem & Sch.⁺	Jogindernagar1	220	July-October
Taxus baccata L.⁺	Winch Camp	2220	September-November
Tecoma grandiflora Loisel.⁺©	Shanan	1100	Summer season
Terminalia arjuna Wt. & Arn.⁺	Jogindernagar	1220	May-September
Terminalia bellirica (Gaertn.) Roxb.⁺	Chauntra	1040	May-September
Terminalia chebula Retz.⁺	Jogindernagar	1220	April-June
Terminalia tomentosa (DC) Wt. & Arn.⁺	Basahi	1350	August-March
Thevetia neriifolia Juss. *ex* Steud.⁺	Jogindernagar	1220	July-September
Thuja orientalis L.⁺©	Jogindernagar	1220	July-October
Thunbergia coccinea Wall.©	Jogindernagar	1220	October-March
Tinospora cordifolia Miers.⁺	Makriri	1290	June-July
Trachelospermum lucidum (D. Don) Schuamann⁺	Harabagh	1470	May-September
Trichosanthes tricuspidata Lour.⁺	Dhelu	1220	May-October
Tropaeolum mazus L.©	Jogindernagar	1220	May-November
Ulmus laevigata Royle⁺	Jogindernagar	1220	May-August
Viburnum cotonifolium Don⁺	Winch Camp	2220	April-September
Viburnum stellulatnm Wall.⁺	Winch Camp	2220	May-July
Vigna unguiculata L.©	Sui	2040	June-October

Contd...

Table 1. Contd...

Vinca rosea L.[+] ©	Jogindernagar	1220	Throughout the Year
Vitex negundo L.[+]	Bassi	930	March-August
Vitis himalayana Brand.*	Winch camp	2220	April-November
Vitis jacquemontii Parker*	Shanan	1100	April-May
Vitis rependa Wt. & Arn.*	Winch Camp	2220	April-June
Wendlandia heynei (Roem. & Schult.) Sant. & Merch.[+]	Basahi Dhar	1350	April-June
Woodfordia floribunda Salisb.[+]	Makriri	1290	March-June
Xylosma longifolium Clos.[+]	Dhelu	1220	February-May
Zanthoxylum armatum DC[+]	Jimjima	1500	April-June
Ziziphus jujuba Mill.[+]	Seri	1080	July-August

©-denotes cultivated plants of tehsil Jogindernagar.

Methodology of Study

The collected plant specimens were dried, preserved and mounted as per the known herborizing practices outlined by Jain & Rao (1977). Botanical identification of the selected species was first done with the help of regional floras (Chauhan, 1999; Chowdhery & Wadhwa, 1984; Collett, 1902; Dhiman, 1976; Hooker, 1872-1897; Nair, 1977; Polunin & Stainton, 1984; Stainton, 1988) and later carefully matched with the authenticated specimens at the herbaria of Botanical Survey of India, Northern Circle and Forest Research Institute, Dehradun. Nomenclature of these taxa was confirmed from Bennet (1986) and Wielgorskaya (1995). Economic valuation of all the presently recorded ethnobotanical species was also carried out to calculate the total importance values (TIV) as per detailed methodology outlined by Belal & Springuel (1996). Voucher specimens of afore-mentioned taxa have been deposited in the Ethnobotanical herbarium of Department of Biosciences, Himachal Pradesh University, Shimla for future reference. Besides, colour photographs of the plant species in their natural habitat was also taken.

Enumeration of Ethnobotanical Data

Systematic enumeration of the plants is in alphabetical order of their botanical names, synonym of the species followed by their respective families in the parentheses. Besides this, information on English, Sanskrit and Regional

Names, distribution, description, reproductive cycle, habitat ecology, material examined, part/s used, folk uses, active constituents, biological activity, uses in literature, major Ayurvedic formulations and trade has also been furnished and presented, if available. The data gathered were screened with help of available literature (Kirtikar & Basu, 1935; Anon., 1948-1976; Chopra *et al.*, 1956, 1969; Maheshwari & Singh 1984, 1990; Ambasta, 1986; Singh *et al.*, 1987; Singh, 1988; Jain, 1991), besides many other books and articles published in different journals. For species marked with asterisk, it indicates that the ethnobotanical information presented is new to the existing knowledge and reported for the first time. Likewise, those marked with '+' denote the additional usage not known for the species.

□□□

Chapter 2

Ethnobotany of Trees, Shrubs and Climbers of Himalaya

Abies pindrow Royle⁺ (Pl. 1A)

Syn.: *A. webbiana* Lindl. var. *pindrow* (Royle) Hook. f., *A. spectabilis* (Don) Spach, *Picea smithiana* Loud., *Pinus webbiana* Wall., *P. striata* Ham., *P. spectabilis* Lamb.

Family: Pinaceae.
Vern.: Tosh.
English, Sanskrit & Regional Names

Eng.	:	Himalayan Silver Fir;
Sans.	:	Manduparni, Talispatra, Dhaatipatra;
H.	:	Talispatra;
Kash.	:	Badar, Budar, Tung;
Tam.	:	Taalispatri;
Tel.	:	Taalispatri.

Distribution : India. Afghanistan-W. Nepal: 2100-3600m.

Description : A slender pyramidal tree with 4-7 cm long distichous leaves which are shining green above and with two faint silvery lines beneath. Male and female cones borne on the same tree.

Reproductive Cycle : April-May.

Habitat Ecology : Often found growing along with *Picea smithiana* forming mix forests.

Material Examined : EBH-446, 03-01-1999.

Parts Used : Wood.

Folk Uses : Turpentine oil considered beneficial in **arthritis**. Wood used for **building work** and **furniture**.

Active Constituents : Chief active constituents of heartwood are friedelin, taraxerol, b-amyrin and urosolic acid.

Uses in Literature: Reported to be used in India as an antiflatulent, antiperiodic, antispasmodic, appetiser, astringent, carminative, contraceptive, diuretic, expectorant, febrifuge, fodder, paper pulp, psychoactive plant, purgative, stomachic, tonic; and for asthma, catarrh of bladder, chest infection of infants, chronic bronchitis, construction, dentition, dyeing, floor boards and water containers, gonorrhoea, haemoptysis, headcahe, hoarseness, making packing cases, pulmonery, neuralgia, phithisis and roofing of houses (**Ambasta, 1986; Chauhan, 1999; Jain *et al.*, 1994; Kapur, 1996; Kapur *et al.*, 1996; Thakur *et al.*, 1989; Watt, 1972**).

Major Ayurvedic Formulations : Kanaksara, Lavanbhaskar-churna, Panchguna-taila, Sudarshan-churna, Talisadichurna, Talisadi-vati.

Trade : Bengali-birmi, Taalispati.

Abrus precatorius L.+ (Pl. 1B).

Syn.: *A. minor* Desv., *A. pauciflorus* Desv.

Family: Fabaceae.

Vern.: Rati.

English, Sanskrit & Regional Names

Eng. : Indian Liquorice, Jeryquity, Papernoster pea, Rosary pea, Weather plant;

Sans.	:	Bhilabhushana, Gunja;
Ass.	:	Gunja, Krishanala, Kakachinchi;
B.	:	Chunhati, Gunj, Kawet;
Bo.	:	Ghungchi;
G.	:	Chanoti, Gunja;
H.	:	Chirmiti, Gnunghachi;
Kan.	:	Guengungi;
Kash.	:	Shangir;
Mal.	:	Irattimadhuram, Shekkunni,
Mar.	:	Chanoti, Kunch;
Oriya	:	Kainsho, Kotibepolo;
P.	:	Labri, Ratak;
Tam.	:	Adisamiyai, Atti, Ghurie-ghenza;
Tel.	:	Kukkutamu, Sinnaguruginja;
U.P.	:	Ujarki ghumchi.

Distribution : China, Pacific Island, Tropical Africa, The West Indies, below 1200m elevation.

Description : A copiously branched woody climber showing 20-40 membranous leaflets, reddish corolla and bright scarlet seeds with a dark spot in hilum.

Reproductive : May-August.

Habitat Ecology : Forests, shrubberies; common.

Material Examined : EBH-709, 15-05-1999.

Parts Used : Roots. Seeds.

Folk Uses : Roots considered **aphrodisiac**. 5g powdered roots given with honey, once daily for 7 days in **whooping cough**, and its decoction in combination with black pepper checks **diabetes**. Seeds **abortifacient**, and considered symbolic for **wealth**. In veterinary, 1-2g powdered seeds given with fodder as remedy for curing **anthrax**.

Active Constituents : Active ingredients are precol, abrol, glycyrrhizin, abrasine, precasine and isoflavon-quinones I, II, III

(Roots), abrin A & B, abricin, abridin, abric acid, hypaphorine, alainine, serine, abralin, abrine and abranin (Seed), and Xyloglucosyldelphinidin and p-coumarylgalloyl-glucosldelphnidin (Seed-coat).

Biological Activity : Antitumourous, anthelmintic, antispermatogenic and spasmogenic activities (+)ve.

Uses in Literature: Known to be used as an abortifacient, alexiteric, analgesic, anthelmintic, antidote to poison, antiperiodic in preventing malaria, antiseptic, antitumourous, aphrodisiac, blood purifier, CNS depressant, colic, diuretic, emetic, febrifuge, laxative (vet), masticatory, measure of weights for goldsmiths, purgative, refrigerant, suppository for causing abortion, symbolic (to ward off evil spirits), tonic; and for affections of hair and nervous system, anthrax (vet), appetite, boils, bone fracture (vet), bronchial problem (vet), bronchitis, convulsions, cough, cure of galls on the neck, decoration of baskets and bracelets, diarrhoea, dropsy, eczema, eye complaints, facilitating delivery, flatulence, gall neck (vet), gonorrhoea, haemorrhoids, hastening labour, headache, hemicrania, impotency, indigestion, inflammation, inhibiting conception, insect bite (vet), itching, killing lice, leprosy, leucorrhoea, liver (vet), menstrual complaints, muscular pain, nervous disorders (vet), paralysis, parturition, peptic ulcers, pnuemonia, retaining placenta (vet), rheumatism, scabies, sciatica, scorpion bite, skin eruptions, snake bite (vet), sore throat, stiffness of shoulder joints, paralysis, stomach complaints, strengthening and toning stomach, swelling, syphilis, ulcer (vet), wounds and ulcers, wounds (vet) (**Ambasta, 1986; Chauhan, 1999; Dagar & Dagar, 1996; Datt, 1996; Gill & Nyawuame, 1994; Goud & Pullaiah, 1996b; Jain, 1991, 1996; Kothary & Moorthy, 1996; Kurian, 1995; Malik, 1996; Mandal & Basu, 1996; Mishra *et al.*, 1996b; Nazarudeen *et al.*, 1996; Painuli & Maheswari, 1996; Prasad *et al.*, 1996; Pullaiah & Kumar, 1996; Rama Rao & Henry, 1996; Ravishankar & Henry, 1992; Saini, 1996b; Satapathy & Brahmam, 1996; Sharma, 1996; Singh & Pandey,**

1996; Singh & Prakash, 1994; Singh *et al.*, 1996b; Siwakoti & Varma, 1996).

Major Ayurvedic Formulations : Gunja-bhadra-rasa, Gunjadi-taila, Gunja-pralepa, Mrtasanjivani-gutika, Nili-bhringadi-taila, Uchatta.

Trade : Gunchi, Rati.

Acacia catechu Willd. (Pl. 1C).

Syn.: *A. polyacantha* Willd., *A. wallichiana* DC, *A.* catechuoides DC, *A. sundra* Bedd., *Mimosa catechu* Roxb., *M. catechuoides* Roxb.

Family: Fabaceae.

Vern.: Khair.

English, Sanskrit & Regional Names

Eng.	:	Black Catechu, Catechu Nigraum, Cutuch, Pegu Catechu;
Sans.	:	Dantadhavana,Jivumangalaya, Vakrakantaka;
Ass.	:	Kat, Koir;
B.	:	Kahyer, Kuth;
Bo.	:	Khaderi, Lalkhair;
G.	:	Kher, Kheribayal;
H.	:	Babul, Katha, Khair;
Mal.	:	Kadaram, Senkainnali;
Mar.	:	Khaderi, Lalkheir;
Oriya	:	Bimbu, Khodira;
Tam.	:	Karangalli, Kasukkalti;
Tel.	:	Khadiramu, Mallasandra.

Distribution : Throughout Peninsula and from Indus eastwards to Assam, 500-1350m.

Description : Trees having recurved prickles, pale yellow flowers and glabrous, oblong stalked pod.

Reproductive Cycle : May-July.

Habitat Ecology : Planted along hilly slopes.

Material Examined : EBH-699, 17-07-1999.

Parts Used : Heartwood ('katha').

Folk Uses : Cutch obtained from its heartwood used with alum for curing **mouth sores** (also in veterinary), **throat diseases**, and its combination along with oil considered a good emollient against **leprosy**; and also as **masticatory**.

Active Constituents : These are catechin, catechu, tannic acid, tannin (Bark) and α, β, γ,-catechin and 1-epicatechin (Wood).

Biological Activity : Antiviral, antiinflammatory hepatoprotective, hypoglycaemic and spasmolytic activities confirmed.

Uses in Literature : So far reported in India as an abortifacient, anticancerous, astringent, aphrodisiac, cutch used in paan, dentrifice, digestive, dye, expectorant, fibre, gum arabic, host for lac insects, plywood adhesives, tonic; and for affection of gum, asthma, bleeding from nose, blood dysentery, bronchial affections, burns, causing impotency if in excess, chest pain, child birth, colic pain, congestion of faeces, conjunctivitis, constipation (vet), consumption, cough, cracked palate (vet), diarrhoea with pyrosis, eczema, gonorrhoea, gravel phthisis, haemoptysis, hoarseness, increasing the secretion of milk after delivery, inflammation, intermittent fevers, itching, leprosy, leucorrhoea, menorrhagia after confinement, menstrual complaints, Mithila paintings, muscular growth (vet), obstinate skin diseases, otorrhoea and tonsils, pain in chest, piles, printing, rheumatism, scurvy, skin diseases, soothening throat, sore gums, sore in mouth, sores of nails and soles (vet), strangulation of intestine, suppressing sexual desire, swelling (vet), swelling of foot, syphilis, tanning, toothache, ulcers, and wounds (vet) (**Ahuja, 1965; Ambasta, 1986; Asolkar *et al.*, 1992; Chauhan, 1999; Jain, 1991; Jain *et al.*, 1991; Jha & Basak, 1994; Maheswari & Singh, 1987; Masih, 2003; Mishra *et al.*, 1996b; Kothary & Moorthy, 1996; Saini, 1996a; Shukla & Verma, 1996; Siwakoti & Varma, 1996; Watt, 1972**).

Major Ayurvedic Formulations : Arimedadi-taila, Chandanadi-vati, Dasnar churna, Khadiraadi-kwatha, Khadiraastaka, Khadiradi-gutika, Khadiradivati, Khadirishta, Khadir-sar, Kutajarishta, Lavangadi-vati.

Trade : Kathaa, Katthaa, Paakhraa.

Acacia concinna DC+ (Pl. 1D).

Syn.: *A. hooperiana* Zippel., *A. sinuata* (Lour.) Merrill, *A. rugata* Ham. *ex* Mentham, *Mimosa concinna* Willd., *M. rugota* Lam., *M. abstergens* Spring.

Family: Fabaceae.

Vern.: Shikakai.

Sanskrit & Regional Names:

Sans. : Aila, Rassaul;
B. : Banritha, Toldang;
Bo. : Sikekdi;
G. : Chikakai;
H. : Kocji, Ritha;
Kan. : Sige;
Mal. : Chikaka;
Mar. : Shikakai;
Tam. : Shikai;
Tel. : Shikaya.

Distribution : Throughout India, China.

Description : A tree possessing abundant, minute, hooked prickles. Leaflets pale green above and glaucous beneath. Pod strap-shaped with broad sutures.

Reproductive Cycle : January-April.

Habitat Ecology : Usually in dry and wet deciduous forests, planted.

Material Examined : EBH-26, 02-04-2000.

Parts Used : Bark. Leaves. Seeds. Flowers.

Folk Uses : Pods used as **shampoo** for washing hair. Bark added in preparation of local wine '**sur**', and ash of burnt bark recommended for the care of **asthma** and **lungs dysinfection**.

Active Constituents : Principle ingredients are lupeol, α-spinasterol, hexacosanol, α-spinasterone, acacic acid, lactone (bark); and a nortiterpene-acacidal, acactic acid lactone 3-acetate, sonunin I and sonunin-II (pods).

Biological Activity : Acacialactam inhibits mitogenic activity and used as immunosuppressant.

Uses in Literature: Reported to be used as deobstruent, detergent, dye, emetic, expectorant, fish poison, mild cathartic, nauseant, detergent, purgative, tonic; and for bilious affections, chutney, flatulence, dandruff, jaundice, malarial fever, skin diseases, tanning, and washing hair (**Ambasta, 1986; Watt, 1972**).

Achyranthes aspera L.+ (Pl. 2A).

Syn.: *Digera muricata* Mart.

Family: Amaranthaceae.

Vern.: Puthkanda, Uptkanda.

English, Sanskrit & Regional Names

Eng. : Prickly Chaff Flower;

Sans. : Adhoganta, Adhvashalya, Aghamargava, Aghata, Apamarga, Apangaka, Dhamargava, Malakanta, Mayuraka, Sthalamangari;

Ass. : Apang;

B. : Apang;

Bo. : Aghada;

G. : Agheda, Aghedi;

H. : Apang, Chichra, Chirchira, Latjira;

Kan. : Uttarane;

M.P. : Apamar;

Mal. : Katalati;

Mar. : Aghada, Aghara;

P.	:	Kutri;
Tam.	:	Nayuviri;
Tel.	:	Antisha, Apamaragamu;
U.P.	:	Latzira, Lahchichra.

Distribution : Throughout India.

Description : An undershrub with leathery leaves. Spikes robust, 30-45 cm long. Seeds inverted.

Reproductive Cycle : May-October.

Habitat Ecology : Waste places, roadsides and forest edges.

Material Examined : EBH-1184, 12-09-2001.

Parts Used : The Whole Plant. Roots. Seeds.

Folk Uses : Plant juice recommended as an **aphrodisiac**. 2g ash of burnt root useful in **asthma** (twice daily for 15-20 days). Root paste **antidote** to **snake bite**, and also massaged over gums to cure **tooth infection**. A gargle of root juice useful for **throat infection**. Seeds considered good for **suppressing hunger**.

Active Constituents : Constituents isolated are ecdysterone (Polypodine A), ecdysone (Roots), linoleic (49.4), oleic (22.6), palmitic (18.6), stearic (4.4), behenic (1.8), arachidic (1.6), myristic (1.2) and lauric (0.4%) acids(seed oil), two saponins; saponin A and saponin B, saponin â-methyl ester (seeds).

Biological Activity : Antibacterial, anthelmintic, anti-inflammatory, cytotoxic, diuretic, cardiotonic and abortifacient activities confirmed.

Uses in Literature: Known to be used as an abortifacient, antidote to snake bite, antiseptic, appetiser (vet), astringent, cardiotonic, cataract, cereals (seeds) in times of scarcity, colic, diuretic, dye, emetic, febrifuge, general tonic, germicide, laxative, magico-theraupetic in labour to achieve easy delivery, manure, oligogalactia, psychoactive plant, spinach, tonic after delivery, tooth brush, toothpowder, vegetable; and also for abortion, anasarca, antifertilty in women, anti-inflammatory activity, ascites, asthma, biliousness, bleeding in delivery, blood dysentery, bodyache, boils, bowel complaints, bronchitis (vet), chest pain, chicken pox, cold, cracks (vet), cutaneous diseases, cuts, dandruff, debility, delivery (vet), dental caries, dog bites, dropsy, ear complaints, eruptions of skin, expelling placenta, expelling worms, external cuts, eye (vet), eye complaints, feeding new born babies, fever (vet), gastric (vet), general dysentery, goitre, gonorrhoea, gout, gum troubles, headache, hydrophobia, improving potency of man during coition, indigestion, injuries (vet), insect bite (vet), jaundice, keeping of evil's spirits, leucoderma, liver complaints, malarial fever, menorrhagia, migraine, muscle sprains, opacity of cornea (vet), pain in ribs, piles, poisonous insect bites, pneumonia (vet), pyorrhoea, reducing hunger, regaining consciousness, relieving fever, removing dead foetus (vet), renal complaints, rheumatism, scorpion bite, sore, stomachache, swelling of eyelid, syphilis, throat irritations, toothache, ulcers, warts of body, and whooping cough (**Ambasta, 1986; Aminnudin & Girach, 1991, 1996; Bhandary *et al.*, 1996; Datt, 1996; Borthakur, 1993; Chauhan, 1999; Chowdhery, 1996; Dwarakan & Ansari, 1996, Gill *et al.*, 1997; Girach, 1992, Girach & Aminnudin, 1995; Goud & Pullaiah, 1996b; Hembrom 1996; Henry *et al.*, 1996; Hosagoudar & Henry, 1996c; Jain *et al.*, 1994, Jha *et al.*, 1991, 1996; Khanna *et al.*, 1996b; Lal *et al.*, 1996; Manadhar, 1989 a, b, 1996; Mishra *et al.*, 1996; Mohanty & Padhy, 1996; Pandey & Verma, 2003; Prasad *et al.*, 1996; Pullaiah & Kumar, 1996; Raghupathy & Mahadevan, 1991; Rajenderan *et al.*, 1996; Rama Rao & Henry, 1996; Rao *et al.*, 1996; Reddy *et al.*, 1996; Saini, 1996b; Saklani & Jain, 1994; Samwatsar & Diwanji, 1996a; Sharma & Borthakur, 2003; Sharma & Sood, 1997; Shome *et al.*, 1996; Shukla & Verma, 1996; Sikarwar, 1996; Singh &**

Maheswari, 1992; Singh & Pandey, 1996; Singh & Prakash, 1996; Singh & Rao, 2003; Srivastava *et al.*, 2003; Sundaravelu *et al.*, 2003; Swami & Gupta, 1996; Tirkey *et al.*, 2003b; Vihari 1995).

Major Ayurvedic Formulations : Abhaya-lavana, Agasti-haritaki-leha, Agnimukha, Apamargaksara, Apamargaksara-taila, Apamarga-siddha-taila, Gudapippali, Jyotismati-taila.

Achyranthes bidentata Blume+ (Pl. 2B).

Syn.: *A. wightiana* Wall., *A. lanceolata* Wall., *A. aspera* Wall.

Family: Amaranthaceae.

Vern.: Puthkanda, Uptkanda.

Regional Names

Ass.	:	Apamarga, Bankhat;
H.	:	Adhajhara, Oga;
Kash.	:	Puthkandiari;
U. P.	:	Latjeera.

Distribution : Temperate & Sub-Tropical Himalaya.

Description : An undershrub with thin, membranous leaves and slender spikes which are rarely more than 10-12.5 cm long.

Reproductive Cycle : June-August.

Habitat Ecology : Ravine slopes.

Material Examined : EBH-1185, 23-07-2001.

Parts Used : The Whole Plant. Leaves. Roots.

Folk Uses : Decoction of plant useful in reducing accumulation of **uric acid** in body, and also it is considered important in the preparation of local wine '**sur**'. Paste of leaves useful as an antidote to **scorpion sting,** and its juice beneficial as **diuretic**.

Active Constituents : Roots yield ecdysterone, ecdysone, emodins and physcion.

Biological Activity : Antimicrobial (root extr.) and abortifacient (stem-bark) activities (+) ve.

Uses in Literature: Used in India as an astringent, diuretic, galactagogue (vet), tooth brush, tooth powder; and for blisters in mouth, cholera, dental caries, edible purposes (leaves and seeds), hemicrania, indigestion, jaundice, pyorrhoea, scorpion sting, scrofula, skin diseases, swelling, toothache and whooping cough (**Chowdhery, 1996; Jain, 1991; Jain *et al.*, 1991; Kapur & Singh, 1996; Lalramnghinglova, 1996; Manandhar, 1990, 1996; Molla & Roy, 1996; Singh *et al.*, 1980**).

Actinidia chinense var. *deliocosa* (A. Chev.) A. Chev. (Pl. 2C).

Family: Actinidiaceae.

Vern. Name : Kiwi.

English Name

Eng. : Chinese Goose Berry.

Distribution : Cultivated upto 500-2100m.

Description : A woody vine. Apices of the leaves at the end of branches sharp pointed. Fruit rusty brown with hairy surface, oblong in shape.

Reproductive Cycle : February-November.

Habitat Ecology : Planted.

Material Examined : EBH-1345, 17-11-2002.

Parts Used : Fruit.

Folk Uses : Fruits **edible** as source of vitamin C.

Uses in Literature : Reported to be used for preparing jams, marmalade, decorating ice creams, wine and as edible (**Ambasta, 1986**).

Trade : Kiwi.

Adhatoda vasica Nees+ (Pl. 2D).

Syn : *A. zeylanica* Medic., *Justicia adhatoda* L.

Family : Acanthaceae.

Vern. : Bansa, Basti, Basunti.

English, Sanskrit & Regional Names

Eng.	:	Malabarnut leaf;
Sans.	:	Bhisangmata, Raktappittaghni;
Ass.	:	Herbuksha;
B.	:	Bakas, Baasaka patra;
Bo.	:	Adalsa, Vasuka;
G.	:	Adsoge, Arduosa, Paandru;
H.	:	Adalsa, Vasaka, Banasaa;
Kan.	:	Adusoge;
Mal.	:	Attalotakam;
Mar.	:	Baksha;
Oriya	:	Rottoomuli;
P.	:	Bhekar;
Tam.	:	Adadoli, Kattumurunghai;
Tel.	:	Addasaramu;
U.P.	:	Adusa, Arusa, Rusa, Roos.

Distribution : Distributed in India from Punjab-Assam. 500-1600m.

Description : A strong smelling gregarious shrub with shortly stalked, ovate, lanceolate leaves and white streaked flowers dotted with pink. Capsule club-shaped.

Reproductive Cycle : April-November.

Habitat Ecology : Wastelands, shrubberies; common.

Material Examined : EBH-171, 21-04-2002.

Parts Used : The Whole Plant. Stem. Leaves. Flowers. Root- bark.

Folk Uses : Plant used in preparation of local wine, '**sur**' and an excellent **gun powder**. Massage of oil obtained by boiling 500g dried powdered plant, 25g 'Ashwagandha', 5g 'Jayaphal', 25g 'Jwati' and 10g 'cloves' with 'Sarson oil' useful in **arthritis** and **rheumatic pains**. 'Gulkand' of flowers beneficial in **asthma**. 2-3g mixture of its powdered root-bark, stem-bark of *Aegle marmelos* and stem of *Tinospora cordifolia* (1:1:1), given with 'Shitopaladi-churan' and honey to cure **asthma**. Leaves hasten **ripening** of banana in storage, and its paste useful in **suppuration of boils**. Also, 20 ml decoction of its leaves along with rind of *Punica granatum* and roots of 'Baraein' (*Acorus calamus*), given twice daily for 10-15 days to cure **cough**, **cold** and **asthma**.

Active Constituents : Chief principles are vascicine, vasinone, vascicone, vascine, β-sitosterol, tritriacontane, vasicinine, vascicinol, 4 new quinazoline alkaloids-vascicoline, adhatodine, vasciolinone and anistone (leaves) and vasicinone (inflorescence).

Biological Activity : Bronchodilatory, uterotonic, hypotensive and abortifacient activities found to be (+) ve.

Uses in Literature : So far known in India as an anthelmintic, antipyretic, antiseptic, antispasmodic, aromatic, beads for rosaries, blood purifier, bronchodilator, electuary, expectorant (vet), febrifuge, fuel for brick-burning, green manure, gun powder charcoal, ingredient of 'chyawanprash', insect repellent for stored grains, psychoactive plant, refrigerant in fever, uterine tonic (vet), vegetable, yellow dye; and for abortion, arthritis, asthma, bleeding from nose, boils, child birth, chronic bronchitis, chronic cold, consumption, cough, curing eczema, diarrhoea (vet), diseases of mouth, dislocation of joints, dropsical affections attended with anaemia, dyeing, dysentery (vet), dysuria or painful and difficult urination, easing delivery, eye pain, flatulence, glandular tumours, gonorrhoea, heart troubles, horses for dysentry, infusion, injuries, intestinal worms, jaundice with chronic liver disorders, joint pains, killing domestic flies, leucoderma, loss of memory, malaria, mastitis (vet), neuralgia, opthalmia, packaging, phthisis, pulmonary affections, relieving body pain after delivery, removal of foetus (vet), respiratory troubles, rheumatism (vet), scabies (vet), skin affections, sores of eye, stomach (vet), stomachache, swelling, tuberculosis, tumours, venereal complaints, vomiting, whooping cough, wrapping paddy in preparation of rice beer 'yeast' and also to stop post partum haemorrhage (**Ambasta, 1986; Aswal, 1996; Borthakur, 1993; Borthakur & Sarma, 1996; Chowdhury *et al.*, 1996; Dixit & Pal, 1996; Girach, 1992; Girach *et al.*, 1996; Gogoi *et al.*, 2003; Huidrom, 1996; Jain, 1968, 1991; Kapur & Singh, 1996; Katewa & Arora, 1997; Kothary & Moorthy, 1996; Kurian, 1995; Lal *et al.*, 1996; Lalramnghinglova, 1996; Mahato & Mahato, 1996; Mao, 1993; Mehrotra *et al.*, 1995; Mishra *et al.*, 1996a, 1997; Negi *et al.*, 1993; Prasad *et al.*, 1996; Raghupathy & Mahadevan, 1991; Rama Rao & Henry, 1996; Rana *et al.*, 1996; Ranjan, 1996; Saini, 1996a, b; Sharma & Borthakur, 2003; Sikarwar, 1996; Siwakoti & Varma, 1996; Srivastava *et al.*, 2003; Swami & Gupta, 1996; Tirkey *et al.*, 2003b; Varma, 1997; Watt 1972**).

Major Ayurvedic Formulations : Asokarista, Chandanadi-taila, Chyawanprasha, Kanakasava, Kasamitra, Kasni, Phalatri-kaadi-kwaatha, Pradarari-rasa, Punarnavarista, Vasaacaa-avleha, Vasaaca-churna, Vasacaandanadi-ghrta, Vasakamadhu, Vasakasava, Vasaputaphaka, Vasarista, Vasavaleha, Viaspanaka.

Trade : Adusa, Baansaa-patti, Vasakapatti,

Aegle marmelos (L.) Correa (Pl. 3A, B).

Syn.: *Feronia pellucida* Roth., *Crateava marmelos* L., *C. reliogosa* Ainstie

Family: Rutaceae.

Vern.: Bil.

English, Sanskrit & Regional Names

Eng.	:	Bael Fruit Tree, Bengal Quince, Holy Fruit, Indian Quince, Stone Apple;
Sans.	:	Atimangaliya, Mahakapithakya, Satyadharma, Tripatra;
Ass.	:	Bel;
B.	:	Bela, Vilva;
Bo.	:	Bila;
G.	:	Bil, Billy;
H.	:	Bel, Siriphal;
Kan.	:	Bilvapatra;
M.P.	:	Bael;
Mal.	:	Kuvulam, Mavilamu;
Mar.	:	Bel;
Oriya	:	Bilwa, Sripholo;
Tam.	:	Aluvigam, Iyalbudi;
Tel.	:	Bilvamu, Sailushamu.

Distribution : Throughout India.

Description : A small deciduous spiny tree. Leaves glabrous, leaflets 3, ovate-lanceolate. Flowers white, sweet-scented. Berry grey or yellow, globose.

Reproductive Cycle : April-July.

Habitat Ecology : Planted, wastelands or near houses.

Material Examined : EBH-1091, 23-07-2001.

Parts Used : Fruit. Leaves.

Folk Uses : Ripe fruits **edible** and **pickled**, and considered **refrigerant** and prescribed for **gastrointestinal ailments, burning sensation** of the soles of the feet and as a **tonic**. Gum used as an **adhesive** and leaves offered to **appease** Lord Shiva.

Active Constituents : Main constituents are aegeline, γ-sitosterol (leaves) and umbelliferone, dictamnine, xanthotoxol, xanthotoxin, scoparone, isopimpinellin, isoimperation, N-2-methoxy-2-(4-Methoxy-phenyl) ethyl cinnamamide, marmelin and its methyl ether, bergapten, marmesin, osthol and auraptin (fruit pericarp).

Biological Activity : Activities exhibited are hypoglycaemic.

Uses in Literature : Known to be used as an adhesive, appetiser, aromatic, astringent, beads, candy, cardiac depressent, cementing mixtures, fish poison, funereal pyre of rich, ingredient of 'chayawanprash', laxative, plaster for walls, post-natal, psychoactive plant, refrigerant, sacred, scorbutic, soaps, soothing agents for intestines, stomachic, sweet drinks, fuel, tree motif on 'Shiva' temples, substitute for tea, tannin and varnish, wish granting tree; and for abdominal accumulation of liquid in the cellular tissue accompanied with constipation and jaundice, abdominal pain for abscess, allaying urinary troubles, anthrax (vet), apthae, asthma, backache, bite of mad dog, blisters (vet), boil (vet), bowel complaints, breast pain, bronchitis, burning sensation of the soles of the feet, catarrhs, chest pain, cholera, chronic dysentery, cold, conjunctiva, convulsions, cough, cramp, dandruff, darkening and shining hairs, diabetes, diarrhoea (vet), dyeing, edible purposes, eye complaints, fever (vet), feverishness, gastric trouble, heart palpitations, hyperacidity, hypochondriasis, indigestion, inflammation of the mucous membrane having a free discharge, injuries, intermittent fevers, intestinal disorders, irritation of alimentary canal, joint pain (vet), keeping evil spirits away (planted near stones with snake sculptures on it), menstrual disorders, mouth sores, nausea, night fever, opthalmia, paintings (mithila), relieving headache, rinderpest (vet), scurvy, snake bite, sore (vet), strengthening and giving tone to the stomach, sunstroke, toys, tympany, ulcer (vet), vomiting and wound (**Ambasta, 1986; Aminuddin & Girach, 1993; Bajpayee & Dixit, 1996; Borthakur, 1996b; Chauhan, 1999; Chowdhury *et al.*, 1996; Datt, 1996; Dwarakan & Ansari, 1996; Goud & Pullaiah, 1996a; Huidrom,**

1996; Jain *et al.*, 1994; Jha & Basak 1994; Katewa & Arora, 1997; Kothari & Moorthy, 1996; Kurian, 1995; Lalramnghinglova, 1996; Manandhar, 1990; Masih, 2003; Mehrotra *et al.*, 1995, Mishra *et al.*, 1996a; Negi & Gaur, 1991; Painuli & Maheswari, 1996; Pandey & Verma, 2003; Parmar & Kaushal, 1982; Prasad *et al.*, 1996; Pullaiah & Kumar, 1996; Rama Rao & Henry, 1996; Ravishankar & Henry, 1992; Sharma & Sood, 1997; Shome *et al.*, 1996; Singh, 1991; Singh & Maheswari, 1992; Singh & Rao, 2003; Singh *et al.*, 1997; Siwakoti & Varma, 1996; Srivastava *et al.*, 2003; Tirkey *et al.*, 2003b; Vihari 1995; Watt, 1972).

Major Ayurvedic Formulations : Amritarista, Bilva-churna, Bilvadi-kwatha, Bilvadi-leha, Bilvadi-majjadi-yog, Bilwa-mool-twak, Brhatgangadhra-churna, Chywanprasha, Dasmula-rista, Dasmula-taila, Dhanya-panchka-kvatha, Gangadhara-churna, Maha-narayan-taila, Pushyanuga-churna, Vatsakadi-kvatha.

Trade : Belgiri, fruits, dried pulp & roots are sold in the drug markets of India.

Aesculus indica Colebr.+ (Pl. 3C).

Family : Hippocastanaceae.

Vern.: Kanhor, Khanor.

English & Regional Names

Eng. : Indian Horse Chestnut tree;
H. : Bankhor, Gugu, Pangar, Pankar;
Kash. : Hane, Hanudun, Kakra;
P. : Bankhor, Khanor.

Distribution : Afghanisan - C. Nepal, 1800-3000m.

Description : Trees with opposite compound leaves and white or yellow flowers. Capsule brown, smooth, ovoid.

Reproductive Cycle : May-September.

Habitat Ecology : Forests, Shady ravines.

Material Examined : EBH-1019, 26-07-2001.

Parts Used : Wood. Fruits. Leaves. Seeds.

Folk Uses : Wood used for **construction work** and leaves as **fodder**. Seeds used as **detergent** and its oil massaged for **arthritis** and **rheumatic tendencies**. Powdered fruits (kernels), 'Pathakar' consumed on fasting days, and also after frying in 'desi ghee', given twice daily for three days to check **excessive bleeding during menstrual period**.

Active Constituents : Active properties are due to aesculuside, aescigenin, escin, protoaescigenin, quercitin-3-xylosido-glucoside, kaempferol-3- xylosido-glucoside, quercitin-3-glucosido-rhamnoside, kaemp-ferol-3- xylosidoglucoside, kaemferol-3- glucosido-rhamnoside, quercitin-3-glucoside, quercitin-3-diglucoside, quercitin-3,4-diglucoside, kaempferol-3-α-L-arabinofuranoside, kaempferol-3-α-L-rhamnofuranoside, avicularin and quercitin-3- α-L-rhamnofuranoside (seeds), castaprenol-11, castaprenol-12 and castaprenol-13 (leaves), quercitin, β-sitosterol, escin (leaves) and rutin and astragalin (stem).

Biological Activity : Hypotensive and cardiotonic activities (+) ve.

Uses in Literature: Known to be used as an anthelmintic, detergent for woollen clothes, diuretic, drinking cups of Tibetans, fodder, galactagogue (vet), flour (kernels edible), platters, psychoactive plant, shoe-heels, edible (vet) and wrapping paper; and for ague, building, colic (vet), cracks, dislocated joints, effusions of blood, frost bite, furniture, haemorrhoids, indigestion (vet), intermittent fevers, leucorrhea, loose motions (vet), match-splints, mathematical instruments, migraine, obstinate constipation, packing cases, phlebitis, pneumonia (vet), preventing thrombosis, rheumatism, skin fissures, stomach diseases (vet), ulcers, variecose veins, water troughs, wounds (vet), and whooping cough (**Ambasta, 1986; Chauhan, 1999; Chopra *et al.*, 1956; Jain, 1991; Jain *et al.*, 1994; Manandhar, 1980a; Negi & Gaur, 1991; Parmar & Kaushal, 1982; Rana *et al.*, 1996; Singh, 1996b; Watt, 1972**).

Major Ayurvedic Formulations : Suntain oil and veterinary formulations. Presently the seeds are being used in preparation of shampoo (**Chauhan, 1999**).

Agave americana L.+ (Pl. 3D).

Syn.: *A. cantula* Roxb.

Family : Agavaceae.

Vern. : Ramban.

English, Sanskrit & Regional Names

Eng.	:	Amercan Aloe, Carata, Century plant;
Sans.	:	Kantala;
B.	:	Banskeora, Bilati-ananash, Bilatipat, Carata;
G.	:	Fangli-kunvara;
H.	:	Banskeora, Barakanwar, Hatisengar, Kantala, Rakas-pattah, Ramkanta;
Kan.	:	Bhuttale, Budukattalenaru;
Mal.	:	Panam, Katrazha;
Mar.	:	Parkand;
Tam.	:	Anaik-kat-razhai, Pithakalabuntha;
Tel.	:	Rakshimatalu.

Distribution : Grown as a hedge plant throughout India.

Description : A woody plant with leaves showing curved tips, and marginal hooked teeth crowded at base, and dull-green flowers.

Reproductive Cycle : March-May.

Habitat Ecology : Open slopes, roadsides, wastelands; common.

Material Examined : EBH-1074, 24-08-2002.

Parts Used : Leaves. Shoot.

Folk Uses : Plant grown as **hedges**, and its juice works as a good **antiseptic**. Tender shoots, 'Maanu' acts as a **blood purifier**. Leaves used as **brush** for **painting**, and its fibres used for **ropes** and **cordage**.

Active Constituents : Hecogenin, chlorogenin, 10 steroidal saponins- agavoside A, B, C, C′, D, E, F, G, H and I- isolated from leaves. While sugars were glucose, galactose, xylose and rhamnose.

Biological Activity : Antibacterial and anticancerous activities found to be positive.

Uses in Literature: So far used as an anthelmintic, antisyphilitic, diaphoretic, diuretic, emmenagogue and laxative for ascites, fire wood, fish poison, fodder, laxative, paper material, poultice, resolvent, substitute for soap; and for anasarca, ascites, boils, bruishes and contusions, brushes, burns, cancer, conjuctivitis, controlling exparasite (vet), cordage, cut, diarrhoea, dropsy, dysentery, fever, fever (vet), goitre, gonorrhoea, mats, paralysis, proof against ravages of white ants, scrofula, scurvy and venereal sores, skin diseases (vet), skin disorders, thatching, warts (vet), and wounds (**Ambasta, 1986; Jain, 1991; Jain *et al.*, 1991; Kapur, 1996; Lalramnghinglova, 1996; Rana *et al.*, 1996; Rao, 1996; Saxena & Vyas, 1983; Singh, 1996b; Watt, 1972**).

Albizzia julibrissin Durazz.+ (Pl. 3E).

Syn.: *A. mollis* Boiv., *Acacia mollis* Wall., *A. nemu* Willd., *Mimosa kalkora* Roxb., *M. arborea* Thunb.

Family : Fabaceae.

Vern. : Kurmur, Lal Sirin.

English & Regional Names

B.	:	Kankora;
H.	:	Barau, Kurmura, Lalsiris;
Oriya	:	Tentuliya;
P.	:	Brind, Mathirshi, Surangru;
Tam.	:	Selaivagai;
Tel.	:	Nallasinduga.

Distribution : Pakistan-Bhutan; Asia, Africa: 1000-2100m.

Description : Small deciduous trees having membranous, pubescent leaflets and pink, fragrant petals. Pods glabrous, membranous, grey-pale brown.

Reproductive Cycle : May-August.

Habitat Ecology : Forests, riversides, cultivated areas, often planted as a shade tree.

Material Examined : EBH-358, 14-07-1999.

Parts Used : Leaves. Flowers.

Folk Uses : Leaves serve as an excellent **fodder**. Decoction of flowers containing cow's urine used externally to cure open **wounds** and various **skin ailments.**

Active Constituents : Flowers rich in cis- and trans-Linalool oxide, α-ocimene, isopentanol and 4-penten-2-0ne.

Uses in Literature: Reported so far in India as a psychoactive plant; and for furniture (**Chowdhery, 1996; Jain *et al.*, 1994; Watt, 1972**).

Albizzia lebbeck Benth.+ (Pl. 4A).

Syn.: *A. latifolia* Boivin, *Acacia speciosa* Willd., *A. sirisa* Ham., *A. lebbeck* Willd., Mimosa speciosa *Jacq.*, M. sirisa *Roxb.*, M. lebbeck *L.*

Family : Fabaceae.

Vern. : Sirin, Siris.

English, Sanskrit & Regional Names

Eng. : Parrot Tree, Siris Tree, Sizzling Tree;
Sans. : Barhapushpa, Bhandi, Bhandika, Kapitaka, Kapitana, Krishnasiris, Lomashapushpaka, Murdhapushpa, Plavaga, Shankiniphala, Vishaghati, Vishahanta;
B. : Siris, Sirisha;
Bo. : Garso, Harreri, Siris;
G. : Kaliosavas, Kalokansakio, Kalosadasadu,Kaloshirish, Pilosarshio;
H. : Garso, Kalshish, Kalsir, Lasrin, Mathirshi, Shiris, Shirish, Sirai, Sirar, Sirin, Tantia;
Kan. : Baggera;
M. : Vagu;
Mal. : Kattuvaka, Nenmani, Vaka;
Mar. : Chichola, Kalashiras, Mothasiras, Shirish;
Oriya : Vaghe, Kotvaghe;
Tam. : Adukkavajai, Kalindi, kattuvagai, pandil, Perunvagai, Piram, Undugam;
Tel. : Dirisanamu, Girisamu, Kapitanamu, Peddadirisanamu, Sirishamu, Siriso.

Distribution : Throughout India. Ascending to 1650m.

Description : Trees with 7-30 cm long glabrous leaves. Flowers white. Pods glabrous.

Reproductive Cycle : April-May.

Habitat Ecology : Frequently planted.

Material : EBH-727, 16-05-1999.

Parts Used : Leaves. Flowers. Root-bark. Wood.

Folk Uses : Wood used for **furniture** and **agricultural implements**. Decoction of flowers prescribed for **allergy**. Leaves serve as good **fodder** and decoction of roots (5-10 ml) taken on empty stomach as an **anthelmintic** and for **bleeding gums**.

Active Constituents : Chief constituents are flavandiols-melacacidin and lebbeacacidin (heartwood), d-catechin, leuco-cyanidin, condensed tannins (bark), vicenin-2, rcynoutrin, rutin, myricitin, robibin, albegenin (leaves), a new saponin characterised as echino-cystic acid-3-O-2-rhamnopyranosyl (1>5)-β-D-xylofuranosyl (1>5)-β-D-arabinofuranosyl (1>4)-β-D-glucopyranoside (root), and crocetin, lebbeck-inins D,F,G and H (flowers).

Biological Activity : Anti-asthmatic and anti-anaphylactic activities (+)ve.

Uses in Literature: So far used as an alterative, antidote to poisons, aphrodisiac, astringent, camel fodder, detergent, expectorant, firewood, poultice for swelling of cervical glands, psychoactive plant, restorative, symbolic (god of love uses flowers on arrow of desire), tannin, tonic; and for agricultural implements, arthritis, asthma, boats, boils, canoes, carbuncle, carriages, cold, cooling, cough, diarrhoea, dyeing, dysentery, edible purposes, eruptions, eye troubles, fever (vet), furniture, gonorrhoea, gum ulcers, house building, infected teeth, interior decoration, leather, leprosy, leucoderma, night blindness, oil-mills, opthalmia, picture frames, piles, railway carriage work, reducing enlargement of cervical glands, retaining seminal fluid, rheumatism (vet), snake bite, sugar cane-crushers, swelling, toys manufacturing, tympany (vet), ulcers, well curbs, and wounds (**Ambasta, 1986; Annonymous, 1948; Chauhan, 1999; Gill & Nyauwame, 1994; Goud & Pullaiah, 1996b; Jain, 1991; Jain *et al.*, 1994; Khanna *et al.*, 1996b; Kurian, 1995; Raghupathy & Mahadevan, 1991; Rama Rao & Henry, 1996; Siwakoti & Varma, 1996; Srivastava *et al.*, 2003; Watt, 1972**).

Major Ayurvedic Formulations : Deshanglepa, Mahasirisagad, Sirisarishta, Sirisa-vleha.

Albizzia stipulata Boiv. (Pl. 4B).

Syn.: *A. chinensis* (Osb.) Merr., *Mimosa chinensis* Obs., *M. stipulata* Roxb., *Acacia stipulata* (Roxb.) DC, *A. smithiana* Wall., *A. marginata* Ham., *Arthrosprion stipulatum* Hassk.

Family: Fabaceae.

Vern. : Oyee.

Regional Names

Ass.	:	Sau, Selcha;
B.	:	Amluki, Chakua;
Bo.	:	Udala;
H.	:	Kanujera, Sam Sundra;
Kan.	:	Hottabage, Kalbage;
Mal.	:	Mottavaka, Pottuvaka;
Mar.	:	Udala;
Oriya	:	Gudanandia, Reyi;
P.	:	Kasir, Oe, Sirin;
Tam.	:	Katturmjil, Silaivogai;
Tel.	:	Chinduga, Kondachiguru.

Distribution : India, China, S.E. Asia: 300-1300m.

Description : A tree. Leaves pubescent or hairy. Stipules pink and large-sized. Flowers yellow. Pods glabrous.

Reproductive Cycle : May-June.

Habitat Ecology : Riversides, Cultivated areas.

Material Examined : EBH-664, 16-05-1999.

Parts Used : Fruit. Leaves. Bark. Stem.

Folk Uses : 2-4g mixture of powdered bark (250g) in combination with (25g) 'Mulathi' (*Glycyrrhiza glabra* L.) given twice daily for 10 days to restore the **normal function of the liver**. Also, its bark and root serve as an important constituents of **local wine** 'sur'. Leaves serve as fodder and wood used for **construction** and **agricultural implements**.

Biological Activity : 50% EtOH extr. of aerial parts spasmogenic and diuretic. Seeds showed marked hypoglycaemic activity in normal rats but not in alloxan-diabetic rats. Aq. extr. of bark acts as uterine consrictor.

Uses in Literature: Used as cattle fodder, fuel, and gum; and for sizing daphne paper, agricultural implements, boxes, building, cart wheels, furniture, tea boxes, turnery articles and wooden bells (**Ambasta, 1986; Watt, 1972**).

Alnus nepalensis Don⁺ (Pl. 4C).

Syn.: *Clethropsis nepalensis* Spach., *Betula Leptostachya* Wall.

Family : Betulaceae.

Vern. : Kharaeun.

English & Regional Names

- **Eng.** : Indian Alder;
- **H.** : Utis;
- **P.** : Koni.

Distribution : Temperate Himalaya. 1000-3000m. (H.P.-S.W. China, Burma, S.E. Asia)

Description : Deciduous trees. Branchlets glabrous. Leaves shortly pointed. Male catkins 10-25cm, numerous. Wing of nutlet-thin, membranous. Fruiting cones 1-2.5cm.

Reproductive Cycle : May-December.

Habitat Ecology : Commonly found growing in damp forests, Riversides, Wet Gullies.

Material Examined : EBH-1226, 16-11-2002.

Part Used : Leaves.

Folk Uses : Leaves used as **fodder** and for rearing **silk worms**.

Biological Activity : Hypoglycaemic and spasmolytic activity and effect on CNS found positive.

Uses in Literature : Reported earlier as an antidiarrhoeal, antispasmodic, fuel, tea leaves (dried bark), and for cholera, cuts, diarrhoea, dyeing, general carpentry, house construction, stomach disorders, tanning, tea boxes, and wounds (**Ambasta, 1986; Jain, 1991; Jamir & Rao, 1990; Lalramnghinglova, 1996; Rao & Jamir, 1982; Watt, 1972**).

Trade : Indian Alder.

Alnus nitida Endl. (Pl. 4D).

Syn.: *Clethropis nitida* Spach.

Family : Betulaceae.
Vern. : Bhuj.
Regional Names

Kash.	:	Chamb;
P.	:	Sharol;
U.P.	:	Kunis, Utis.

Distribution : Kashmir -W. Nepal: 300-1500m.

Description : Large deciduous trees having long pointed, toothed leaves and ovoid, oblong fruits. Wings of nutlet thick and leathery.

Reproductive Cycle : August-October.

Habitat Ecology : Riversides; common.

Material Examined : EBH-428, 11-10-2002.

Parts Used : Bark. Leaves.

Folk Uses : Leaves used as **fodder** and bark for **dyeing**.

Active Constituents : Bark contains tannins.

Uses in Literature : Used in India as an antidiarrhoeal, antispasmodic, fodder, tea leaves (dried bark), and for bedsteads, cholera, construction of baskets, cuts, diptheria, dyeing, hooked-sticks of rope-bridges, rheumatism, tanning, and tying loads (**Ambasta, 1986; Kapur, 1996; Watt, 1972**).

Aloe vera (L.) Webb. & Berth (Pl. 5A).

Syn.: *A. barbadensis* Mill., *A. perfoliata* var. *vera* L.

Family : Liliaceae.

Vern. : Ghi kawar.

English, Sanskrit & Regional Names

Eng. : Aloe, Barbados Aloe, Common Indian Aloe, Curcago Aloe, Jaffarabad Aloe;

Sans. : Adala,Ajara, Amara, Ambudhisrava, Aphala, Atipichhila, Bahupatri, Brahmaghni, Grihakanya, Kantakapravrita, Sthuladala, Sukantaka,Tarani, Vipulsrava;

Ass. : Chai-kuwari;

B. : Ghritakumari;

G. : Kadvikunwar, Kunvar, Kuvara;

H. : Ghiguvara, Ghikumari, Gvarapatha, Kumari, kuvarapatha;

Kan. : Loli-sara;

Mal. : Chenninayakam, Cherukattala, Chuvannakattala, Ghritakumari, Kaltaavala, Kaltala, Kaltavaya, Kumari;

Mar. : Koraphada, Koraphanta, Korkand, Kunvarpatha;

Oriya : Ghikunri, Ghritokumari, Kumari;

Tam. : Angani, Angini, Ettakalavanda, Kalavanda, Kattalai, Kodiyan, Manjaikattalai, Musamvaramu, Sirukattalai, Sivappusottukkattalai, Sirappukkattalai, Sottukk-attalai.

Distribution : Kerala, Tamilnadu and other warm regions of India, below 1500m.

Description : Acaulescent, stoloniferous shrubs. Leaf rosettes arising from the ground. Flowers vermillion-coloured.

Reproductive Cycle : March-April.

Habitat Ecology : Usually along paths of field borders, slopes.

Material Examined : EBH-184, 13-04-2000.

Parts Used : Latex. Leaves.

Folk Uses : Leaves cooked for **discomforts** in **stomach** and **jaundice**, and its gel applied as an emollient for **burning sensation** of soles of the feet, and also in treatment of burns. Also, useful in band-aid for **suppuration of boils** and healing **external injuries**.

Active Constituents : Chief ingredients are barbaloin aloesin, aloesone β-sitosterol (leaves) and 4 partially acetylated glucomannans, β-barbaloin and aloe-anodin (jelly).

Biological Activity : Barbaloin inhibits the growth of *mycobacterium tuberculosis*. Plant extr. (100 mg/kg) prevents implantation in 50 and 100% rats at oral dose of 100 and 200 mg/kg. Supernatant of gel cytotoxic against C3HIOTI/2 cells at initiation stage of transformation.

Uses in Literature: Reported to be used as an aperient in apoplexy, aromatic, carminative, cathartic, coolant in fever, diphoretic, diuretic, emmenagogue, fibre, flavouring agent, refrigerant, stimulant, stomachic, tonic, vermicidal; and for acute indigestion, asthma, blisters, boil (vet), breast tissue hardening, bronchitis, burns (vet), checking growth of tumours, chest pain, colic, combining with tonics, curing bone spavin and in ostisis of horses, cutaneous leishmaniasis, delivery (vet), dermatitis, digestion (vet), dyeing, easy shedding of placenta during delivery (vet), enlarged lymphatic glands, eye diseases, falling hair, flatulence, giddiness, gin preparation, gonorrhoea, headache, healing wounds and fresh cuts, hepatomegaly, indigsetion (vet), injuries, itching (vet), leprosy, liver and spleen ailments, mastitis (vet), menstrual complaints, miscarriage (vet), *Mycobacterium tuberculosis*, obstinate constipation, pickling, piles, prevention of miscarriage, relieving pains in muscles, rheumatism, ringworm (vet), sexual vigour and vitality, skin diseases, softening swollen parts, soothing burns, sore (vet), sores of bullocks, swelling (vet), swelling of foot, typhoid, washing hairs (vet), wounds, and X-ray burns (**Ambasta, 1986; Borthakur**

& Sarma, 1996; Chauhan, 1999; Datt, 1996; Geetha *et al.*, 1996a, b; Girach & Aminuddin, 1995; Gogoi *et al.*, 2003; Hosagoudar & Henry, 1996a; Jain, 1991; Kurian, 1995; Mishra *et al.*, 1996b; Rama Rao & Henry, 1996; Saini, 1996b; Samwatsar & Diwanji, 1996a; Satapathy & Brahmam, 1996; Sharma & Borthakur, 2003; Shome *et al.*, 1996; Singh & Pandey, 1996; Watt, 1972).

Major Ayurvedic Formulations : Aloes compound, Bhringamaladi, Cukkum-tippalyadi-gutika, Kumaryadi-vati, Kumaryasava, Loha-rasayana, Rajahpravartini-vati, Rajapra-vartain-vati.

Andrachne cordifolia (Decne.) Muell.-Arg.+ (Pl. 5B).

Syn.: *A. decaisneana* Baill., *Arachne cordifolia* (Decne.) Hurus., *Leptopus cordifolia* Decne., *Phyllanthus cordifolia* Wall., *P. hoffmeisteri* Klotzsch, *P. glauca* Wall.

Family : Euphorbiaceae.

Vern. : Dudh-ki-Buti, Pini.

Regional Names

P. : Gurguli, Kurkni, Kurkuli;

U.P. : Bhatula.

Distribution : In most parts of Himalayas: 1650-2600m.

Description : Small shrubs. Leaves alternate, long-stalked, thin, ovate-oblong with lower surface pale and hairy. Flowers green. Capsule globose.

Reproductive Cycle : April-October.

Habitat Ecology : Common in ravine slopes.

Material Examined : EBH-1027, 16-10-2002.

Parts Used : The Whole Plant. Leaves. Fruits.

Folk Uses : Decoction (10 ml) of whole plant **diuretic** (twice a day for seven days).

Active Constituents : Active principles isolated from aerial parts are glut-5(10)-en-3-one and new triterpene-glut-5(10)-en-1β-OL.

Uses in Literature : Used so far as poison; and for scorpion and snake bites. The plant is poisonous to cattle when browsed by them in early morning on empty stomach (**Jain, 1991; Kapur & Nanda, 1996; Watt, 1972**).

Araucaria columnaris (Forst.) Hook. (Pl. 5C).

Syn.: *A. cookii* R. Br. *ex* Lindl.

Family: Araucariaceae.

Vern.: Araucaria

English Name

Eng : Monkey Puzzle Tree.

Description : A tree with lower branches usually abscising. Short shoots arise from the adventitious buds at the base. Male inflorescences 2.5-6 cm long. Scales about 2-4 cm wide. Seeds winged, with a terminal thorny tip.

Reproductive Cycle : September-October.

Habitat Ecology : Planted.

Material Examined : EBH-1248, 12-10-2001.

Plant/Parts Used : The Whole Plant.

Folk Uses : Plant grown for **ornamental** purposes.

Uses in Literature : Used in general carpentary work (**Ambasta, 1986**).

Arundinaria falcata Nees (Pl. 5D).

Syn.: *A. utilis* Cugh.

Family: Bambusaceae.

Vern. : Banji, Nagral.

English & Hindi Names

Eng. : Himalayan Bamboo;

H. : Gurwa, Spiag, Spikso.

Distribution : W. Himalaya: 1200-2100m.

Description : Shrub-like bamboos. Leaves not net-veined and having glabrous sheath. Flowers occuring on leafless stem. Glumes ribbed.

Reproductive Cycle : May-November; flowering sporadically.

Habitat Ecology : Common in oak forests and forms thick inpenetrable under wood-thickets; Winch camp, 2230m.

Material Examined : EBH-400, 15-05-2003.

Parts Used : Stem. Leaves.

Folk Uses : Plant serves as an excellent material for **baskets**, **brooms**, **'kalam'** (pen), **agricultural implements** and **umbrella** (for working in fields), and also as **fodder**.

Uses in Literature : Reported earlier to be used as base material for house, fibre, fodder; and for agricultural implements, baskets, brooms, cattle shed, containers for compost fertilizer, grains, roofing, storage wings, mats and fishing rods and hookas (**Ambasta, 1986; Negi *et al.*, 1993; Watt, 1972**).

Asclepias currasavica L.⁺ (Pl. 6A).

Family: Asclepiadaceae.

Vern.: Ghamera, Parvini.

English, Sanskrit & Regional Names

Eng.	:	Curassavian Swallow-Wort, West Indian Ipeca-cuanha, Wild Ipecacuanha;
Sans.	:	Kakatundi;
Bo.	:	Kakatundi, Kuraki;
Mar.	:	Kurki
P.	:	Kaktundi.

Distribution	:	Throughout India. Naturalised throughout Himalaya to 1500m.
Description	:	An undershrub with opposite-oblong, lanceolate leaves and orange-red corolla. Follicles long, smooth, swollen, pointed.
Reproductive Cycle	:	Flowers throughout the year.
Habitat Ecology	:	Common in wasteplaces, roadsides, forest edges, fields.
Material Examined	:	EBH-35, 12-05-2000.
Parts Used	:	Flowers.
Folk Uses	:	Plant grown as **hedges** and its flowers considered **sacred** for **offering** to **dieties**.
Active Constituents	:	Hesperidin isolated from stem.
Biological Activity	:	Hypoglycaemic, anti-tumorous, CNS-stimulant and cardiotonic activities found (+) ve.
Uses in Literature	:	So far used in India as an adulterant, anthelmintic, antidysenteric, astringent, cathartic, emetic, fish poison,

purgative, styptic, sudorific; and for abdominal pain, cancer, corns and treatment of sores and wounds, dysentery, eradicating warts, haemorrhages, piles, stufffing, treatment of gonorrhoea, ulcer, vomitting, and warts (**Ambasta, 1986**; **Jain, 1991**; **Jain *et al.*, 1997**; **Kurian, 1995**; **Watt, 1972**).

Asclepias physocarpa Schlecht (Pl. 6B).

Family: Asclepiadaceae.

Vern. Name: Gol-gappa Phool.

Distribution	:	Cultivated.
Description	:	A shrub with narrow upright leaves. Corolla deeply lobed, and often reflexed. Flowers yellowish-white.
Reproductive Cycle	:	April-July.
Habitat Ecology	:	Planted.
Material Examined	:	EBH-559, 24-07-2001.
Parts Used	:	Flowers.
Folk Uses	:	Planted as ornamental.
Uses in Literature	:	A severe gastro-intestinal irritant; may prove fatal (**Ambasta, 1986**).

Atyloshia platycarpa Benth.+ (Pl. 6C).

Syn.: *A. geminiflora* Dalz., *Canthospermum distans* Royle.

Family: Fabaceae.

Vern.: Pahadi-Rongi, Ban-Sem.

Distribution	:	N.W. Himalaya. 300-2400m.

Description : Trailing shrubs with orbicular leaflets and yellow flowers. Pods flat. Seeds ovoid, black.

Reproductive Cycle : May-August.

Habitat Ecology : Grassy fields, roadsides, agricultural fields.

Material Examined : EBH-602, 28-08-2002.

Parts Used : Roots. Seeds.

Folk Uses : Seeds consumed as **vegetable**. Powdered roots (1-3g) taken with honey (once daily after 4th day of menstrual cycle for 10 days) to check **conception**.

Uses in Literature : Used for eye sores (**Jain, 1991**).

Bambusa nutans Wall.+ (Pl. 6D).

Syn.:*Bambusa falconeri* Munro

Family: Bambusaceae.
Vern.: Ddang.
Regional Name

B. : Mahlbans.

Distribution : H.P.-Sikkim. Commonly cultivated upto 1700m.

Description : Solitary stems, arising from a creeping rhizome and without spiny branches. Internodes 38-45cm long. Leaves linear-lanceolate, pubescent beneath.

Reproductive Cycle : Flowering sporadically.

Habitat Ecology : Cultivated; Seinthal, 750m.

Material Examined : EBH-105, 14-03-2000.

Parts Used : Stem. Leaves.

Folk Uses : Leaves used as **fodder**, and stem good for **building works** and **wooden utensils**.

Uses in Literature : Reported to be used for house building, and as pulp (**Ambasta, 1986; Jain, 1991**).

Major Ayurvedic Formulations : Shitolashadi-churn.

Trade : Bans.

Bauhinia vahlii Wight & Arn.+ (Pl. 7A).

Syn.: *B. racemosa* Vahl., *Phanera vahlii* Benth.

Family: Fabaceae.

Vern.: Tor.

English & Regional Names

- **Eng.** : Camel's Foot Climber;
- **B.** : Chehur;
- **G.** : Chambelli;
- **H.** : Jallur, Maljhan;
- **Mal.** : Motanvally;
- **Mar.** : Chambal, Chambura;
- **Oriya** : Shiali,
- **P.** : Taur;
- **Tam.** : Mandarai;
- **Tel.** : Adattige, Madapu.

Distribution : Kashmir-Sikkim: upto 1500m.

Description : A large climber having branches terminating in a pair of opposite tendrils and cordate-orbicular leaves, whitish yellow petals, and woody pods with velvet hairs.

Reproductive Cycle : April-June.

Habitat Ecology : Forests, Rocky slopes.

Material Examined : EBH-693, 25-05-2001.

Parts Used : Leaves. Stem. Wood.

Folk Uses : **Ropes** made from the plant. Wood considered to form an excellent **gun powder**. Leaves used for making '**pattal**' and '**duna**' (meal plates). Extr. of its tender leaves (5-10 ml) kills **intestinal worms** (once daily for 5 days).

Active Constituents : Chief constituents are quercitroside, iso-quercitroside, rutoside (plant); tannin (17%) (bark) and campesterol, stigmasterol, β-sitosterol, agathisflavone and its mono- and d- mrthyl derivatives, (kaempferol, quercitin) its 3-glucoside and betulinic acid (leaves).

Uses in Literature : Used so far as an adhesive for fixing the hair, aphrodisiac, demulcent, finger rings- worn as a charm against-dropsy, rough table clothes, tonic, vegetable, vemifuge; and for amenorrhoea, antifertility, bone fracture (vet), dysentery, local cigarretes, making meal plates, malaria, mucilaginous remedies, rain hats and caps, ropes and cordage, seed pulp (eaten raw), stomachache, syphilis, thatching, and umbrellas (**Ambasta, 1986; Banerjee, 1996; Goud & Pullaiah, 1996a; Jain, 1991, 1996; Kumar & Pullaiah, 2003; Rama Rao & Henry, 1996; Rao, 1996; Sahu, 1996; Saini, 1996b; Singh, 1996a, b; Singh & Prakash, 1994; Tosh, 1996; Varma, 1997; Watt, 1972**).

Bauhinia variegata L.⁺ (Pl. 7B).

Syn.: *B. purpurea* Wall., *B. candida* Roxb., *Phanera variegata* Benth.

Family: Fabaceae.

Vern.: Karale.

English, Sanskrit & Regional Names

Eng.	:	Buddhist Bauhinia, Mountain Ebony, Orchid Tree;
Sans.	:	Kovidara;
B.	:	Raktanchan;
Bo.	:	Segapa-Munthari;
G.	:	Kovindara;
H.	:	Kachnar;
Kan.	:	Kanchavala;
Mal.	:	Chuvannamandaram;
Mar.	:	Raktakanchan;
Oriya	:	Boroda;
Tam.	:	Sagapumanchori;
Tel.	:	Devakanchanamo.

Distribution : Common in India, Burma, China upto 1800m.

Description : A tree with deeply 2-lobed glabrous leaves. Petals with purple veins and fragrant. Pods glabrous.

Reproductive Cycle : March-August.

Habitat Ecology : Forests, planted in villages.

Material Examined : EBH-460, 26-07-2001.

Parts Used : Flowers. Leaves. Bark.

Folk Uses : Flower buds **cooked, pickled** and used as **vegetable** and in **culinary preparations**. 1-2g powdered flowers in combination with 'saunf' helpful in **epilepsy** and for regaining lost **sexual vigour**. Leaves used as **fodder** for enhancing **lactation**. Decoction of bark (15-20 ml) alone or with black pepper taken as tea for **promoting flow of urine, leucorrhoea**, easy **shedding of placenta** after delivery and **glandular adenytis.**

Active Constituents : Active properties are due to isoquercitin, astragalin quercitin, hentriacontene octacosanol reducing sugars, stigmasterol, β-sitosterol and β-p-glucopyranoside (flowers), kaempferol glycosides quercitroside, isoquercitroside and rutoside (plants), myricetol glycosides (seeds), and 5,7-dihydro-xyflavanone-4′-O-α-L-rhamnopyranasyl-β-D-gluc-opyranoside (I) (stem).

Biological Activity : Seed extr. exhibits marked agglutinating for human blood. Antibacterial and CNS depressant activity found (+) ve.

Uses in Literature : So far reported to be used as an alterative, anthelmintic, antidote to snake poison, antiinflammatory, astringent, carminative (roots), culminative, fodder, gentle laxative, pot herbs (flower bud), sacred, tonic, vermifuge (vet); and for agricultural implements, beedi manufacturing (leaves), body/ muscular swelling, cuts (vet), cutaneous troubles, diarrhoea, dyeing, dysentery, dyspepsia, enhancing lactation, epistaxis, expelling placenta (vet), fatness, flatulence, gastropathics (vet), goitre, gripping pains from the stomach and bowels, indigestion, leprosy, malaria, mumps, oilseeds, paintings (Mithila), petals used in a curd preparation, pickling, piles, preventing obesity and decomposition of the blood, reducing corpulence, scrofula, skin diseases, snake bites, suppuration, syphilis, tanning, tuberculosis, tumours, ulcer (vet), uterine bleeding, ulcers, and wounds (vet) (**Ambasta, 1986; Banerjee, 1996; Chauhan, 1999; Chhetri, 1994; Gill & Nyawuame, 1994; Hosagoudar & Henry, 1996a; Jain, 1991; Jain *et al.*, 1991; Jha & Basak, 1994; Jha *et al.*, 1996; Khanna *et al.*, 1996a, b; Kurian, 1995; Lal *et al.*, 1996; Lalramnghinglova, 1996; Masih, 2003; Negi *et al.*, 1985; Rao, 1996; Saini, 1996b; Singh, 1991, 1996b; Sinha, 1996; Watt, 1972**).

Major Ayurvedic Formulations : Chandanasava, Kanchanaaradi-kwath, Kanchanara-guggulu, Kanchanara-gutika, Kanchanar-kwath, Ushirasava, Vidanga-sava.

Trade : Stem bark sold in the drug market as 'Kachnar chhal'.

Berberis aristata DC⁺ (Pl. 7C).

Syn.: *B. aristata* var. *floribunda* (G. Don) Hook.f. & Thoms.

Family: Berberidaceae.

Vern.: Kashmal, Panjolu.

English, Sanskrit & Regional Names

Eng.	:	Indian Barberry, Tree Turmeric;
Sans.	:	Daruharidra, Kantakati, Kateri;
B.	:	Daruhaldi;
G.	:	Daruharidra, Daruhuldur;
H.	:	Chitra, Darhald, Kashmal;
Kan.	:	Daruhaladi, Maradarishana, Maradarishina;
Kash.	:	Rasvat;
Mal.	:	Maradarisina, Maramanjal;
Mar.	:	Daruhald;
Oriya	:	Daruhalidi, Daruharidra;
P.	:	Chitrak, Kashmal, Simlu;
Tam.	:	Mullukala, Usikkala.
Tel.	:	Manupusupu;
U.P.	:	Darhald.

Distribution : H.P.-Nepal: 1800-3500m.

Description : A shrub with red young branches, few teethed obovate leaves. Flowers yellow. Young fruits white, blue when ripe.

Reproductive Cycle : May-September.

Habitat Ecology : Shrubberies, common; Winch Camp, 2270m.

Material Examined : EBH-531, 31-07-2001.

Parts Used : Roots. Stem. Leaves.

Folk Uses : Ripe fruits, '**kashmal**' considered as **blood purifier**. Wood used as **fuel** and leaves an excellent **fodder**. A yellow **dye** extracted from its roots and stems. And its powdered roots given with honey as **mild laxative** to children.

Active Constituents : The principle alkaloids of root bark are berberine, karachine, taxilamine, oxycanthine and umbellatine.

Biological Activity : Antibacterial, antidepressant and antitubercular activities have been attributed to berberine.

Uses in Literature : Reported to be used as an alterative, antiperiodic, antiscorbutic, blood purifier, cooling laxative to children, deobstruent, diaphoretic, fuel, psychoactive plant, stomachic, tonic; and for affections of the eye, ammenorrhoea, anorexia, apthous sores, bites of rat and snake, boils, cholera, diarrhoea, dyeing, dyspepsia, edible purposes, enlargement of spleen, eye complaints, infection in horses, intermittent and remittent fevers, intestinal catarrh, jaundice, liver (vet), malarial affections, menorrhagia, painful micturition, piles, rheumatism, skin diseases, stomach disorders, sun-blindness, tannins, ulcers, uterine troubles, and also in cicatrization, and vomitting during pregnancy (**Ambasta, 1986; Anonymous, 1948-1976; Chauhan, 1999; Jain, 1968, 1991; Jain *et al.*, 1994; Kirtikar & Basu, 1935; Parmar & Kaushal, 1992; Sensarma, 1991; Shome *et al.*, 1996; Singh 1996b; Watt, 1972**).

Major Ayurvedic Formulations : Ashwagandhyarista, Bhrargaraja-taila, Chandana-adivati, Darbhyadi tel, Darbhyadi-leha, Darbyadi-kwath, Darbyadi-taila, Dashanglepa, Jatyadi-taila, Khadira-gutika, Khadirarishta, Laghu-vishgarbha-taila, Punarnava-mandoor, Rasanjanadi-churna, Sudershan-churna, Triphala-ghrita.

Trade : Traded in Indian markets as 'Rasaunt'.

Berberis lycium Royle (Pl. 7D).

Syn.: *B.insignis* Hook.f.

Family: Berberidaceae.

Vern.: Kashmal.

English, Sanskrit & Regional Names

Eng. : Raisin Berberry;
Sans. : Daruharidra;
Bo. : Kasmal Rasout;
H. : Kushmal.

Distribution : Pakistan - E. Nepal. India:1500-3000m.

Description : Shrubs with angular branches, narrowely lanceolate leaves, having a few, small teeth. Flowers pale-yellow. Berries ovoid and violet, and covered with blume.

Reproductive Cycle : April-July.

Habitat Ecology : More or less common in cultivated areas of W. Himalayas.

Material Examined : EBH-345, 12-04-2001.

Parts Used : Roots. Stem. Leaves. Fruit.

Folk Uses : Plant provides **forage** for sheep and goats. Ripe fruits edible. Leaves chewed as an **appetiser**. Stem used as **fuelwood**, and chewed for relief from toothache. Decoction of root beneficial in treatment of **diarrhoea, piles, opthalmia** and **jaundice**. 3-4g mixture of its powdered roots and stems with those of 25g 'Ashwagandha', 20g 'Black pepper' and 10g 'Jayphal', given twice a day with cow's milk for 20-25 days as **blood purifier**.

Active Constituents : Chief principles isolated from roots are umbellatine, berberine, berbericine hydrochloride, berbericinine hydroiodide, oxyberberine and berbamine.

Biological Activity : Antianotic, antisecretory, anticonvulscent and spasmolytic activities confirmed (Rastogi & Mehrotra, 1991).

Uses in Literature : Used in India as an alterative, antidote for poisoning, antimalarial, antiseptic, aperient, blood purifier, carminative, coagulent, febrifuge, mild laxative, sedative, tonic; and for allaying irritability and thirst, bleeding piles, boils, chest and throat troubles, cholera, chronic diarrhoea, cough, diarrhoea, disorders of urinary tract, dysentery, ear problems, edible purposes, enlargement of liver and spleen, epigastrium, eye diseases (chronic opthalmia), increasing appetite, jaundice, leprosy, malaria, menorrhagia, oral ulcers, oriental sores, periodic neuralgia, piles, promoting digestion, skin diseases, sores, spleen troubles, stomach disorders, throat pain, and urogenital disorders (**Ambasta, 1986; Chauhan, 1999; Jain, 1991; Kapur *et al.*, 1996; Lal *et al.*, 1996; Negi & Gaur, 1991; Rana *et al.*, 1996; Singh, 1996b, Watt, 1972**).

Major Ayurvedic Formulations : Ashwagandhyarista, Bhrargaraja-taila, Chandana-adivati, Darbhyadi-tel, Darbhyadi-leha, Darbyadi-kwath, Darbyadi-taila, Dashanglepa, Jatyadi-taila, Khadira-gutika, Khadirarishta, Laghu-vishgarbhataila, Punarnava-mandoor, Rasanjanadi-churna, Sudershan-churna, Triphala-ghrita.

Trade : Traded in Indian markets as 'Rasaunt'.

Boehmeria platyphylla Don (Pl. 8A).

Syn.: *B. macrophylla* Horn., *B. macrostachya* Wedd., *Urtica macrostachya* Wall.

Family : Urticaceae.

Vern.: Chamrala, Handa.

Regional Name

H. : Gargela.

Distribution : India. S.W. China. Burma: 800-2700 m.

Description : A shrub with opposite leaves; small nearly white coloured flowers; compressed fruit and achenes enclosed in dry perianth.

Reproductive Cycle : July-September.

Habitat Ecology : Very common in wastelands, secondary forests.

Material Examined : EBH-1107, 11-08-2002.

Parts Used : Young leaves. Stem.

Folk Uses : Paste of the plant along with a pinch of 'Black pepper' applied as an emollient for quick **healing of cuts** and **wounds**.

Active Constituents : Cryptopleurine, 5α-stigmastan-3,6-dione, pomolic acid, oleanolic acid, lupeol & β-sitosterol, secophenanthroquinolizidine isolated from leaves and twigs.

Biological Activity : Hypotensive and spasmolytic activities and effect on CNS found +ve.

Uses in Literature : Known to be used as fodder, galactagogue; and for cuts, diarrhoea (vet), dress making, dysentery, eczema, stomach disorders, wounds, and yielding a fibre used for cordage and fishing lines (**Ambasta, 1986; Jain, 1991; Kharkonger & Joseph, 1981; Manandhar 1987a, 1989a; Rama Rao & Henry, 1996; Rao, 1996; Uniyal & Chauhan, 1973**).

Bombax ceiba L.+ (Pl. 8B).

Syn.: *B. heptaphylla* Cav. Diss., *B. heptaphyllum* Roxb., *B. malabaricum* DC, *Salmalia malabarica* (DC) Schott.& Endl. Melet., *Gossampinus malabarica* (DC) Merrill., *G. rubra* Ham.

Family: Bombacaceae.

Vern.: Simbal.

English, Sanskrit & Regional Names

Eng. : Cotton Tree, Red Silk Cotton Tree;
Sans. : Bahuvirya, Nirgandhapuspi;
B. : Roktosimul, Tula;
Bo. : Saer, Shembal;
G. : Ratoshemalo, Shimul;
H. : Kantishembal, Simal;
Mal. : Mullilavu, Unnamurrika;
Mar. : Kanterisamar, Shevari;
Oriya : Buro, Mochorso;
Tam. : Ilavam, Pougar, Selavagu;
Tel. : Kondaburaga, Pinnaburaga.

Distribution : India, Bhutan, S. China. S. E. Asia to 1400m.

Description : Tall trees with prickly stems; shortly-stalked, 5-7 glabrous leaflets, flowers whitish red and ovoid, downy capsules.

Reproductive Cycle : February-May.

Habitat Ecology : Common in lower valleys and Sub-Himalayan regions.

Material Examined : EBH-319, 13-04-1999.

Parts Used : Leaves. Flowers. Fruits. Gum.

Folk Uses : Unripe fruits consumed as **vegetable** and ripe ones to restore the **sexual potency**. Sweet infusion of its leaves taken as a substitute to **chyawanprash**. Decoction of flowers used as **febrifuge** and for curing **piles**. Cotton good for stuffing **pillows** and **mattresses** and gum as an **adhesive**. Wood considered good for **construction work**.

Active Constituents : Hydrolysis of gum yields arabinose, galactose, galacturonic acid, catechu-tannic acid and traces of rhamnose.

Biological Activity : Antiviral, hypoglycaemic and oxytocic activities found (+)ve.

Uses in Literature : Used in India as an alterative, aphrodisiac, astringent, contraceptive, cooling tonic, filling material for jackets, cushions and pillows, demulcent, diuretic, emetic, expectorant, fibre, firewood, fodder, haemostatic, laxative, restorative in excessive loss of vital fluid, spermator, stimulant, stomachic, styptic, tonic, tubercular, upholstery; and for abdominal pain, acne pimples, anaemia, asthma, atrophy, bacillary dysentery, bleeding gums, blisters, boils, bone fracture, calculus infections, canoes, catarrhal affections, chicken pox, cholera, chronic cystitis, coffins, cordage, cough, delivery (vet), diarrhoea, diarrhoea (vet), dislocation of bone (vet), doors, edible purposes (buds, seeds), emaciation, enlarged spleen, facilitating delivery, fever, fishing float, fistula, flatulence (vet), glandular swellings, gonorrhoea, haemorrhoids, hematuria, impotency, indigestion (vet), inflammation, intercostal neuralgia, leprosy and rinderpest diseases, leucorrhoea, lining of wells, mattresses, menorrhagia, night-pollution, oilseeds, packing fragile material, pain in legs, painful mensturation, pickling, planking, pleurisy, pneumonia, relieving giddiness, rheumatism, rutting (vet), scab boards, scabies, scorpion bite, shingles, skin disorders, sleeping bags, small pox, snake bite, sore in mouth, spermatorrhoea, spleen complaints, styptic, syphilis, tea boxes, tonsilitis, toothache, toys, ulceration of bladder, ulcers, urinary troubles, vitality, water troughs, and window-shutters (**Ambasta, 1986; Amminudin & Girach, 1991; Aswal, 1996; Bajpayee & Dixit, 1996; Banerjee, 1996; Brahmam & Saxena, 1990; Chauhan, 1999; Girach, 1992; Girach & Amminudin, 1995; Gogoi & Borthakur, 1991; Katewa & Arora, 1997; Lalramnghinglova, 1996; Manandhar 1989a, b; Mishra *et al.*, 1996b; Mukherjee & Namhata, 1990; Rama Rao & Henry, 1996; Rao, 1996; Saini 1996b; Shome *et al.*, 1996; Sikarwar, 1996; Sikarwar & Kaushik, 1992; Singh 1996b; Singh & Zaheer, 1996; Singh & Maheswari, 1992; Singh & Pandey, 1996; Siwakoti & Varma, 1996; Watt, 1972**).

Major Ayurvedic Formulations : Abhaya-dirista, Laghugangadhar-churna, Mocarasa-siddha-ghrita, Salmali-ghrita.

Trade : Mochavasa, Supari-ka-phool.

Bougainvillea spectabilis Willd. (Pl. 8C).

Family: Nyctaginaceae.
Vern.: Garden-Glory.
English & Regional Name

Eng.	:	Garden Glory;
H.	:	Bogenvillea.

Distribution : Often cultivated.

Description : A woody, straggling thorny shrub; with ovate to orbicular-ovate leaves; small, yellow coloured flowers; and red coloured showy bracts with number exceeding the flower.

Reproductive Cycle : January-June.

Habitat Ecology : Generally planted as ornamental.

Material Examined : EBH-157, 11-01-2000.

Parts Used : The Whole Plant.

Folk Uses : Grown as **ornamental.**

Active Constituents : Glycosides of isorhamnetin and quercetin isolated from fresh flowers bearing ivory-white bracts and pinitol isolated from leaves.

Biological Activity : Pinitol (0.01g/kg p.o.) exhibits significant hypoglycaemia in normal fasted mice, maximum being at the end of 2 hr. alloxan-induced diabetic mice on treatment with pinitol for 72 hr (5 doses).

Uses in Literature : Known to be used as ornamental; and for paintings (Mithila) and ulcers (**Gogoi & Borthakur, 1991; Jha & Basak, 1994; Lalramnghinglova, 1996**).

Broussonetia papyrifera Vent.[+] (Pl. 8D).

Family: Urticaceae.

Vern.: Karyun, Paper mulberry.

English & Regional Names

Eng. : Paper Mulberry;

H. : Jangli-toot;

Kan. : Kaagda.

Distribution : Punjab-Arunanchal Pradesh, Meghalaya to 1000m.

Description : A tree having 7.5-20cm long membranous, toothed leaves and red coloured fruits.

Reproductive Cycle : April-May.

Habitat Ecology : Hilly slopes, forests.

Material Examined : EBH-711, 14-04-2001.

Parts Used : Fruit. Leaves. Wood.

Folk Uses : Fruits **edible**. Wood suitable for **building works, furniture** and **handle of guns**, and leaves as **fodder**.

Uses in Literature : So far used as diaphoretic, diuretic, fibre for paper, fodder, tonic; and for furniture, hard boards, indigestion, lens paper, packing cases, paper handkerchiefs, ply wood, rearing silk worms, sport goods, and umbrella covers, (**Ambasta, 1986; Rao, 1996**).

Buddleia crispa Benth.* (Pl. 9A).

Syn.: *B. tibectica* Sm. var. *grandiflora* Marquand

Family: Loganiaceae.
Vern.: Butter fly bush.

Distribution : India. Pakistan to S. W. China: 1200-4000m.

Description : A large-sized shrub with lanceolate-ovate leaves; purple, fragrant flowers and tomentose capsule.

Reproductive Cycle : March-May.

Habitat Ecology : Shrubberies, open slopes; common in W. Himalaya.

Material Examined : EBH-533, 14-04-2001.

Parts Used : Buds. Dried Stem.

Folk Uses : Buds kept in houses for **fragrance** and its stem suitable for **walking sticks.**

Buddleia paniculata Wall.* (Pl. 9B).

Syn.: *B. nepalensis* Colla

Family: Loganiaceae.
Vern. Name: Shivra.

Distribution : India. W.Nepal-Bhutan: 900-2400m.

Description : Tall shrubs with ovate or oblong leaves; corolla tube pink having orange throat and tomentose capsule.

Reproductive Cycle : March-April.

Habitat Ecology : Forests, shrubberies; common.

Material Examined : EBH-292, 22-03-2003.

Parts Used : Root. Leaves. Flowers.

Folk Uses : Flower used in preparation of local wine, '**sur**'. Leaves given as **fodder**. Decoction of roots useful in **hernia** (1 teaspoonful, twice daily for 5 days).

Butea monosperma Kuntze (Pl. 9C).

Syn.: *B. frondosa* Koen. *ex* Roxb., *Erythrina monosperma* Lamk., *Plaso monosperma* Kuntze

Family: Fabaceae.

Vern.: Palah, Palash.

English, Sanskrit & Regional Names

Eng.	:	Bastard Teak, Bengal Kino, Flame of the Forest, Butea Gum;
Sans.	:	Bijasneha, Ksharashreshtha, Vatapotha;
B.	:	Palas, Polasi;
Bo.	:	Khakra, Palasa;
G.	:	Khakra, Palasso;
H.	:	Chichra, Kankrei, Palas;
Kan.	:	Muttuga, Muttuga mara, Thoras;
Mal.	:	Brahmavriksham, Palasinjamata, Palasin-samatha;
Mar.	:	Kakradha, Phalas;
Oriya	:	Kinjuko, Polaso;
Tam.	:	Kutumurukku;
Tel.	:	Tellamoduga, Vatapodhamu;
U.P.	:	Dhak.

Distribution : Common throughout the greater parts of India:1300-1400m.

Description : A tree with brown silky pubescence on shoots; pinnately 3-foliate leaves; bright, orange red coloured petals; and argenteo-canescent pod which narrowed suddenly into a stalk.

Reproductive Cycle : March-May.

Habitat Ecology : Common throughout dried areas, borders of agricultural fields.

Material Examined : EBH-712, 18-04-2003.

Parts Used : Leaves. Flowers. Seeds. Bark. Wood.

Folk Uses : Plant considered **sacred**. 10 ml decoction of its leaves, gum, flowers and bark (1:1:1:1) considered beneficial in **enlarged spleen**, **diabetes**, and **skin diseases** (twice daily for 15-20 days). Seeds **anthelmintic**.

Active Constituents : These are butrin, isobutrin-5, coreopsin, isocoreopsi, sulphurein, monospermoside and isomonospermoside (flower), tetramers of leucocyanidin (bark), a nitrogenous acidic compound along with palasonin; α-amyrin, β-sitosterol, its glucoside and sucrose and a new lactone-n-hemicosanoic acid-δ-lactone (seeds).

Biological Activity : Anthelmintic activity found positive due to palasonin.

Uses in Literature : Known earlier to be used as an abortifacient, alterative, anthelmintic (vet), aperient, aphrodisiac, astringent, beedi wrapper, colouring material, contraceptive, cooling, depurative, diabetic, diuretic, dye (flowers), febrifuge, fibre for paper making and cordage, fish poison, fodder, fuel, haemostatic, insecticidal, lac host, laxative, ornament (ear top), psychoactive plant, rubiefacient (vet), sacred (associated with Buddha's life; born from the body of Brahma a curse of Parvati turned Brahma into this tree), steriliser, stomachic (vet), stomatitis, tonic, vermifuge (vet); and for amenorrhoea, ascaris, asthma, bladder, blood pressure, boils, bone fracture, bowels, butea gums, catarrh, cold, colic, controlling parasite (vet), cough, delirium, diarrhoea, diuresis, dog bite, dysentery, dyspepsia, elephantiasis, expelling roundworms and ringworms, eye complaints, flatulence, gout, gum, haemorrhage affections, heart-burn, hydrocele, inflammation in the innerside of the mouth, jaundice, kidney trouble, leprosy, leucorrhoea, meal plates for serving food, menorrhagia, menstrual disorders, news print manufacture, night blindness, offerings, orchitis, paintings as colouration (Mithila), phthisis, piles, pimples, promoting menstrual flow, pterygium, relaxed sore throat, removing dhobie's itch, retains saline soils, septic and congested throat, skin diseases, snake bite, sore (vet), structural works, sun stroke, swellings, tanning, tubercular, tumours, ulcers, urine complaints, water scoops, well curbs, worshipping, wounds, and dummy corpse (**Ambasta, 1986; Bajpayee & Dixit, 1996; Banerjee, 1996; Bhandary *et al.*, 1996; Chopra *et al.*, 1956; Dey, 1980; Gill & Nyawuame, 1994; Girach & Aminuddin, 1995; Goel *et al.*, 1984; Jain, 1968, 1991, 1992, 1996; Jain *et al.*, 1994; Jha & Basak, 1994; Jha & Varma, 1996; Kothari & Moorthy, 1996; Kurian, 1995; Lalramnhinglova, 1996; Masih, 2003; Mohanty & Padhy, 1996; Rama Rao & Henry, 1996; Ranjan, 1996; Rao, 1996; Ravishankar & Henry, 1992; Reddy *et al.*, 1996; Saini, 1996b; Satapathy & Brahmam, 1996; Shah & Singh, 1990; Sharma, 1996; Sikarwar & Kaushik, 1992; Singh 1991; Singh & Prakash, 1994; Singh & Zaheer, 1996; Siwakoti & Varma, 1996; Srivastava *et al.*, 2003; Varma, 1997; Ved Prakash & Mehrotra, 1987; Watt, 1972**).

Major Ayurvedic Formulations : Ayaskrti, Janma-ghutti, Krimighna-churn, Krimi-kuthara-rasa, Krimimudgar-ras, Mahanarayana-taila, Nyagrodhadi-kwatha-churna, Palasabijadi-churna, Palsa-ksharaghrita, Plasa-puspasava, Plash-ghrit.

Trade : Traded in Indian markets as 'plashapapra' (seeds), 'dhaak phool' or 'tesh' (flowers) and 'bengal kino' (gum).

Buxus wallichiana Baill.+ (Pl. 9D).

Syn.: *B. sempervirens* L., *B. emarginata* Harb., *B. longifolia* Jacq.

Family: Buxaceae.

Vern.: Shamshad.

English, & Regional Names

Eng.	:	Box-Wood Tree;
H.	:	Chikri, Papri;
Kash.	:	Chikri;
P.	:	Shamshd.

Distribution : India. Afghanistan-C.Nepal: 1800-2700m.

Description : A small tree having yellow-green, strongly scented flowers and leaves which are shining above and pale-green beneath. Capsule ovoid, wrinkled.

Reproductive Cycle : April-November.

Habitat Ecology : Often gregarious: in shady rock ravines.

Material Examined : EBH-788, 12-11-2001.

Parts Used : Wood. Leaves.

Folk Uses : Leaves used as **cattle feed** and wood for making **agricultural implements.**

Active Constituents : Active properties of wood and leaves are due to buxine.

Uses in Literature : Used as diaphoretic, febrifuge, purgative; and for bone fracture, boxes, cabinet work, carving, combs, mathematical and other precision instruments, rheumatism, syphilis, toys, and turnery (**Ambasta, 1986; Jain, 1991; Watt, 1972**).

Caesalphinia decapetala (Roth.) Alston (Pl. 10A).

Syn.: *C. crista* Thunb., *C. sepiaria* Roxb., *C. ferox* Hassk., *C.japonica* Sieb & Zucc., *C. bonduc* (L.) Roxb., *C. bonducella* (L.) Flem., *Reinchardia decaptela* Roth., *Bianacaea scadens* Todero., *Guilandina bonduc* (L.)

Family : Fabaceae.

Vern.: Drogaadi, Durghara.

English, Sanskrit & Regional Names

Eng.	:	Bonduc-Nut, Fevernut, Indian Filbert, Niekar, Physic Nut, The Mysore Thorn;
Sans.	:	Kantakini, Putikaranja, Tinagachhika;
B.	:	Natakanja, Notukoranja;
Bo.	:	Sagurghota;
G.	:	Galga, Kachhki;
H.	:	Kanja, Kanthe Karanja;
Kan.	:	Gajagakayi;
Kash.	:	Kando;
Mal.	:	Kazanchik-kuru;
Mar.	:	Kanchaki, Karbath;
Oriya	:	Kokokolenja;
P.	:	Chillur;
Tam.	:	Kachukkai, Shriyinda;
Tel.	:	Gacha, Sukajambuka;
U.P.	:	Kuchaitha, Karjee.

Distribution : India. Pakistan - Bhutan. China. Japan. S. E. Asia to 2200m.

Description : Climbing prickly shrubs having pinnate leaves; yellow coloured flowers and glabrous, flat pods.

Reproductive Cycle : April - July.

Habitat Ecology : Stony open slopes, ravines.

Material Examined : EBH-1113, 24-07-2001.

Parts Used : Leaves. Seeds.

Folk Uses : Poultice of seeds useful against **burns** and that of leaves good for **swellings** and **hydrocele**.

Active Constituents : Seeds rich in a-caesalpin, β-caesalpin, γ- caesalpin, δ-caesalpin, aspartic acid (675.0 μg/g), arginine (224.0 μg/g), citrulline (3.64%), a new diterpine- î-caesalpin and bonducellin.

Biological Activity : Plant extr. (500 mg/kg p.o.) administered on days 1-8 of post-coitum exhibits significant contraceptive activity in female hamsters.

Uses in Literature : Reported earlier as an anthelmintic, antiperiodic (vet), antipyretic, febrifuge, insecticidal (vet), laxative, pulmonary tuberculor, soap nuts, symbolic (seeds strung to red silk worm as an abortion amulet collar), uterine tonic; and for asthma, boils, burns, cholera (vet), colic, cosmetics, diarrhoea (vet), discharges from ears, emmenagogue, epilepsy, essential oil, expelling intestinal worms (vet), fencing, fever (vet), filariasis. general debility, hydrocele, indigestion and liver trouble, inflammation of umblicis (vet), inflammed piles, inflammed swellings, lac, malarial fever, menstrual complaints, nervous complaints, ornamental bracelets, remittent fever, removal of placenta (vet), rheumatism (vet), scrofula, skin diseases, snake bite, tanning, ulcer, wounds, and for yellow dye obtained from flowers (**Ambasta, 1986**; **Gill & Nyawuame, 1994**; **Girach *et al.*, 1996, 1997**; **Henry *et al.*, 1996**; **Jain, 1991**; **Kothari & Moorthy, 1996**; **Ragupathy & Mahadevan, 1991**; **Rama Rao & Henry, 1996**; **Reddy *et al.*, 1996**; **Saini, 1996b**; **Singh & Maheswari, 1992**; **Singh *et al.*, 2003**; **Singh *et al.*, 1996b**; **Sinha & Pandey, 1996**; **Watt, 1972**).

Major Ayurvedic Formulations : Karaanjadi, Kubesaaksha-vati, Lata-karanja-bija-taila, Latakaranja-vati, Visamajvaraghni-vati, Vishamjvaraantaka-churna.

Trade : Katkaranja, Natkaranja, Saagargola.

Callicarpa macrophylla Vahl.+ (Pl. 10B).

Syn.: C. *incana* Roxb., C. *tomentosa* Koenig *ex* Vahl., C. roxburghii Wall.

Family: Verbenaceae.

Vern.: Dahimal, Dussah.

Regional Names

Ass.	:	Priyangu;
B.	:	Mathara, Mattranja;
G.	:	Lata-priyangu;
H.	:	Daya, Shiwali;
Kan.	:	Gandhapriyangu, Priyangu;
Mal.	:	Chimpompil;
Mar.	:	Gavhala, Priyangu,;
Oriya	:	Priyangu;
P.	:	Priyangu;
Tam.	:	Gnazalpoo, Gnazhal;
Tel.	:	Prenkanamu, Prkhanam;
U.P.	:	Daia.

Distribution : Throughout India; ascending to 2000m.

Description : Shrubs having branches and inflorescence densely clothed with wool-like tomentum; shortly stalked crenate leaves; pink coloured flowers and succulent, globose, white coloured droop.

Reproductive Cycle : July-September.

Habitat Ecology : Commonly occurring in shady localities of ravines and scrub jungles.

Material Examined : EBH-1093, 18-09-2001.

Parts Used : Inflorescence. Aerial Plant Parts. Seeds.

Folk Uses : A gargle of decoction of its seeds helpful in checking **mouth blisters**, and inflorescence one of the constituents of local wine, '**sur**'. Paste of the plant parts applied externally for healing **wounds**.

Active Constituents : Main ingredients of leaves are luteolin, apigenin, luteeolin-7-O-glucusonide and apigenin-7-O-glucuronide.

Uses in Literature : Used in India as an astringent, blood purifier, refrigerant; and for abdominal trouble, agricultural implements, boils, cuts, dysentery, emetic (vomiting), fever, headache, indigestion, rashes on tongue, rheumatism, skin diseases, stomach disorders, throat, ulcers, and wounds (**Ambasta, 1986; Jain, 1991; Manandhar, 1980a, b, 1987a, 1989b; Megoneitso & Rao, 1983; Negi & Gaur, 1991; Rama Rao & Henry, 1996; Singh, 1991; Watt, 1972**).

Major Ayurvedic Formulations : Ashwangandharista, Chandaasava, Dasamoolarista, Drakshrista, Eladi-churna, Kanaka-taila, Khadiradi-gutika (Mukharoga), Kumkumadi-taila, Nilikadya-taila, Priyangradi-churna, Priyanu-churna.

Trade : Flower buds traded as 'Phool priyangu' in the drug markets of North India.

Callistemon lanceolatus DC (Pl. 10C).

Syn.: *C. citrinus* Skeels, *C. cirinus* Satpf.

Family: Myrtaceae.

Vern.: Bottle brush.

English Name

Eng. : Bottle Brush Tree.

Distribution : Cultivated in Himalaya to 1500m.

Description : A small evergreen tree having glossy, lanceolate. leaves and 7-15 cm long flowers with very long red stamens.

Reproductive Cycle : March-April.

Habitat Ecology : Planted.

Material Examined : EBH-328, 11-04-2001.

Parts Used : Flower spikes.

Folk Uses : Planted as an **ornamental**.

Calotropis procera R. Br.⁺ (Pl. 10D).

Syn.: *C. hamiltonii* Wall., *C. wallichii* Wight, *C. heterophylla* Wall., *Asclepias procera* Willd.

Family: Asclepiadaceae.

Vern.: Ak.

Sanskrit & Regional Names

Sans. : Adyitapushpika, Alarka, Arka, Bimbora, Dirghpuspa, Ganarupaka, Kashthila, Mandara, Shamblau, Sharakarapushpa, Shitakarka, Shuklaphala, Vasuka, Vritamallika;

Ass. : Ag, Ak, Akada, Madar, Safedak;

Mar. : Mandara;

Oriya : Orko;

P. : Ak, Shakarallighal, Shakarulusar;

Tam. : Vellerukku.

Distribution : Throughout India. Ascending to 1350m.

Description : An erect, white-downy shrubs with milky latex, pink coloured purple spotted flowers; and thick, wrinkled, follicles covered with woolly pubescence.

Reproductive Cycle : February-May.

Habitat Ecology : Wastelands, river-beds, roadsides, forest clearings; common.

Material Examined : EBH-1198, 14-04-2002.

Parts Used : Flowers. Leaves. Roots. Seeds. Latex.

Folk Uses : Flowers chewed for **flatulence** and **migraine**. Seed oil massaged for curing **rheumatoid arthritis**. Latex considered useful for **toothache**, **ringworm** and **piles**. Decoction of root and leaves beneficial in **rheumatoid pains**.

Active Constituents : Active principles isolated are uzarigenin, syriogenin, proceroside, calotropin, calotoxin, uscharin, uscharidin, α- and β-amyrin, β-sitosterol, calotropagenin; uzarigenin and syriogenin (latex), a new triterpene calotropenyl acetate, cyanidin-3-rhamnoglucoside (flowers), quercitin-3-rutinoside [roots (1.66), stem (5.01), flowers (7.63) and latex (9.74%)], benzoyllineolone, benzoylisolineolone and a new compound voruscharin (root-bark).

Biological Activity : Calotropin exhibits digitalis like action on the heart.

Uses in Literature : Reported to be used as an abortifacient, alterative, antidote in scorpion sting, antiseptic, bow strings, camel fodder, cardiac, diaphoretic, emetic, gun powder charcoal, infanticide, laxative, magico-therapeutic (put on forehead of women in labour to achieve easy delivery), masticatory, psychoactive plant, symbolic (abode of god and taboo to offer to godesses), tonic, tooth brush; and for asthma, boils, bronchitis, bronchitis (vet), cholera, chronic rheumatism, cold, constitutional syphilis, corn, cuts (vet), dental caries, diarrhoea, dropsy, dysentery, eczema, elephantiasis, epilepsy, expelling worms, fever, fibre, gastric troubles of children, guinea worm, idiopathic ulceration, insect bite, leprosy, malarial fever, menstrual disorders, mercurial cachexia, mouth diseases (vet), neck injury (vet), pain, piles, pyrexia (vet), rheumatoid arthritis, ringworm, scorpion bite, skin diseases (vet), small pox, sores, spleen complaints, stomach disorders, swellings, syphilis, toothache, whooping cough, and wounds (vet), **(Ambasta, 1986; Bhattacharya, 1996; Datt, 1996; Huidrom, 1996; Jain, 1991, 1996; Jain *et al.*, 1994; Kapur *et al.*, 1996; Khanna *et al.*, 1996a; Maheshwari & Singh, 1984; Masih, 2003; Mishra *et al.*, 1997; Pandey & Verma, 2003; Saini, 1996a, b, Sharma & Borthakur, 2003; Sharma & Sood, 1997; Shome *et al.*, 1996; Sikarwar, 1996; Singh & Prakesh, 1996; Singh & Maheswari, 1983; Singh & Prakash, 1996; Srivastava *et al.*, 2003; Watt, 1972)**.

Major Ayurvedic Formulations : Arkadi-churna, Arka-lavana, Arkaolakanda, Arka-taila, Arkeswar, Dhanvantara-ghrita, Ekangavira, Mahavata-vidhavamba, Mahavisagarbha-taila, Pravala-pancamrta, Sutikabharana.

Camellia sinensis (L.) Kuntze (Pl. 11A, B).

Syn.: *C. theifera* Griff., *C. thea* Link., *Thea sinensis* L., *T. assamica* Masters

Family: Theaceae.

Vern.: Cha, Chai.

English & Regional Names

Eng.	:	Assam Tea, China Tea, Indian Tea, Tea;
Ass.	:	Hilkat;
B.	:	Chha
Mal.	:	Chaya;
Mar.	:	Chaha;
Oriya	:	Cha;
P.	:	Cha;
Tam.	:	Karupputeyilai, Pachai Heyilai, Teyilai;
Tel.	:	Nallateyaku, Teyaku, Tiyaku.

Distribution : Extensively cultivated in Darjeeling, Assam, in and around Palampur in W. Himalaya to 2300m.

Description : An evergreen shrub with white flowers and light green coloured pods.

Reproductive Cycle : June-September.

Habitat Ecology : Cultivated.

Material Examined : EBH-1062, 15-04-2002.

Parts Used : Leaves.

Folk Uses : Tea leaves chewed for removing **bad smell** from mouth, and its beverage considered **refreshing, stimulant**, and

anti-hypertensive. Also, these are used as a **colouring material**.

Active Constituents : Leaves possess a new proanthocyanidin gallate epigallocatechin (4β>8)-3-O-galloylepicatechin (I) along with two hydrolyzable tannine-1,4,6-tri-O-galloyl-β-D-glucose and 1-O-galloyol-4,6-(-) hexahydroxy-diphenoyl-β-D-glucose, procya-nidins B-2, B-4, C-1, three new red pigments-epitheaflagallin, epitheaflagallin-3-O-gallate and theaflagallin together with theaflavin and its 3-O-gallate, 3′-O-gallate and 3,3′-di-O-gallate; a growth inhibitor identified as ABA; 2 new flavonoid glycosides-camellianin A and and camellianin B along with epigenin and triacontanol.

Biological Activity : Anticoagulant and antithrombic properties of green tea polysaccharides demonstrated *in vitro* and *in vivo* in mice, rabbit and guinea pig.

Uses in Literature : Used in India as an abortifacient, antihypertensive, antiinflammatory, astringent, diuretic, folk beverage, nervine tonic, psychoactive plant, sedative, stimulant; and for conjuctivitis, cuts and injuries, diarrhoea, dysentery, removing bad smell from mouth, skin burns, and as colouring material in Mithila paintings (**Ambasta, 1986; Jain, 1991; Jain *et al.*, 1994; Jha & Basak, 1994; Lalramnghinglova, 1996; Shome *et al.*, 1996**).

Canna indica L.⁺ (Pl. 11C).

Syn.: *C. chinensis* Willd., *C. indica* var. *orientalis* Rosc., *C. orientalis* Rosc.

Family: Cannaceae.

Vern.: Aeran, Hardarshan, Sudarshan.

English, Sanskrit & Regional Names

Eng. : Indian Bead, Indian Reed, Indian Shot;

Sans. : Devkili, Kamkshi, Krishnamantara, Sarvajaya, Shilarambha, Vanakadali;

B. : Kamakshi, Lalsarbojaya, Sarbajaya;

G. : Akalabera;

H. : Sabbajaya, Sarvajjaya;

Kan.	:	Hudigana, Kalahu;
Mal.	:	Kattuvala;
Mar.	:	Devakali;
Oriya	:	Sorobojoya;
P.	:	Hakik;
Tam.	:	Kalvalai, Kalvalaimani, Kundimani, Puvalai, Siramalai;
Tel.	:	Guruginja.

Distribution : Widely planted in India.

Description : Erect undershrubs with a creeping, subterranean tuberous rhizome and fleshy thick roots. Leaves sub-sessile, oblong, acuminate. Corolla bright red. Capsule red.

Reproductive Cycle : January-February.

Habitat Ecology : Oftenly cultivated in gardens as an ornamental.

Material Examined : EBH-1117, 11-01-2000.

Parts Used : Flower. Leaves.

Folk Uses : Flowers used in **worshipping**, and poultice of leaves heated in oil used against **stomachache** and **internal swellings.**

Uses in Literature : So far used in India as an antidote to poisonous fodder (vet), demulcent, diaphoretic, diuretic, febrifuge; and for dropsy, dyeing, headache, meal plates, necklaces and ornaments, rope twine, sack clothes, swelling of abdomen in cattles, worshipping, wrapping up goods, edible purposes (rhizomes), and thatching houses (**Ambasta, 1986; Bajpayee & Dixit, 1996; Jain, 1991; Lalramnghinglova, 1996; Rama Rao & Henry, 1996; Watt, 1972**).

Major Ayurvedic Formulations : Aragvahadi-kvatha-churna.

Carica papaya L.+ (Pl. 11D).

Family: Caricaceae.

Vern.: Papaya, Papita.

English, Sanskrit & Regional Names

Eng.	:	Melon Tree, Papaw, Papaya, Paw-paw;
Sans.	:	Chirbita, Eranda, Naikadala;
B.	:	Papai;
Bo.	:	Chibhado, Katha, Paputa;
G.	:	Chibda, Eranda Kakdi, Thadchibadi;
H.	:	Anda Kharbuja, Papita;
Mal.	:	Kappalam, Nalikadala;
Mar.	:	Papaya;
Oriya	:	Omrytobhondo, Popoya;
P.	:	Popai;
Tam.	:	Pappli, Parangiyananakki;
Tel.	:	Madananaba, Madhurnakamu.

Distribution : Cultivated mostly throughout India.

Description : A rapidly growing sub-herbaceous, branchless tree. Trunk succulent with milky sap. Leaves palmately lobed and flowers creamy yellow.

Reproductive Cycle : Greater part of the year.

Habitat Ecology : Cultivated; often met as an escape.

Material Examined : EBH-729, 26-11-2000.

Parts Used : Milky Latex. Fruit. Seeds.

Folk Uses : Unripe fruits cooked as a **vegetable** and ripe ones **edible** and good for chronic **liver ailments, expelling worms**, and as **cardiac tonic**. Also, paste of young fruits considered useful against various **skin diseases**. 3-5g powdered seeds prescribed with milk, once daily till cure for curing **diabetes**. Latex and seed oil applied in the treatment of **eczema** and **leprosy**. Also, seeds used as an **adulterant** for black pepper.

Active Constituents : These are hentricontane, b-sitosterol, benzoylthiourea, carpaine (seeds), myristoleic, palmitoleic arachidic acids, pectin and free sugars (fruits).

Biological Activity: Carpaine exhibits antitumour activity *in vitro* against mouse lymphoid leukamia (1-210), lymphocytic leukamia (P-338) and *Ehlrich ascites* tumour cells, and also as anti-microbial against gram (+)ve and gram (-)ve bacteria.

Uses in Literature : Reported earlier to be used as an abortifacient, adulterant, anthelmintic, antidote for cancer, blood anticoagulant, carminative, diuretic, face creams, febrifuge, galactagogue, hair shampoos, purgative, stomachic, source of vitamins and papain, substitute of soap, tonic, vegetable; and for amoebic dysentery, asthma, beri-beri, bleeding piles, boils, bone fracture, candies, cardiac diseases, chest problems, cooling, corns and warts, removing freckles and other blemishes, cough, cuts and wounds, dental caries, diabetes, diptheria, dressing of foul wounds, dysentery, dyspepsia, eczema, emmenagogue, excessive bleeding of the kidneys, expelling intestinal worms, flatulence, flavouring meat, gastric, gonorrhoea, haemoptysis, heart depressions, hoarseness, inflammation, itching, jaundice, kidney stone, malaria, marmalades, pain in gums, pickling, promoting menstrual flow, reducing elephantoid growth, rheumatism, ringworm, skin diseases, soft drinks, spleenic and hepatic enlargements, stomach and bowel disease, stomach and duedonal ulcers, syphilis, tonsils, tooth and gumache, tuberculosis, ulcers, urinary troubles, warts, and wounds, (**Ambasta, 1986; Borthakur, 1993; Dagar & Dagar, 1996; Ghosh *et al.*, 1996; Gill *et al.*, 1993, 1997; Gogoi *et al.*, 2003; Hosagoudar & Henry, 1996a; Jain, 1991; Jain *et al.*, 1997; Kulkarni & Kumbhojkar, 1993; Kurian, 1995; Lalramnghinglova, 1996; Rama Rao &**

Henry, 1996; Rao *et al.*, 1996; Saini, 1996b; Singh & Zaheer, 1996; Siwakoti & Varma, 1996; Watt, 1972).

Carissa congesta Wight+ (Pl. 12A).

Syn.: *C. carandus* L.

Family: Apocynaceae.

Vern.: Garna, Garnae.

English, Sanskrit & Regional Names

Eng.	:	Bengal Currents;
Sans.	:	Bokkarambuka, Jalipushpa, Krishna Pakaphala, Vanalaya;
B.	:	Karamcha, Kurumia;
Bo.	:	Hartundi, Karinda;
G.	:	Karamadan, Karamarda;
H.	:	Karaunda, Karrona;
Kan.	:	Karekayi;
Mal.	:	Kalavu, Karakka;
Mar.	:	Boronda, Haradundi, Karavandi;
Oriya	:	Khyirokoli, Korondo;
Tam.	:	Kalakkay, Perungala;
Tel.	:	Paddavaka, Peddakalive, Vaka.

Distribution : Throughout India, ascending to 1750m.

Description : An erect spinous shrub. Leaves leathery, ovate. Flowers white. Fruit an ovoid berry, green at first, then red, nearly black when ripe.

Reproductive Cycle : April-July.

Habitat Ecology : Open slopes.

Material Examined : EBH-469, 25-05-2002.

Parts Used : Fruit. Stem. Wood.

Folk Uses : Leaves used as sheep-goat **fodder**. Fruits **edible** and considered good for **anaemic**, and also as **blood**

purifier. Wood used for making **combs**, and its ash applied to cure **pain in joints**.

Active Constituents : A new triterpine carissol isolated from fruits along with glucose, galactose, serine, glutamine, alamine, valine, phenylalanine, glycine and carissic acid.

Biological Activity : Hypotensive effect and cardiotonic activity (+) ve.

Uses in Literature : So far used in India as an anthelmintic, antiscorbutic, astringent, expectorant, laxative, antidote for snake bite, stomachic; and for anasarca, bite of rabbid dog, blood dysentery (vet), checking bleeding of internal organs, chutney, combs, convulsions, cooling in bilious complaints, curry, diarrhoea (vet), dropsy, dyeing and tanning, edible purposes, epilepsy, fencing, fever, indigestion, itching, jellies, pickling, madness, postnatal care, puddings, remittent fevers, rheumatism, scurvy, skin infections, sore (vet), spoons, stimulating appetite, strengthening and giving tone to the stomach, tarts and puddings, tussar silk worms and wounds (vet) (**Ambasta, 1986; Bhandary *et al.*, 1996; Jain, 1991; Khanna *et al.*, 1996a; Kurian, 1995; Sahu, 1996; Upadhye & Kumbhojkar, 1996; Watt, 1972**).

Major Ayurvedic Formulations : Kusthahara.

Carpinus viminea Wall. *ex* Lindl. (Pl. 12B).

Family: Corylaceae.

Vern.: Chaeyr.

English Name

Eng. : Indian Hornbean.

Distribution : Temperate Himalaya. S. E. Asia 1500-2700m.

Description : A small tree having glabrous, pendulous drooping branches and leaves with a long tail like tip. Flowers solitary. Male catkins appear before the leaves.

Reproductive Cycle : April-August.

Habitat Ecology : Forests, shady ravines, water courses.

Material Examined : EBH-1231, 11-07-2001.

Parts Used : Leaves. Stem. Wood.

Folk Uses : Leaves relished by **cattles**. Wood used for **furniture**.

Uses in Literature : Used for carpentry work, and also as galactagogue (vet) (**Watt, 1972**).

Carya illenoesis (Wang.) Koch. (Pl. 12C).

Syn.: *C. pecan* Engl & Graebn.

Family: Juglandaceae.
Vern.: Kagzi Akhrot.
English Names
Eng. : Hickory Nut, Pecan Nut Tree.

Distribution : Native of South-East USA; introduced in India.

Description : Tree.

Reproductive Cycle : May-October.

Habitat Ecology : Cultivated, often met with an escape.

Material Examined : EBH-455, 12-07-2001.

Parts Used : Roots. Fruit. Stem-bark.

Folk Uses : Fruits **edible**, also 1-2 fruits recommended for children in **bed-wetting**. A **black dye** extracted from its rind. Bark useful for **scourring** teeth, and it is boiled in mustard oil along with *Terminalia chebula* and applied externally for imparting **black dye** to hair. Paste of leaves applied on affected parts for curing **skin diseases**.

Active Constituents : Quercitin-3-glucoside, 3-galactoside, 3- rhamnoside and 3-arabinoside isolated from leaves.

Caryopteris wallichiana Schauer[+] (Pl. 12D).

Syn.: *C. odorata* (D. Don) Robinson, *Volkameria odorata* Roxb., *Clerodendro odoratum* Don.

Family : Verbenaceae.
Vern.: Bharmeda, Chingari.

Distribution : Sub-Tropical Himalaya. 300-2100m.

Description : Shrubs having shortly-stalked, toothed leaves with white, fragrant flowers tinged with blue, dark blue fruits.

Reproductive Cycle : April-August.

Habitat Ecology : Open slopes, foot hills.

Material Examined : EBH-252, 11-07-2001.

Parts Used : Stem Wood.

Folk Uses : Plant considered **sacred** and planted in courtyard of houses. Stem used for making **walking sticks** and branches used as **fuel wood**.

Uses in Literature : Reported in India for walking sticks (**Ambasta, 1986**).

Cassia fistula L.[+] (Pl. 13A).

Syn.: *C. rhombifolia* Roxb., *Cathartocarpus fistula* Pers.

Family: Fabaceae.
Vern.: Aahali, Amaltas.

English, Sanskrit & Regional Names

Eng. : Drum Stick, Golden Shower, Indian Laburnum, Purging Cassia;
Sans. : Aragvadha, Chaturangula, Kritamalluka;
Ass. : Soanru;
B. : Amultas, Sonali;
G. : Balla, Garmala;
H. : Amaltas, Bandarlauri;
Kan. : Kakke;
Mal. : Saturangulam, Svarnnakam;
Mar. : Bhawabaya, Chimkani;
Oriya : Soturongulo, Sunari;
P. : Alash, Karangal;
Tam. : Arogaram, Irayavinnadagam;
Tel. : Aragvadhamu, Sampakamu.

Distribution : Common in Himalaya upto 1400m.

Description : A small tree with 8-16 leaflets, yellow flowers and cylindric pod.

Reproductive Cycle : March-July.

Habitat Ecology : Forests, shrubberies, open hill sides.

Material Examined : EBH-651, 07-04-2001.

Parts Used : Leaves. Seeds. Fruit. Stem Wood.

Folk Uses : Wood good for **agricultural implements**. Fruit extr. checks **constipation**. 5-10 ml decoction of seeds, given twice a day for 3-5 days as **diuretic**, **stomachic**, **anthelmintic**, **laxative** and for **promoting appetite**. Young leaves cooked and taken for **menstrual disorders**.

Active Constituents : Various constituents are fistucacidin (bark), leucoanthocyanidin- 5-4'-dihydroxy flavon-3, 4-diol (I) (sap wood), rhein, glucose, sucrose and fructose (fruit pulp), galactomannan (seed), rhein glucoside, fistulic acid (pods), vernolic (6.1), sterculic (2.0) and malvalic (1.5%) acids (seed oil), seven new biflavinoids β-sitosterol (I-VII) and two triflavonoids (VIII-IX), (-) epicatechin, procyanidin B-2, (-) epiafzelechin and its 3-O-glucoside (leaves).

Biological Activity : An aq. extr. of fruit pulp exhibits antibacterial activity.

Uses in Literature : So far used as an antibacterial (vet), antinarcotic to revive nervous system paralysed under the influence of liquor, antiseptic, blood purifier, cooling agent, edible (vet), emetic, febrifuge, insectiside (vet), laxative (vet), lenitive, mild cathartic, ornamental, psychoactive plant, purgative (vet), symbolic abode of god, tonic (roots), vegetable, vermicide; and for abdominal pain, adust bile, agricultural implements, amenorrhoea, anthrax (vet), antifertilty, asthma, biliousness, black water fever, blindness, body swellings, boils, burns, cancer, carts, cataract, chest infection, chest pain, cold (vet), colic, confectionary, cough, diabetes, diarrhoea (vet), dog bite, dropsical swellings, dyeing, dysentery, dyspepsia, dysuria, eczema, epilepsy, expelling hookworms, flatulence, flavouring coffee and tobacco, gout, heart diseases, house construction, intestinal worms(vet), jaundice, leprosy, liver complaints, malarial fever, paralysis, pimples, ploughs, pustules, rheumatic pains, rice pounders, ring worm, scorpion bite, skin diseases, snake bite, stomach disease (vet), swelling, swelling in throat (vet), syphilis, tanning, thoracic obstructions, tongue sore (vet), tool handles and wheels, toothache, urinary troubles, whooping cough, and wounds (vet) **(Ambasta, 1986; Balasubramaniam & Prasad, 1996; Bhandary *et al.*, 1996; Chauhan, 1999; Dixit & Pal, 1996; Gill & Nyawuame, 1994; Girach & Amminudin, 1995; Gogoi & Bortahkur, 1991; Goud & Pullaiah, 1996b; Jain, 1968, 1991, 1996; Jain *et al.*, 1994; Kothari & Moorthy, 1996; Kumar, 1996; Kurian, 1995; Lalramnghinglova, 1996; Manandhar, 1989a; Masih, 2003; Molla & Roy, 1996; Pullaiah & Kumar,**

1996; Rama Rao & Henry, 1996; Sahoo & Mudgal, 1995; Saini, 1996a; Sharma & Sood, 1997; Sharma, 1996; Sikarwar & Kaushik, 1992; Sikarwar, 1996; Singh, 1991; Siwakoti & Varma, 1996; Srivastava *et al.*, 2003; Tiwari & Tiwari, 1996; Tosh, 1996; Watt, 1972).

Major Ayurvedic Formulations : Aargavadhaadi-arishta, Aargavadhaadi-taila, Aragvadhadi-Kwath, Aragvadhadi-leha, Aragvadhadi-taila, Aragvadharistaz. Agnikumar-churn, Arishta, Kankayan-gutika, Leha and taila, Mahamanjisth-yadyarista, Mahamarichayadi-taila, Rasnadi-Kwatha-Yoga.

Trade : Cassia. Dry fruit and pulp is marketed.

Cassia laevigata Willd. (Pl. 13B).

Syn.: *C. floribunda* Cavan.

Family: Fabaceae.

Vern.: Raili.

Distribution : Throughout India: naturalised in Himalaya to 2200m.

Description : An erect shrub having ovate-elliptic long-pointed leaflets; yellow flowers and cylindric pods, with a thick stripe.

Reproductive Cycle : June-December.

Habitat Ecology : Scrub forests, waste grounds.

Material Examined : EBH-498, 21-07-2001.

Parts Used : Leaves. Flowers. Seeds.

Folk Uses : Leaf paste applied as an **antidote** in snake bite. Seed extr. given to check **throat infection**. For **indigestion**, **chutney** prepared by pounding its flowers with those of *Rhododendron arboreum,* black pepper and a pinch of rock salt considered beneficial.

Active Constituents : Main constituents are chrysophanol, physcion and quercetin-3,7-dirhmnopyranoside (seeds), rhamnetin-3-galactosyl(1→4) galacatopyranoside, rhamnetin-3-galactosyl (1→6) galactopyranoside and oleanolic-3-galactosyl (1→4)galactopyranoside along with quercetin, docosyl alcohol, carnaubyl alcohol, ceryl alcohal and octacosanol (flowers) and floribundone 1 and floribundone along with N1, N8-di-benzoyl spermidine & physicion (leaves).

Cedrella serrata Royle+ (Pl. 13C).

Syn.: *Toona serrata* (Royle) M. Roem.

Family: Meliaceae.

Vern.: Darlu.

English & Regional Names

Eng. : Hill Toon;

P. : Drawa;

U.P. : Darlu.

Distribution : W. Himalaya: 1000-2650m.

Description : Tree. leaves odd pinnate, flowers pinkish white, petal glabrous. Capsule ovoid, seeds winged only at the upper end.

Reproductive Cycle : June-October.

Habitat Ecology : In moist, shady places.

Material Examined : EBH-1014, 11-10-2002.

Parts Used : Leaves. Stem. Wood.

Folk Uses : Twigs used as **cattle feed**. Wood considered useful for **agricultural implements**, **furniture** and **building work**.

Uses in Literature : Used for bridges, furniture, packing cases and plywood manufacturing, poles, sieves, hoops, ulcers, and in chronic infantile dysentery (**Ambasta, 1986; Kapur, 1996; Watt, 1972**).

Cedrella toona Roxb. *ex* Rottl.+ (Pl. 13D).

Syn.: *T. ciliata* Roem,
C. febrifuga Blume, *C. hexandra* Wall., *C. longifolia* Wall.,
C. australis F. Muell.

Family: Meliaceae.

Vern.: Tooni.

English, Sanskrit & Regional Names

Eng.	:	Happy Tree, Indian Mahogany, Mountain Cedar, Red Cedar, Sandal;
Sans.	:	Kubheraka, Nandivriksha;
Ass.	:	Henduripoma, Jia;
B.	:	Lud, Tunna;
Bo.	:	Deodari, Tundu, Tupa;
H.	:	Tun, Tuna, Tunkajhar;
Mal.	:	Devabaram, Malarveppu;
Mar.	:	Deodari;
Oriya	:	Mona Limba;
P.	:	Bisru, Darab, Khanam, Tun;
Tam.	:	Madagirvembu, Sandavanembu;
Tel.	:	Gali, Nandi;

Distribution : Tropical Himalaya. Throughout the hilly districts of C. & S. India upto 1000m.

Description : Trees with even pinnate leaflets varying from 10-20. Panicles longer than leaves. Flowers pink coloured. Capsule ovoid.

Reproductive Cycle : May-June.

Habitat Ecology : Wastelands, roadsides, open slopes.

Material : EBH-343, 11-05-2001.

Parts Used : Leaves. Bark. Stem Wood.

Folk Uses : 5-10 ml decoction of bark given to check **fever**, **boils** and also for **regulating the menstruation** (twice a day for 5 days). Wood considered best suitable for **carpentry** and **building works**.

Active Constituents : Constituents isolated are cedrelone (heartwood), geranylgeranoil (trunk wood), cedrelone, 1, 2-dihydrocedrelone, bergapten and â-sitosterol, a new tetranorterpinoid-6-deoxy cedrelone, a coumarin-siderin (wood), two new B-secotetranortiterpenoids- toonacilin and 6-acetoxytoonacilin (bark).

Biological Activity : Activities found positive are anticancerous, antiamoebic, hypoglycaemic, spasmolytic and CNS depressant.

Uses in Literature : Used in India as antiperiodic, antipyretic, antiseptic, astringent, cattle feed, dye (yellow), emmenagogue, hoops for sieves, mild febrifuge, tonic; and for bridges, bronchitis, ceilings, chronic dysentery of infants, cigar cases, diarrhoea, doors, dyeing, dysentery, floors, furniture, gastric trouble, medicine chests and surgical instruement cases, plywood manufacture, tea boxes, ulcers, windows, wood carvings, and wounds (**Ambasta, 1986; Jain, 1991; Kapur & Singh, 1996; Siwakoti & Varma, 1996; Watt, 1972**).

Cedrus deodara (Roxb.) Louden⁺ (Pl. 14A).

Syn.: *C. libani* Barrel. var. *deodara* Hook. f., *Abies deodara* Lindl., *Pinus deodara* Roxb.

Family: Pinaceae.

Vern.: Deaar, Devdar.

English, Sanskrit & Regional Names

Eng. : Himalayan Cedar;

Sans. : Bhadravata, Indravriksha, Svehavriksha;

B.	:	Gunduguragi, Pitadaru;
G.	:	Vanseodeodar;
H.	:	Deodar, Kilankaper;
Kash.	:	Dadar, Dar, Deowar;
Mal.	:	Devataram;
Mar.	:	Devadaru;
P.	:	Geyar, Kairval, Keolikelmang;
Tam.	:	Tevadaram, Vandugolli;
Tel.	:	Devadaravu.

Distribution : Himalaya: 1800-3000m.

Description : Robust lofty tree with oblong cones and winged seeds which are broadly triangular, and with rounded sides.

Reproductive Cycle : June-July.

Habitat Ecology : Forming forests.

Material Examined : EBH-139, 26-07-2001.

Parts Used : Roots. Stem Wood. Resin.

Folk Uses : Deodar oil massaged for **repelling leeches** and to check **skin dryness**. Wood excellent for **furniture** and **building works**. Decoction of bark (5 ml) given twice daily for 3 days as **febrifuge**.

Active Constituents : Active constituents are cedeodarin, cedrin, dihydromyricetin, cedrinoside, taxifolin meso-secoisolariciresinol, cedrusinin (wood).

Biological Activity : Stem-bark extr. exhibits significant anti-inflammatory activity in rat.

Uses in Literature : Reported to be used as an alterative with turmeric and guggul, anthelmintic, carminative, diaphoretic, diuretic, stomachic, symbolic (abode of gods); and for ascites, bilious fevers, bridges, chronic skin diseases, construction, costiveness, diarrhoea, doors, dropsy, dysentery, intermittent fevers, flatulence, furniture, gonorrhoea, gout, headache, heart palpitations, inflammation, leprosy, material requirements, packing cases, paralysis, piles, pulmonary complaints, railway sleepers, rheumatism, ring worm (vet), shingles, skin diseases, syphilis, ulcers, urinary disorders, window frames; and also in indolent swellings (**Ambasta, 1986; Jain, 1991; Kapur & Singh, 1996; Chauhan, 1999; Singh, 1991; Watt, 1972**).

Major Ayurvedic Formulations : Chandra-prabha-vati, Devdarvadi-kwath, Khadi-rarishta, Laghu-vishgarbha-taila, Punarnava-man-door, Rasnaadi-kwath, Sudershan churna.

Trade : Deodar.

Celtis australis L. (Pl. 14B).

Syn.: *C. caucasia* Willd., *C. acata* Ham.

Family: Ulmaceae.

Vern.: Khirak.

English & Regional Names

- **Eng.** : European Nettle Tree, The Honey Berry Tree;
- **Kash.** : Brimji;
- **P.** : Khark, Khirk, Ku, Batkar;

Distribution : Temperate Himalaya, 1350-2650m.

Description : Trees. Leaves 2-nerved, alternate. Flowers pale yellow. Fruit drupe and long stalked.

Reproductive Cycle : April-October.

Habitat Ecology : Planted near villages.

Material Examined : EBH-634, 11-10-2001.

Parts Used : Fruit. Stem Wood.

Folk Uses : Fruits **edible**. Wood used for **agricultural implements** and leaves as **fodder**.

Active Constituents : Bark yields betulin, 3,3′-di-O-metyl-ellagic acid, gallic acid and querbrachilol.

Uses in Literature : Known to be used as cattle fodder, seeds for fatty oil, vegetable; and for agricultural implements, ammenorrhoea, blow instruments, coach whips, colic, cups, edible purposes (fruits), hay-forks, leprosy, oars, ram-rods, spoons, turnery, wagons, walking sticks, and whip handles (**Ambasta, 1986**; **Jain, 1991**; **Jain** ***et al.*****, 1991**; **Kapur** ***et al.*****, 1996**; **Maheswari** ***et al.*****, 1981**; **Watt, 1972**).

Trade : Nettle Wood.

Cestrum nocturnum L.[+] (Pl. 14C).

Family: Solanaceae.
Vern.: Rat-Ki-Rani.
English & Regional Names

Eng.	:	Lady of Night, Night Blooming Jasmine;
B.	:	Hasna,
H.	:	Rat-Ki-Rani.

Distribution : Throughout India; commonly cultivated upto 1500m.

Description : An evergreen shrub with ovate-lanceolate, long pointed, shining leaves. Flowers greenish yellow; scented at night.

Reproductive Cycle : August-October.

Habitat Ecology : Wastelands, border areas of agricultural fields; common.

Material Examined : EBH- 12-08-2001.

Parts Used : Flowers.

Folk Uses : Flowers considered **sacred** for offering to deities, and its powder (2-3g) given with cow's milk on empty stomach for **sinusitis**.

Active Constituents : Chief active principles of flowers are benzaldehyde, benzyl alcohol, benzyl acetate, methyl benzoate, phenyl acetaldehyde, phenyl ethyl alcohol, linalool and α-farnesene.

Biological Activity : Aerial parts spasmolytic, diuretic and hypotensive.

Uses in Literature : Recorded in India for epilepsy (**Uphof, 1968**).

Cinnamomum camphora (L.) Nees & Eberm+ (Pl. 14D).

Syn: *Laurus camphora* L.

Family: Lauraceae.

Vern.: Kapur.

English, Sanskrit & Regional Names

Eng. : Camphor, Camphor Cinnamomum, Official Camphor;
Mal. : Ghanarasam, Himavaluka;
Oriya : Chondrongyon, Romokorpuro;
Tam. : Indu, Kirasemiran, Pugalvisurandiram;
Tel. : Chandramu, Ghanasaramu, Karpuramu.

Distribution : Cultivated in Himalaya to 1500m.

Description : Trees with alternate leaves and buds enclosed in silky, caducous, imbricating scales. Flowers hairy within. Fruits globular, with thickened club-shaped base.

Reproductive Cycle : April-August.

Habitat Ecology : Planted.

Material Examined : EBH-1108, 15-08-2003.

Parts Used : Wood. Bark. Fruit.

Folk Uses : Fruit juice **antidiarrhoeal**. 10 ml decoction of inner bark (500g) mixed with 50g black pepper and 25g dried ginger powder, given thrice daily for 3-5 days to bring down **high fever**. Wood used for performing '**havan**'.

Active Constituents : Two new alkaloids acyclic monoterpenediols, 3,7-dimethylocta-1,7-dien-3,6-diol and 3,7-dimethylocta-1,5-dien-3,7-diol, a new lignan- (-) trans-2-(3,4-dimethoxybenzyl)-butrolactone (I)- and (-) trans 3-hydroxy-5,7-dimethoxy –3′,4′-methylenedioxyflavan along with dimethyl matairesinol and kusunokinin (heartwood) and sesquiterpene hydrocarpone-a-ylangene, b-santalene, δ-guanine, δ-cadinene, calamenene, calacorrone g-patchoulene and 1,6-dimethyl-4-isopropyl-7,8-dihydronaphthalene (camphor oil) isolated.

Uses in Literature : So far reported as acro-narcotic poison, analeptic in cardiac depressions, anthelmintic, antidote to strychnia, anti-leprotic, anti-phlegmatic, antiseptic, antispasmodic, aphrodisiac, carminative, counter irritant in muscular strains, diaphoretic, natural camphor, psychoactive plant, sedative, stimulant; and for affections of the genito-urinary system, angina pectoris, anodyne properties, asthma, bed-sores, camphor, cancer, curing wounds in hoofs, delirium, dysmenorrhoea, fever, gangrene, gout, hysteria, impotency, inflammation, inflammatory conditions of the eye, insanity, malaria, muscular strains, muttering, nervous headaches, nervous-ness, nymphomania, painful affections of the uterus, perfumery, rheumatic conditions, sleeplessness, small pox, spermatorrhoea, whooping-cough, and wounds (vet) (**Ambasta, 1986; Jain *et al.*, 1994; Mishra *et al.*, 1996b; Sharma & Sood, 1997; Vihari, 1995; Watt, 1972**).

Cinnamomum tamala Nees & Eberm+ (Pl. 15A).

Syn.: *C. albiflorum* Nees, *C. cassia* Don, *C. pauciflorum* var. *â* Tazia, *Persea tamala* Spreng., *Laurus cassia* Roxb., *L. albiflora* Wall., *L. tamala* Ham., *L. tazia* Ham., *L. somcaurium* Ham., *L. sailyana* Ham.

Family: Lauraceae.

Vern.: Gudpatraj, Mithpatra, Tejpatra.

English, Sanskrit & Regional Names

Eng.	:	Cassia cinnamomum, Indian Cassia Lignea;
Sans.	:	Dalavhaya, Patrakhya, Vasanavhaya;
Ass.	:	Dopati;
B.	:	Tejpat, Chota sinkoli;
Bo.	:	Darchini, Tamala;
G.	:	Tamalpatra;
H.	:	Dalchini, Tezpat;
Kan.	:	Thit-kya-bo;
Kash.	:	Dalchini-Pan, Tajpatra;
Mal.	:	Karuvapatta Patram;
Mar.	:	Dalchinitiki, Sambharpana;
Oriya	:	Tejapatra;
P.	:	Tajpater;
Tam.	:	Talishappaalttiri;
Tel.	:	Aakupatri, Tallishapatri.

Distribution : Tropical & Subtropical Himalaya 450-2100m.

Description : Trees with leathery leaves and white flowers in pubescent panicles. Drupe black when ripe.

Reproductive Cycle : June-August.

Habitat Ecology : Forests.

Material Examined : EBH-753, 03-07-2001.

Parts Used : Leaves.

Folk Uses : Leaves used for **flavouring** and its decoction (10-20 ml) alone or in equal combination with those of 'mulathi'

(*Glycyrrhiza glabra* L.), 'black pepper' (*Piper nigrum* L.), and 'soa' (*Anethum sowa* Roxb.) given once daily for 5 days for **cough, cold, bronchitis, asthma, tuberculosis, flatulence, menstrual disorders,** and as **carminative**. Also, its powder good for **scouring teeth.**

Active Constituents : Essential oil from leaves contains cinnamic aldehyde, phellandrene, eugenol, linalool, cinnamaldenyde, 3,4',5,7-tetrahydroxy-flavone, 3,3',4',5,7-pentahydroxy-flavone, kaempferol-3-O glucopyranoside, kaempferol-3-0-sophoroside, kaempferol-3,7-di-O-rhamno-pyranoside and quercetin-3-O-rutinoside.

Uses in Literature : So far used as an adulterant of true cinnamon, carminative, condiment (leaves), expectorant, ingredient of 'chywanprash', narcotic, 'paan' stimulant, stomachic, sudorific; and for affection of the nerves or heart, asthma, bites of serpents, bronchitis, calico printing, cold, colic, cough, debility of stomach, diarrhoea, dropsy, dyeing, enlargement of spleen, essential oil, fever, flatulence, gonorrhoea, headache, heart and throat complaints, indigestion, manufacturing of soap, nausea, oil in perfumery, pain in the womb, piles, poisoning by opium, retention of urine and catamenia, rheumatism, suppression of lochia after child birth, and scorpion sting (**Ambasta, 1986; Chauhan, 1999; Ghosh *et al.*, 1996; Jain, 1991; Lalramnghinglova, 1996; Mehrotra *et al.*, 1995; Negi & Pant, 1990; Rana *et al.*, 1996; Sharma & Sood, 1997; Singh, 1991, 1996b; Sinha, 1996; Watt, 1972**).

Major Ayurvedic Formulations : Amrtabindu, Amrtadhara, Ashwagandharista, Chandraprabhavati, Citrakadi-taila, Dasamoolarista, Karpura-sundaru-vatika, Karpuradi-vati, Karpura-rasa, Kasisadi-taila, Khadira-rishta, Lavangadi churna, Lodhrasava, Sitopaladi-churna, Sudarshan-Churna, Talisadi-Churna, Vajraka-taila, Vasasava, Yog-Raj-Guggulu.

Trade : Leaves collected and sold in the markets by the name of 'tejpat'.

Cissampelos pareira L.[+] (Pl. 15B).

Syn.: *Cissampelos pareira* L. var. *hirsuta* (Buch.-Ham. *ex* DC) Forman, *C. caapeba* L.,*C. convolvulacea* Willd., *C. hirsuta* Buch.-Ham. *ex* DC, *C. orbiculata* Buch. DC, *C. heradiflora* Wall., *C. obectata* Wall.,*C. discolor* Buch., *C. diversa* Miers, *C. elata* Miers, *C. grallatoria* Miers, *C. eriantha* Miers, *C. subpeltata* Thwaites, *C. delicatula* Miers, *C. nepalensis* Rhodes, *Cocculus orbiculatus* DC, *C. villous* Wall., *Menispermum orbiculatum* L.

Family: Menispermaceae.

Vern.: Bhatindu, Bhindu, Pakhtundu.

English, Sanskrit & Regional Names

Eng.	:	False Periera, Periera Brava, Velvet Leaf;
Sans.	:	Ambashthika, Ruchaishya, Vidhakarni;
Ass.	:	Saruthubukilata;
B.	:	Akanadi, Nimuka;
Bo.	:	Pahadmul, Pahadvel;
G.	:	Venivel;
H.	:	Akauadi, Dekhnirbissi;
Kan.	:	Padavali;
M.P.	:	Karepar;
Mal.	:	Kattuvalli, Patuvalli;
Mar.	:	Padavalli, Padavel;
Oriya	:	Okanibindhi;
P.	:	Bat, Batindu Path;
Tam.	:	Ap patta, Puttutiruppi, Vatattirupi;
Tel.	:	Pata.

Distribution	:	Usually in Tropical & Sub-Tropical parts of India upto 2000m.
Description	:	A climbing shrub with orbicular-ovate leaves; pale-green flowers and red drupe.
Reproductive Cycle	:	April-September.
Habitat Ecology	:	Rocky slopes, screes.
Material Examined	:	EBH-1001, 12-09-2002.

Parts Used : Leaves. Roots. Seeds.

Folk Uses : Extr. of leaves given for **spermatorrhoea** (1 teaspoonful twice daily for 3-5 days), and its paste with 'bantulsi' (*Ocimum basilicum* L.) heals **wounds.** Roots chewed for **inflammation of skin**, and decoction considered **abortifacient** and **stomachic**. Decoction of seeds useful for **phlegm** and **kidney stones**.

Active Constituents : Active principles are menismine, cissamine, parierine, 1-bebeerine, hayatinin, hayanin, d-isochendrodendrine, 4"-0-methyleurine (root bark), isoguinoline, pelosine, berbenine, reserpine, cissampetive, hyatin, hyatinin, quercitel and sterol (roots).

Biological Activity : CNS depressant activity confirmed.

Uses in Literature : Reported earlier to be used in different parts of India as an anthelmintic, antidote against poisons, antilithic, antiperiodic, astringent, cardiac, carminative, diuretic, expectorant, febrifuge, fibre yielding, fish poison, health tonic, psychoactive plant, purgative used in dyspepsia, sedative, stomachic, suppurative, symbolic (wearing in neck cures headache), tonic; and for abscess, toothache, antifertility, asthma, bite of dog, bladder, bodyache, boils, bone fracture, bowel complaints, bronchitis, burns, carbuncle, catarrhal affections of the bladder, chest pain, chronic liver disorders associated with alcoholism and dyspepsia, cold, colic, cough, country wine, cystitis, diabetes, diarrhoea, diphtheria, dropsy, dysentery, enlarged spleen, epilepsy, eye complaints, fermentation, fibre, gastric complaint, headache, heart burn, inflammation of kidney, insect bites, itching, jaundice, labour pain, leprosy, leucoderma, leucorrhoea, local drinks, malaria, pains, peptic ulcer, piles, pimples, prolapsed uterus, recovering energy after child birth, regularizing menstruation, rheumatism, scorpion bite, sedative action on the mucous membranes of the genito-urinary organs, sinuses, skin diseases, small pox, stomach ulcer, stone, swelling of gum, ulcer, unhealthy sores, urinary troubles, venereal complaints, weakness of stomach, and wounds (**Ambasta, 1986; Aminuddin & Girach, 1991; Borthakur, 1993, Brahmam & Saxena, 1990; Brahmam *et al.*, 1996; Chauhan, 1999; Datt, 1996; Dwarakan & Ansari, 1996; Gogoi & Boissya, 1984; Gogoi *et al.*, 2003; Hosagoudar & Henry, 1996 a, c; Islam, 1996; Jain, 1991, 1996; Jain *et al.*, 1991, 1994, 1997; Jha & Varma, 1996; Kumar *et al.*, 1987; Manandhar, 1980a, 1996; Masih, 2003; Painuli & Maheswari, 1996; Rama Rao & Henry, 1996; Rana *et al.*, 1996; Rao, 1996; Reddy *et al.*, 1996; Saini, 1996a; Satapathy *et al.*, 1996; Saxena *et al.*, 1981; Shah & Joshi, 1971; Sharma & Sood, 1997; Singh & Rao, 2003; Singh, 1991; Sinha, 1996; Siwakoti & Varma, 1996; Varma, 1997; Watt, 1972**).

Major Ayurvedic Formulations : Agnimukh-churn, Brhat-gangadhar-churna, Gangadhar-Churna, Kutajastak-kwath, Mahayograj-guggulu, Pathadi-kwath, Pippalaya-sav, Prada-rantaka-lauha, Pushayug-churn, Sarasvata-ghrita, Stanyasodhana-kasaya-churna.

Trade : Roots sold in the drug market as 'Patha'.

Citrus aurantium L. (Pl. 15C).

Syn.: *C. aurantium* var. *Bigardia* Hook. f.

Family: Rutaceae.

Vern.: Sangtra, Santra.

English, Sanskrit & Regional Names

Eng. : Bigarde Orange, China Orange, Orange Tree, Portugal Orange, Seville, Sour, Bitter,;

Sans. : Chakradhivasi, Tvagagandha;

B. : Kamlanembu, Narungi;

Bo. : Cantara , Naranghi;

G. : Narangi;

H. : Amritphal, Narunge, Sangatra;

Kan. : Hearlay;

Mal. : Madhuranaranna;

Mar.	:	Nargungasala, Sakulimba;
Oriya	:	Nagoronga, Narongi;
P.	:	Naranj, Santara;
Tam.	:	Narangam, Puttansini;
Tel.	:	Battavinarinja, Narinja.

Distribution : Hot valleys of Himalaya.

Description : A small tree. Leaves elliptic. Flowers pure white. Fruit globose.

Reproductive Cycle : October-March.

Habitat Ecology : Planted.

Material Examined : EBH-132, 11-01-2000.

Parts Used : Fruit. Rind.

Folk Uses : Fruits **edible** as an excellent source of **vitamin-C**. Burnt ash of fruit rind given with honey to check **cough**.

Active Constituents : These are 5-hydroxyauranetin, 6,7-dimethoxycoumarin, 5-o-demethylnobiletin, aurantin, hesperidin (peel), hesperidin, naringin and poncirin (fruits) a-pinene, b-pinene, limonene, a-phellandrene, decanal, citrohellal, a-terpineol, geranyl acetate, geraniol, and citral (oil)

Biological Activity : a, b-pinenes, limonene, octanol, citronellal inhibit seed germination and seedling growth in *Amaranthus retroflexus*. Naingen and naringenin selectively inhibit *in vitro* growth of cancer cells in human (breast, colon, liver).

Uses in Literature : Reported earlier to be used as allxiphasmic, antispasmodic, blood purifier, carminative, disinfectant, flavouring agent, psychoactive plant, refreshing drink, tincture, sedative in nervous and hysterical cases, stimulating, stomachic, tonic; and for acne, allaying thirst in fevers, bergamot oil, bilious affections, catarrh, colds, coughs, dyspepsia, edible purposes, eczema, flatulence, foetid ulcers, general debility, gout, improving the appetites, infusion, intestinal worms, lime oil, liqueurs and other drinks, marmalades, perfumery, preparation of confections, psoriasis, and vomiting (**Ambasta, 1986; Jain *et al.*, 1994; Srivastava *et al.* 2003; Watt 1972**).

Citrus grandis (L.) Osbeck+ (Pl. 15D).

Syn.: *C. maxima* (Burm.) Merr., *C. decumana* L., *Aurantium maximum* Burm.

Family: Rutaceae.

Vern.: Chakotra.

English, Sanskrit & Regional Names

Eng.	:	Forbidden Fruit, Paradies Apple, Pommel-Moose Shadock Pummelo;
Sans.	:	Madhukarkati;
B.	:	Batavinebu, Botornebu;
Bo.	:	Papanass;
G.	:	Obakotru;
H.	:	Batavinebu, Chakotra, Sadaphal;
Kan.	:	Chaoktre, Sakkota;
Mal.	:	Bambitinarakam;
Mar.	:	Panis;
Oriya	:	Batabilembu;
P.	:	Chakotra;
Tam.	:	Pambalimasu;
Tel.	:	Pampanpanasa.

Distribution : Cultivated.

Description : A tree with large, ovate-oblong leaflets. Flowers white. Fruit large-sized, pale globose with thick rind.

Reproductive Cycle : November-March.

Habitat Ecology : Planted near houses.

Material Examined : EBH-61, 19-02-2000.

Parts Used : Fruit. Rind.

Folk Uses : Fruit **edible**, and checks **vomiting**. Its rind taken along with a pinch of 'black pepper' for expelling **intestinal worms**.

Active Constituents : Main ingredients isolated are naringin, poncirin (peels), nerolidol (29.3), limonene (18.2), linalool (16.4) and farnesol (15.7%) (flower essential oil), 5-geranyloxy-7-methoxycoumarin, isoimperatorin, bergapten, and 5-geranyloxypsoralen (bergamottin) (fruits).

Uses in Literature : Used as nutritive, refrigerant, edible; and for cholera, convulsive coughs, constipation and inflammed nerves, deserts, diabetes, diarrhoea (vet), epilepsy, flatulent pains, heart disease, indigestion (vet), jams and marmalades, insanity, morbid cancer, nervous disorders, offerings, painful swellings, rheumatism, scurvy, ulcer, vomiting, waist and back pains, (**Ambasta, 1986; Borthakur, 1996; Gogoi & Borthakur, 1991; Kurian, 1995**).

Citrus jhambiri Lush.* (Pl. 16A).

Family: Rutaceae.
Vern.: Jhambirdi, Jhamirdi.
English Name
Eng. : Rough Lemon

Distribution : Hot valleys of Himalaya.

Description : Tree.

Reproductive Cycle : October-March.

Habitat Ecology : Oftenly planted near border of the fields.

Material Examined : EBH-58, 11-03-2000.

Parts Used : Fruit. Rind.

Folk Uses : Rind of fruits recommended for expelling **intestinal worms**, and the juice of its **edible** fruits in combination with black pepper and a pinch of salt useful in the **disorders of liver** and **gall bladder**. Sacred vermillion, **'kungu'** is made by mixing its fruit juice with 'turmeric' (*Cucuma longa* L.) and 'suhaga'.

Active Constituents : A new flavone-tangeretin along with β-sitosterol, hesperidin isolated from peel oil.

Citrus limon (L.) Burm. f.+ (Pl. 16B).

Syn.: *C. medica* var. *lomonum* L. (Risso) Hook. f., *C. limonum* Risso. *C. medica* L. var. *limon* L.

Family: Rutaceae.

Vern.: Khatta.

English & Regional Names

Eng.	:	Lemon;
B.	:	Bara Nebu, Goranebu;
G.	:	Motalimbu;
H.	:	Bara nibu, Jambhira, Paharinembu, Paharikagzi;
Kan.	:	Bijapura, Bijuri;
Mar.	:	Idalimbu, Thoralimbu;
Oriya	:	Periya Yelumichai;
Tel.	:	Bijapuram.

Distribution : Valley of the Himalayas.

Description : A small thorny tree having pinkish young foliage and flower buds. Leaves ovate; petioles marginate or winged. Fruits medium-sized, ovoid-yellow with thin rind.

Reproductive Cycle : November-February.

Habitat Ecology : Cultivated in gardens.

Material Examined : EBH-114, 12-01-2001.

Parts Used : Fruit. Leaves.

Folk Uses : Fruits used as a **pickle** and in preparation of **refreshing drinks**, and also for **culinary purposes**. Heated half portion of its fruit applied on affected part for relief due to **pain** from **twisted ankle**.

Active Constituents : Fruit peel yields apigenin, luteolin, chrysoseriol, quercitin, isorhamnetin, new flavonols- limocitrin and limocitrol- sinapic acid, p-coumaric acid, scopoletin, umbelliferone, hesperidin, citrusins A, B, C and D, coniferin syringin and dehydro-diconeferyl alcohol-4β-glucoside.

Biological Activity : Coniferin and syringin (1.0 mg/100 g, *i. v.*) found to raise blood pressure of SHR-Sp rats whereas dehydrodiconiferyl alcohol -4β-glucoside (*i. v.*) lowers it in spontaneously hypertensive rats.

Uses in Literature : Used earlier as an anthelmintic (vet), carminative, disinfectant, juice (vet), nervine tonic; and for burns, citric acid, colds, corns, coughs, culinary purposes and in the preparation of beverages, dandruff, febrifuge, flavouring liqueurs, headache, hypertrophy of spleen, indigestion, lemon oil (obtained as bye-products), lowering body heat, pectin, pickles, preventing dandruff in hairs, rheumatism, scabies, scurvy, stomachache, throat diseases, vitality to the body, and warts (**Ambasta, 1986; Dagar & Dagar, 1996; Kurian, 1995; Mishra *et al.*, 1996a; Shome *et al.*, 1996; Tirkey *et al.*, 2003a; Vishwanathan & Singh, 1996**).

Citrus medica L.+ (Pl. 16C).

Family: Rutaceae.

Vern.: Nimbu.

English, Sanskrit & Regional Names

Eng.	:	Adam's Apple, Cedrat, Citron, Melon Lime;
Sans.	:	Amlakshera, Bijaphalaka, Rochanaphala;
B.	:	Baranimbu, Nimbu;
Bo.	:	Bijapura, Limbu;
H.	:	Baranimbu, Nimbu;
Kan.	:	Madala, Mahaphala, Rusaka;
Mal.	:	Gilam, Matalanarakam;
Mar.	:	Mahalunga;
P.	:	Bajauri;
Tam.	:	Kogilachan, Kommattimadalai;
Tel.	:	Lungamu.

Distribution : Valley along the foothills of Himalaya.

Description : A spiny tree with pinkish white flowers. Fruit oblong or ovoid and usually mammilate at the apex.

Reproductive Cycle : April-September.

Habitat Ecology : Cultivated.

Material Examined : EBH-339, 13-04-2001.

Parts Used : Leaves. Fruit. Bark.

Folk Uses : Bark and leaves useful for **scouring teeth** and **against foul breath**. Fruit juice best for checking **jaundice**, **vomiting**, **flatulence** and **indigestion**, and oftenly pickled.

Active Constituents : Fruit affords a new triterpenoid-citrusin-isolated along with palmitic and succinic acids, citropten, b-sitosterol and daulosterol.

Biological Activity : Anthelmintic activity found positive.

Uses in Literature : So far reported as an anthelmintic, antidote to acro-narcotic and animal poisons, antiscorbutic, carminative, febrifuge, flavouring agent, refreshing drinks, stomachic, tumbagotain in hip joints, vermifuge (vet); and for abdominal pain, acnes, acute rheumatism, agricultural implements, biliousness or congestion the liver, bleeding gums, cardiac diseases, checking constipation, diarrhoea, dysentery, dyspepsia, essential oil, fatigue, feckles, foul breath, haemorrhage, hypertrophy of the spleen, improving general weakness and trembling of the limbs, indigestion, inflammatory affections, jaundice, kidney diseases, lemon syrup, malaria, marmalades, mental disorders, mucous and pus from mouth (vet), nausea, pickling, scabies, sciatica, scurvy, stimulating appetite, stomach ulcer, strengthening and giving tone to the stomach, urinary calculus, urinary diseases, veterinary-mucus and pus from mouth, and vomiting (**Ambasta, 1986; Borthakur & Sarma, 1996; Chowdhery, 1996; Jain, 1991; Kurian, 1995; Lalramnghinglova, 1996; Mohanty & Padhy, 1996; Negi & Gaur, 1991; Singh *et al.*, 1997; Tiwari & Tiwari, 1996; Watt, 1972**).

Major Ayurvedic Formulations : Gandhak-vati, Hingol-shodhan, Nimbuka-phala-rasa, Sankh-vati.

Clematis buchananiana DC+ (Pl. 16D).

Syn.: *C. buchananiana* Hook. f. *et* Thomas

Family: Ranunculaceae.

Vern.: Bald Khuja, Dhudki Bel.

English Name

Eng. : Old Man Beard.

Distribution : Throughout Temperate Himalaya 1800-3000m.

Description : Woody climbers. Leaves pinnate, hairy. Flowers pale-yellow and in panicles. Achenes hairy.

Reproductive Cycle : July-October.

Habitat Ecology : Common in forests; shrubberies.

Material Examined : EBH-1206, 12-08-2001.

Parts Used : Aerial Parts. Leaves. Flowers.

Folk Uses : Plant yields **fibres** for **cordage**. 5-10 ml extr. of flowers useful in curing **ulcer** (twice a day for 3 days). Paste of leaves applied as poultice on affected area for relief due to **pain from sprain**.

Biological Activity : Aerial parts antiviral and spasmolytic.

Uses in Literature : So far recorded for cuts, eczema, pain in stomach due to tumour, poison (vet), scabs, skin diseases, sores, stomach disorders, toothache, wounds; and as head wash (**Jain, 1991; Kapur, 1996; Manandhar, 1987a, 1996; Megoneitso & Rao, 1983; Rao & Jamir, 1982b**).

Clematis grata Wall.+ (Pl. 17A).

Family: Ranunculaceae.
Vern.: Chubad, Dhagli Bel.
Hindi Name
H. : Biliri, Ghantiali.

Distribution : Sub-Tropical & Temperate W. Himalaya; 700-2650m.

Description : A climbing shrub with pinnate leaves having three hairy leaflets. Flowers cream-coloured, fragnant. Achenes densely pubescent.

Reproductive Cycle : July-September.

Habitat Ecology : Shrubberies, cultivated areas.

Material Examined : EBH-1168, 14-08-2001.

Parts Used : Leaves. Flowers. Stem.

Folk Uses : Paste of leaves and flowers applied to cure **skin eruptions** and **itching**. Plant used for making **ropes**. A piece of stem is kept in dental caries for relief from **toothache.**

Active Constituents : A new saponin-clematoside S isolated from roots.

Uses in Literature : Used in India for spleen affections (**Kapur, 1996**).

Clerodendrum philippinum Schauer⁺ (Pl. 17B).

Syn.: *C. fragrans* Enum., *C. japonica* (Thunb.) Sweet, *Volkamannia japonica* Jacq.

Family: Verbenaceae.

Vern.: Gandharaj.

Hindi Name

H. : Madan Mohan.

Distribution : Sub-Tropical Himalaya. Grown in Himalaya as a medicinal plant: 2000m.

Description : Shrubs with opposite, broadly-ovate, coarsely toothed leaves; fragrant flowers and red calyx.

Reproductive Cycle : August-October.

Habitat Ecology : Cutivated.

Material Examined : EBH-1094, 11-08-2001.

Parts Used : Flowers. Root.

Folk Uses : Flowers **fragrant** and offered to **deities**. 3-5 ml decoction of roots, given once in the morning for 5 days to cure **cough**. Leaves **refreshing**.

Active Constituents : Active principle of roots are due to triacontane, α-amyrin, β-sitosterol and its glucoside, (24S)-ethylcholesta-5, 22, 25-trien-3β-ol, clerosterol and clerodolene.

Uses in Literature : Used earlier for bodyache and headache (**Jain, 1991**).

Colebrookia oppositifolia Sm.+ (Pl. 17C).

Syn.: *C. ternifolia* Roxb., *Elsholtzia oppositifolia* Poir.

Family: Lamiaceae.

Vern.: Basay, Dhudpa.

Regional Names

Bo. : Bahmani, Bhamini, Darai, Dasari, Dussarica;

H. : Binda, Bindu, Pansra;

Mar. : Darigopa, Merota;

Oriya : Bosili, Dangoti;

P. : Barmera, Briali, Dashane, Duss, Phisbekkar, Sampru, Shakardana, Sulai.

Distribution : Tropical & Sub-Tropical parts of India:ascending to 1350m.

Description : Shrubs. Leaves oblong-lanceolate, crenate. Flowers white in terminal panicles. Nutlet usually one with hairy tip.

Reproductive Cycle : April-November.

Habitat Ecology : Wasteplaces, roadsides, forest clearings, Sal forests scrub jungles, common.

Material Examined : EBH-36, 12-04-2000.

Parts Used : Leaves. Wood.

Folk Uses : Paste of leaves heals **wounds** and a gargle of its decoction checks **mouth ulcers**. Roots one of ingredients of a local wine '**sur**', and wood used as **fuel**.

Active Constituents : Active properties are due to flavones-5-hydroxy-4′, 6, 7, 8,-tetramethoxy flavone and hydroxy-4′, 6, 7,-trimethoxy flavone (aerial parts) and triacontane, triacontanol, β-sitosterol, palmitic, stearic and oleic acids and 5,6,7-trimethoxy flavone and 5,6,7-tetramethoxy flavone (roots).

Uses in Literature : Reported earlier for asthma, bruishes, burns, catarrah (vet), conjuctivitis (vet), cough, cuts, digestion, epilepsy, extracting worms from sores on the leg, eye complaints, giddinesss, gunpowder charcoal, hysteria, sorcery, sores, toothache, wounds; and as anthelmintic (**Ambasta, 1986; Jain, 1991; Jha & Varma, 1996; Kulkarni & Kumbhojkar, 1996; Rama Rao & Henry, 1996; Shreshta, 1985; Siwakoti & Varma, 1996; Varma, 1997; Watt, 1972**).

Cordia dichotoma Forst.[+] (PL. 17D).

Syn.: *C. obliqua* Willd., *C. latifolia* Roxb., *C. myxa* Thwaites, *C. lowriana* Brandis, *C. indica* Lamk.

Family: Cordiaceae.

Vern.: Lasura, Lasyada.

English, Sanskrit & Regional Names

Eng. : Sebastan Plum;
Sans. : Kshudrashleshmataka, Madhubhutadruma;
B. : Bahubara, Chhotabohnaari;
Bo. : Bargund;
G. : Rayagundo, Vadagundo;
H. : Chhotalasura, Lasura;
Kan. : Chikka Challe;
Mal. : Madaviriyasam, Vidi;
Mar. : Chokri, Goden;
Oriya : Domolo, Kophokoturi;
P. : Laswara;
Tam. : Naruvilli, Viriyam;
Tel. : Chinnanakkera, Nekkara.

Distribution : Throughout India: upto 650m.

Description : Trees. Leaves ovate or orbicular, tomentose young and glabrous at adult stage. Corolla white. Fruit a yellow berry.

Reproductive Cycle : April-July.

Habitat Ecology : Common in moist forests.

Material Examined : EBH-1055, 17-07-2001.

Parts Used : Inflorescence. Fruit.

Folk Uses : Fruits pickled, and cooked as vegetable for **stomach disorders, nocturnal emmisions, epilepsy, constipation** and as **demulcent**. Also, given to live stock mixed with feed for **expelling worms**.

Active Constituents : Taxifolin-3-rhamnoside isolated from seeds.

Uses in Literature : Used in India as an anthelmintic, antidysenteric, astringent, blood purifier, demulcent, diuretic, expectorant, fibre yielding, vegetable; and for agricultural implements, anasarca, boats, bone fracture (vet), burning of throat, chest diseases, cholera, chronic fever, cold, colic, constipation, cough, diseases of chest, dropsy, dysentery, dyspepsia, edible purposes (fruits), expelling retained placenta on child birth, expelling worms (vet), gonorrhoea, headache, jaundice, joint pains, lung disease, pickling, spleen diseases, stomachache, syphilis, ulcers, urinary diseases, urticaria, and well-curbs (**Ambasta, 1986; Dagar & Dagar, 1996; Girach *et al.*, 1997; Hosagoudar & Henry, 1996a; Jain, 1991; Kumar & Pullaiah, 2003; Lalramnghinglova, 1996; Negi & Gaur, 1991; Parmar & Kaushal, 1982; Rama Rao & Henry, 1996; Rao, 1996; Saini, 1996b; Singh, 1996b; Watt, 1972**).

Coriaria nepalensis Wall.* (Pl. 18A).

Family: Coriariaceae.

Vern.: Kachog.

English & Regional Names

Eng.	:	Mussoorie Berry;
H.	:	Makola, Masuri;
Kash.	:	Balel, Tadrelu.

Distribution : Himalaya. Pakistan-S.W. China. 1000-2700m.

Description : Glabrous shrubs with quadrangular branches and opposite leaves. Stamens coral-red and black, small fruits enclosed by succulent petals in fruit.

Reproductive Cycle : March-June.

Habitat Ecology : Forests, shrubberies; common.

Material Examined : EBH-287, 12-04-2002.

Parts Used : Fruits. Stem Wood.

Folk Uses : Fruits though **edible** but not relished. A **yellow dye** obtained from its wood.

Cornus macrophylla Wall.[+] (Pl. 18B).

Syn. : *C. brachpoda* C. A. Mey, *C. controversa* Homsley *ex* Prain, *Swida macrophylla* (Wall.) Soja'k.

Family : Cornaceae.

Vern. : Nang.

Regional Names

Ass.	:	Dieng-Phait;
H.	:	Kachur, Kandar, Kasis.

Distribution : Throughout Himalaya. 1350-2650m.

Description : Trees. Flowers pale to yellowish-white. Fruits globose, black, pubescent.

Reproductive Cycle : May-October.

Habitat Ecology : Common, riverine and shady moist places.

Material Examined : EBH-1101, 11-10-2002.

Parts Used : Leaves. Stem Wood.

Folk Uses : Wood used for valuable **furniture, agricultural implements, walking sticks, gun powder charcoal.** Leaves used as **fodder.**

Active Constituents : Betulinic acid (11.2%) and halleridone isolated from dried roots and bark.

Uses in Literature : Recorded earlier for bobbins, edible purposes (fruits, wood), gun powder charcoal, mallet heads, pulleys, shuttles; and as fodder (**Ambasta, 1986**; **Watt, 1972**).

Costus speciosus Smith (Pl. 18C).

Syn. : *C. arabicus* Jacq., *Hellenia grandiflora* Retz., *Banksia speciosa* Koenig *ex* Retz.

Family : Zingiberaceae.

Vern. Name : Keun, Sap-Challi.

Sanskrit & Regional Names

Sans. : Bramhatirtha, Kashmira, Kushtha, Kusthabeda, Padmakarna, Padmapatramula, Padmapunya, Paushkara, Pushkarajata, Pushkarshifa, Subandhu, Vira, Virapushkaravhaya, Vriksharuha;

Ass. : Jamlakhuti;

B. : Keu, Kust;

Bo. : Gudurichakanda, Kemuka;

G. : Pakaramula;

H. : Keu, Kust;

Kan.	:	Chengalvakoshtu;
Mal.	:	Anakkuva, Anappu, Channa, Channakuva, Kottam, Marujanna, Narikkurampu, Patimukam, Pushkarmulam;
Mar.	:	Penva, Pinnaga, Pushkarmula;
Oriya	:	Chittorokudho, Ghanapohora, Kudho;
Tam.	:	Kottam, Kudaram, Kugaimanjal, Kuravam, Kuttaiyidukki, Kuttam, Malaiva-Sambhu;
Tel.	:	Chengalvakostu.

Distribution : U.P.-Sikkim: upto 1500m.

Description : An undershrub. Leaves spirally arranged with funnel-shaped white flowers having contrasting large, bright red bracts. Capsule red.

Reproductive Cycle : June-August.

Habitat Ecology : Forests, shrubberies.

Material Examined : EBH-1216, 08-07-2003.

Parts Used : Rootstock.

Folk Uses : **Planted** in the courtyard of houses with a popular belief to keep away snakes. Rootstock considered **aphrodisiac**.

Active Constituents : Active ingredients of rhizome are tetradecyl-11-methyltridecanoate, 14-oxotricosanoic acid, 14-oxoheptacosanoic acid, 15-oxo-octacosanoil acid, methyl 3-(4-hydroxyphenyl)-2E-propeonate, tigo-genin, and diosgenin.

Biological Activity : Antifertility, antispasmodic, anti-inflammatory, antiarthritic, and CNS depressant activities of alkaloidal fraction from rhizome have been experimentally demonstrated.

Uses in Literature : Described earlier as an anthelmintic, antidote to snake venom, antispasmodic, aphrodisiac, astringent, cardiotonic, cooling, depressant on CNS, depurative, digestive in catarrhal fevers, diuretic, ear drops, edible, feeble astringent, galactagogue, hair lotion, hydrocholeratic, maggoticide, purgative, stimulant, stomachic, tonic, vegetable; and for anaemia, anasarca, asthma, backache, bite of rabid dog, blisters, body pain, boils, bronchitis, burning sensation on urination, burns, catarrhal fever, chicken pox, cholera, cold, constipation, cough, cuts, delirium, dropsy, dysentery, ear and eye complaints, gastric troubles, giddiness, headache, hiccough, increasing lactation after delivery, indigestion, infective hepatitis, inflammation, injuries (vet), internal injuries, jaundice, leprosy, local drinks, nausea, phthisis, piles, postnatal, respiratory troubles, rheumatism after delivery, scabies, snake bites, stomach disease, thirst, tuberculosis, vomiting tendency during labour, white leprosy, worms, wound (vet), wounds; and also to keep snakes away (**Ambasta, 1986; Aminuddin & Girach, 1993; Brahmam & Saxena, 1990; Brahman *et al.*, 1996; Chauhan, 1999; Chowdhery, 1996; Gogoi *et al.*, 2003; Goud & Pullaiah, 1996a, b; Islam, 1996; Jain & Saklani, 1992; Jain, 1968, 1991; Jain *et al.*, 1991; Jamir & Rao, 1990; Mahato & Mahato, 1996; Mandal & Basu, 1996; Masih, 2003; Mishra *et al.*, 1996b; Rama Rao & Henry, 1996; Rao, 1981; Reddy *et al.*, 1996; Saini, 1996b; Shah & Singh, 1990; Sharma & Sood, 1997; Singh *et al.*, 1996; Sinha, 1996; Siwakoti & Varma, 1996; Varma, 1997; Watt, 1972**).

Major Ayurvedic Formulations : Eladiganum employed for corticosteroidal operation in modern pharamceutical preparations.

Cotoneaster bacillaris Wall. *ex* Lindl.+ (Pl. 18D).

Syn.: *C. affinis* Lindl., *C. ignoyus* Klotz, *C. confusus* Klotz, *C. obtusa* Wall.

Family : Rosaceae.

Vern. : Simbli.

Regional Names

Kash.	:	Linu.
P.	:	Khariz, Reush, Sichu;
U.P.	:	Riuns.

Distribution : W. Himalaya. Afghanistan-Bhutan: 1800-3000m.

Description : Shrubs. Leaves ovate-lanceolate. Flowers white, hairy. Fruits black when ripe.

Reproductive Cycle : May-October.

Habitat Ecology : Shrubberies, open slopes; common.

Material Examined : EBH-407, 11-05-2002.

Parts Used : Twigs. Wood.

Folk.Uses : Twigs used as **cattle feed** and wood for making **walking sticks**.

Uses in Literature : Known earlier for agricultural implements, axe handles, golf clubs, jampan poles, tent pegs, turnery, umbrella handles, walking sticks; and as an astringent (stolons) (**Ambasta, 1986**; **Kapur & Nanda, 1996**; **Watt, 1972**).

Cotoneaster microphylla Wall. *ex* Lindl.[+] (PL. 19A).

Syn. : *C. emarginata* Hoffm., *C. buxifolia* Baker

Family : Rosaceae.

Vern. : Khilgu.

Regional Name

Kash. : Khariz Luni;

Distribution : Temperate Himalaya: 1350-3500m.

Description : Procumbent shrubs having upper surface of leaves shining and lower ones pubescent. Flowers white with bright red globose fruit..

Reproductive Cycle : May-June.

Habitat Ecology : Rocks, banks, alpine slopes; common.

Material Examined : EBH-121, 15-05-2000.

Parts Used : Stolons. Branches.

Folk Uses : Branches excellent for making **walking sticks** and stolons considered **astringent** and for **constipation**.

Active Constituents : Sorbitol and hydrocyanic acid isolated from young twigs.

Uses in Literature : So far used as an astringent; and for baskets, turnery and walking sticks (**Ambasta, 1986; Jain, 1991; Kapur & Nanda, 1996; Watt, 1972**).

Cryptolepis buchanani Roem. & Sch.+ (Pl. 19B).

Syn. : *C. reticulata* Wall., *Nerium reticulum* Roxb., *Echites reticulata* Roth., *E. cuspidata* Heyne

Family : Asclepiadaceae.
Vern. : Khumb, Thundra.
Sanskrit & Regional Names

Sans. : Kashmari;
Ass. : Gamar;
H. : Dudhi, Karanta;
Mal. : Kalipalvalu, Kattupalvalli;
Oriya : Gopikonioro, Maloti;
Tam. : Umi-Thekku;
Tel. : Adavipalatige, Gurgupaltige, Rokalliphala, Rytuphala.

Distribution : Throughout India, ascending to 1350m.

Description : Twining shrubs with milky latex; yellow-green flowers; oblong-ovate seeds which are contracted below the tip.

Reproductive Cycle : April-November.

Habitat Ecology : On trees, shrubs; wastelands, forests.

Material Examined : EBH-1078, 26-07-2002.

Parts Used : Fruit. Leaves. Latex.

Folk Uses : Fruits **edible**. Decoction of leaves used as a gargle for **mouth ulcers** and **throat infection**. Latex mixed with mustard oil applied for **urticaria**.

Active Constituents : Active properties of leaves are due to cardenolide-cryptosin and crytanoside B.

Uses in Literature : Recorded earlier as fibre yielding, galacatagogue (vet); and for abdominal pain, anasarca, ascites, babesiosis, bleeding, bodyache, bone fracture, cholera, cuts, dropsy, dysentery, fever, fissures on sole, fits, lactation, making ropes and cordage, paralysis, rheumatism, rickets in children, skin diseases, snake bites, sores, stomachache, syphilis, venereal diseases, and worshipping (twigs worshipped in the harvesting place to get good yield) (**Ambasta, 1986; Brahmam *et al.*, 1996; Henry *et al.*, 1996; Hosagoudar & Henry, 1996a, b, c; Jain, 1991; Rama Rao & Henry, 1996; Rana *et al.*, 1996; Rao, 1996; Ravishankar & Henry, 1992; Singh, 1996a; Singh & Prakash, 1996; Singh & Rao, 2003; Watt, 1972**).

Major Ayurvedic Formulations : Saarewaadi-kwaatha, Saari-vaadhyasava, Saari-waadi-avleha

Cucumis sativus L. (Pl. 19C).

Syn.: *C. muricatus* Wall., *C. hardwickii* Royle

Family: Cucurbitaceae.

Vern.: Kheera.

English, Sanskrit & Regional Names

Eng : Cucumber, Musk Melon, Sweet Melon;

Sans. : Bahpuspha, Kandalu, Kantakilaka, Kantakiphala, Pitapushpa, Sudhavasa, Sushitala, Trapukarkati, Trapusha, Tundilaphala;

B. : Khira, Sasa;

Bo. : Kakri, Kankari;

G.	:	Khira;
Mar.	:	Kakdi, Khira, Tavase;
Oriya	:	Kaknai;
P.	:	Khira, Khiyar;
Tam.	:	Piping Kay;
Tel.	:	Dozakaya.

Distribution : Cultivated throughout India.

Description : Climbing herbs with hirsute stem and simple tendrils. Leaves hispid on both sides. Flowers yellow. Fruit oblong, cylindric.

Reproductive Cycle : May-September.

Habitat Ecology : Grown locally.

Material Examined : EBH-1158, 27-08-2003.

Parts Used : Fruit. Seeds.

Folk Uses : Fruits **edible** and much used in **salads** and as **facial cosmetics**. Seeds ingredient of '**sund**', given to ladies to overcome **post delivery weakness**. Also, its decoction considered as **tonic**.

Active Constituents : Active properties are due to palmitic, stearic and oleic acids, their glycerides, sterols, squalene (fruits), sitosterol and stigmasta-7, 22,25-trien-3β-ol codisterol, 25(27)-dehydroporiferasterol, clero-sterol, avenasterol, isofucosterol, stigmasterol, campesterol, sitosterol, 25(27)-dehydro-chondrillasterol, 24β-ethyl-25(27) dehydro-lathosterol, 24æ-methyllathosterol, spinasterol and 22-dihydrospinasterol (seeds).

Uses in Literature : Used in India as diuretic, facial cosmetics, fruit salads, refrigerant, tonic; and for confectionary, fever, gastric (vet), headache, locust swarm on nagali, perfumery, pnuemonia, sunstroke, sweets, throat affections, urinary infections, and whooping cough (**Ambasta, 1986; Jain,**

1991; Kulkarni & Kumbhojkar, 1993, 1996; Negi & Gaur, 1991; Satapathy & Brahmam, 1996; Vihari, 1995; Watt, 1972).

Cucurbita hispida Thunb.+ (Pl. 19D).

Syn.: *Benincasa cerifera* Savi.

Family: Cucurbitaceae.

Vern.: Dhuda petha.

English, Sanskrit & Regional Names

Eng.	:	Ash Pumpkin, Tallow Gourd, White Gourd Melon.
Sans.	:	Gramya Karkati, Kushmanda, Nagapuspa Phala;
B.	:	Chalkumra;
Bo.	:	Kohala;
H.	:	Golkaddu, Petha;
Kan.	:	Budagumbala Kayi;
Mal.	:	Kumpalanna;
Mar.	:	Kohala;
P.	:	Chalkumra, Golkaddu;
Tam.	:	Kalytanppushinikkay;
Tel.	:	Burdagumuda, Pendligummadikaya.

Distribution	:	Cultivated.
Description	:	Climbers with rigid petiole having pungent hairs and large yellow flowers. Fruits flesh white.
Reproductive Cycle	:	June-September.
Habitat Ecology	:	Grown in house gardens.
Material Examined	:	EBH-610, 26-07-2001.
Parts Used	:	Fruit.
Folk Uses	:	Fruits used for preparing '**Petha**' in **confectionary** and in **sacrificial rites** during religious **wor-shipping**. Ash of burnt fruit applied on head to change all hair into **white**.

Active Constituents : Chief active principle of fruits are β-sitosterol, lupeol, n-tricontanol, mannitol, arginine, aspartic acid, glutamic acid, asparagine, glutamine, proline, hydroxy proline, isoleucine, cysteine, L-leucine, glucose and rhamnose.

Uses in Literature : Reported earlier as an anthelmintic, aphrodisiac, diuretic, general tonic for horses, tonic in insanity, vegetable (flowers); and for asthma, epilepsy, haemoptysis, intestinal worms, preserving corpse, and tuberculosis (**Ghosh *et al.*, 1996; Jha *et al.*, 1996; Lalramnhinglova, 1996; Mahato & Mahato, 1996; Rama Rao & Henry, 1996; Sharma & Borthakur, 2003; Siwakoti & Varma, 1996**).

Major Ayurvedic Formulations : Kalyanaka-ghrta, Kushamanda-ghrta, Kushmanda-avaleha, Kushmanda-churna, Kushmanda-khanda, Kushmanda-paka, Kushmanda-rasayana.

Cucurbita melo L.* (Pl. 20A).

Family : Cucurbitaceae.
Vern.: Phunt.

Distribution : Throughout India, cultivated.

Description : A climbing herb with hirsute stems, reniform leaves, yellow flowers and globose fruits.

Reproductive Cycle : July-September.

Habitat Ecology : Planted.

Material Examined : EBH-586, 27-07-2001.

Parts Used : Fruit.

Folk Uses : Fruits edible as **vegetable**.

Cucurbita pepo L.⁺ (Pl. 20B).

Syn.: *C. pepo* Roxb.

Family : Cucurbitaceae.

Vern.: Kaddu.

English, Sanskrit & Regional Names

Eng.	:	Pumpkin, Vegetable Marrow;
Sans.	:	Karkaru, Kurkaru;
B.	:	Kadimah, Kumra;
Bo.	:	Kaula;
G.	:	Dudia;
H.	:	Kadimah, Safed Kaddu;
Mar.	:	Kohala;
Oriya	:	Panikakhara;
Tam.	:	Suraikayi;
Tel.	:	Budadegummedi.

Distribution : Cultivated.

Description : Large climbers with 5-lobed leaves, sinus between the lobes broad. Fruit large, peduncle woody, strongly grooved.

Reproductive Cycle : June-September.

Habitat Ecology : Planted.

Material Examined : EBH-640, 03-07-2001.

Parts Used : Fruit. Rind.

Folk Uses : Fruits cooked as **vegetable**, and generally used as a **utensil** after drying. Seeds **eaten** as such or after roasting.

Active Constituents : These are cholesterol, ergosterol, campesterol, β-sitosterol (seeds), p-hydroxybenzaldehyde, anisyl alcohal, p-hydroxybenzyl alcohal, veratryl alcohal isolated, isovaniyyl alcohal, p-coumaric acid, phloretic acid, benzyl-β-D-glucoside, 4-methoxybenzyl-β-D-glucoside and 3, 4-dimeth-oxybenzyl-β-D-glucoside (flowers).

Biological Activity : Seeds exhibit anthelmintic activity *in vitro* and *in vivo* against *Hymenolepis nana* and *Dicrocrelium denddriticum.*

Uses in Literature : Known so far as an anthelmintic, pshycoactive plant; and for burns, diarrhoea, edible purposes (leaf, fruit), utensils, poison (vet), worshipping, taenicide (seeds) (**Ambasta, 1986**; **Jain, 1991**; **Jain** ***et al.*****, 1994**; **Jha** ***et al.*****, 1996**; **Watt, 1972**).

Cuscuta reflexa Roxb.+ (Pl. 20C).

Syn.: *C. grandiflora* Wall., *C. verrucosa* Sweet., *C. hookeri* Sweet, *C. macrantha* G. Don, *C. elatior* Chois

Family : Cuscutaceae.

Vern.: Akashbel, Amarbel,Gaasbel.

English, Sanskrit & Regional Names

Eng.	:	Dodder;
Sans.	:	Akshspavana, Amaravallika;
Ass.	:	Amarlati;
B.	:	Hadialgusilutta;
G.	:	Amarbel;
H.	:	Akashabel, Amarabela;
Mar.	:	Nirmulia, Kashavela;
Oriya	:	Kolanirmuli;
P.	:	Amil , Nilathari, Niradhar, Zarbuti;
Tam.	:	Kodiya Gundal;
Tel.	:	Savarapukada;
U.P.	:	Kas-banwar, Banwar.

Distribution : Afghanistan - S.W. China. 600-3300m.

Description : Climber with succulent stem, densely overlacing small trees or shrubs. Flowers waxy white and fragrant.

Reproductive Cycle : July-October.

Habitat Ecology : On shrubs; covering many of the bushes in late monsoon.

Material Examined : EBH-476, 14-10-2002.

Parts Used : Stem. The Whole Plant.

Folk Uses : Paste of stem used as an emollient for suppuration of **boils**. Plant extr. **anthelmintic**, and its decoction considered good as an **abortifacient** and for regulating **menstruation, easy shedding** of placenta after delivery (in live stock), and also in **itching**. Massage of oil boiled with the powdered climber helpful against **arthritis**.

Active Constituents : Principle ingredients are dulcitol, luteolin, quercitin, a glucoside of luteolin (stem), and luteolin, quercetin, a glycoside of luteolin, amerbelin, cuscutin and wax (plant).

Biological Activity : Aq. and Alc. extrs. have relaxant, cardiotonic, spasmolytic and acetylcholine- like action in atropinised rabbit auricles and frog's heart.

Uses in Literature : Known to be used as alterative, anodyne, anthelmintic, astringent, blood purifier, cardiac, carminative, demulcent, depurative, diaphoretic, psychoactive plant, purgative, symbolic (sorcery), tonic; and for antifertility, bilious disorders, bodyache (vet), bone fracture (vet), bronchitis, burns, cold, cough, cuts, dermatitis, diarrhoea, diptheria, displaced knee cap, dyeing, epilepsy, eye diseases, fall of hairs, fodder, haematuria, headache, hydrocele, indigestion, itching, killing lice (vet), melancholy madness, nervine weakness, oedema, poultry disease (vet), protracted fevers, rheumatism, sores, spermatorrhoea, sprain, stomachache, swelling of legs and testicles, tongue ulcers, warts, and wounds **(Ambasta, 1986; Chauhan, 1999; Datt, 1996; Dixit & Pal, 1996; Girach, 1992; Girach & Aminuddin, 1995; Girach *et al.*, 1996; Hembrom, 1991; Jain, 1991; Jain *et al.*, 1991; Katewa & Arora, 1997; Mahato &**

Mahato, 1996; Manandhar, 1985, 1987a; Sahoo & Mudgal, 1995; Saini, 1996a, b; Satapathy & Brahmam, 1996; Singh 1996a; Singh & Prakash, 1996; Singh *et al.*, 1996a, 1997; Siwakoti & Varma, 1996; Srivastava *et al.*, 2003; Tirkey *et al.*, 2003b; Tiwari & Tiwari, 1996; Watt, 1972).

Major Ayurvedic Formulations : It also finds its extensive use in Unani system of medicine as "Itrifal Ustakhuddus".

Cyathula tomentosa (Roth.) Moq.* (Pl. 20D).

Syn.: *C. sequax* Moq., *Achyranthes tomentosa* Roth., *A. sequax* Wall., *Desmochaeta tomentosa* Roem. & Sch., *Polyscelis saquax* Wall.

Family : Amarantaceae.

Vern.: Kathurla.

Distribution : Temperate Himalaya. 700-2650m.

Description : Densely woolly undershrubs with opposite, long pointed leaves; white, shining flowers in long- stalked spikes; awned sepals and a single seeded utricle.

Reproductive Cycle : July-December.

Habitat Ecology : Roadsides, wasteplaces, dry areas.

Material Examined : EBH-1215, 25-11-2002.

Parts Used : Leaves. Inflorescence.

Folk Uses : Leaves serve as **cattle feed**. Inflorescence combined to form **lid** for covering utensils during cooking.

Biological Activity : EtOH (50%) extr. of plant spasmolytic.

Cycas circinalis L.+ (Pl. 21A).

Syn.: *C. sphaerica* Roxb., *C. thuarsii* Br., *C. madagascariensis* Miq.

Family : Cycadaceae.

Vern.: Cycas.

Sanskrit & Regional Names

Sans.	:	Varaguna;
H.	:	Jangali-Madan-Mast-Ka-Phul;
Kan.	:	Mundichalu;
Oriya	:	Maddu;
Tam.	:	Canningay, Madanagama;
Tel.	:	Intalappana.

Distribution : Planted.

Description : A palm tree-like plant with heavy trunk having an elegant rossette of stiff, shining fronds divided to the midrib into 80-100 long, narrow segments.

Reproductive Cycle : February-May.

Habitat Ecology : Cultivated.

Material Examined : EBH-1241, 26-02-2001.

Parts Used : Leaves. Fruits.

Folk Uses : Planted as an **ornamental.** Fruits **edible** and juice of leaves given to check **indigestion**.

Active Constituents : These are cyasin (roots), amentoflavone, podocarpus flavone A, 2,3-dihydroamentoflavone, 2,3-dihydro-hinokiflavone, isoginkgetin and bilobetin (leaves).

Uses in Literature : Recorded earlier as food in time of scarcity (seeds), psychoactive plant, vegetable; and for eradicating rats, flatulence, rheumatic pains, sago starch and edible fruits (**Ambasta, 1986; Jain, 1991; Jain *et al.*, 1994; Nazarudeen *et al.*, 1996; Rama Rao & Henry, 1996; Watt, 1972**).

Cyphomandra betacea (Cav.) Sendt. (Pl. 21B).

Syn.: *Solanum betaceum* Cav.

Family : Solanaceae.

Vern.: Desi Tamatar, Tree Tomato.

English Name

Eng. : Tree Tomato.

Distribution : Cultivated.

Description : Armed shrub. Fruit reddish-black.

Reproductive Cycle : May-September.

Habitat Ecology : Planted near houses.

Material Examined : EBH-158, 16-07-2000.

Parts Used : Fruits.

Folk Uses : Fruits eaten **raw** or cooked as **vegetable.**

Active Constituents : Pelarigonidin-3-glucosylglucoside isolated from reddish-brown peel of fruits while peonidin-3-glucosyl glucoside and malvidin-3-glucosylglucoside isolated from seeds.

Uses in Literature : Used in India for edible purposes (fruit), and jams and preserves (**Ambasta, 1986**).

Dalbergia sissoo Roxb.[+] (Pl. 21C).

Syn.: *D. pendula* Tenore

Family : Fabaceae.

Vern.: Shisham, Tahli.

English, Sanskrit & Regional Names

Eng. : South Indian Red Wood;

Sans. : Agurushinshupa, Mandalapatri;

Ass. : Sissu;

B.	:	Shishu;
Bo.	:	Sissu,Tanach;
G.	:	Sissom, Tanach;
H.	:	Shisham, Sissai;
Kan.	:	Agaru, Biridi;
Mal.	:	Iruvil;
Oriya	:	Simsapa, Sisu;
P.	:	Nelkar, Shewa, Shishai, Tali;
Tel.	:	Etta sissu, Sinsupa.

Distribution : Throughout India ascending to 1650m.

Description : Trees with pale-green, odd pinnate leaves and 3-5, broadly ovate leaflets. Flowers yellow-white. Pods long-stalked, thin and glabrous.

Reproductive Cycle : May-August.

Habitat Ecology : On grassy slopes.

Material Examined : EBH-659, 25-07-2001.

Parts Used : Roots. Leaves. Stem Wood.

Folk Uses : Wood good for **agricultural implements, building works** and as **fuelwood** (for coke); and also its oil considered useful against **skin ailments**. Poultice of leaves useful to check **inflammation**. 1-2g powdered roots prescribed with milk for **flatulence** (twice a day for 5-7 days).

Active Constituents : Principle ingredients are biochanin A-7-glucoside (sissortin), isoflavone-0-glycoside (leaves), 5,7-4'-Trihydroxy-8-methoxyisoflavone (isotectorigenin) (bark), and a new allylphenol of latifolin type-dalbergiphenol along with dalbergenone, dalbergin, methyldalbergin (heartwood).

Biological Activity : Spasmolytic.

Uses in Literature : Known so far as an alterative, antiemetic, astringent, excellent fuel wood, fodder, ornamental (ear pendent), sacred, spasmolytic (vet); and for agricultural implements, allaying vomiting, building works, boils, cabinet wood, carved work, charcoal, charpai legs, diarrhoea (vet), dysentery, eruptions, framing of carriages, gonorrhoea, headache, high class furniture, leprosy, musical instruments, naves of wheel, railway sleepers, scurf, skin diseases, and sleepers (**Ambasta, 1986; Borthakur & Sarma, 1996; Jain, 1991; Mandal & Basu, 1996; Siwakoti & Varma, 1996**).

Trade : Sissoo, Shisham.

Daphne cannabina Wall.⁺ (Pl. 21D).

Syn.: *D. papyracea* Wall. *ex* Steud., *D. odora* Don., *D. bholua* Don., *D. papyrifera* Ham.

Family : Thymelaeaceae.

Vern.: Chamar-Chumb, Raabal.

English & Regional Names

Eng. : Nepal Paper Plant;

H. : Satpura, Setburusa, Setburwa;

P. : Mahadeo Kaphul, Niggijeku.

Distribution : Temperate Himalaya and Khasia Hills: 1500-2100m.

Description : Erect shrubs. Leaves shortly-stalked, more than 5cm long. Flowers white-lilac. Fruit ovoid, succulent and red coloured.

Reproductive Cycle : April-October.

Habitat Ecology : Forests; on dry stony slopes.

Material Examined : EBH-278, 12-10-2003.

Parts Used : Leaves. The Whole Plant.

Folk Uses : Leaves prescribed as **bitter tonic** and 5-10 ml decoction of the plant taken orally, twice daily for 3-5 days to cure **boils**.

Active Constituents : Constituents isolated are sterol and three glycosides G-1, G-2 and G-3, daphnin (roots), daphnetin and daphenetin-8-â-glucoside, taraxerol and its â-sitosterol-D-Glucoside (aerial parts).

Uses in Literature : Reported earlier for cordage fibres from bark for ropes and writing paper, fever, genital swellings, intestinal complaints, ornamental garlands, and urinary disorders (**Ambasta, 1986; Aswal, 1996; Jain, 1991; Lal *et al.*, 1996; Watt, 1972; Yonzone *et al.*, 1996**).

Datura metel L.⁺ (Pl. 22A).

Syn.: *D. fastuosa* L., *D. fastuosa* var. *alba* (Nees) Cl., *D. alba* Rumph.

Family : Solanaceae.

Vern.: Dhatura, Kala Dhatura.

English, Sanskrit & Regional Names

Eng. : Black Dhatura, Purple Dhatura;
Sans. : Kanakonmatta, Shivashekwa;
B. : Kaladhutura;
Bo. : Kaladhatura;
G. : Kalodhaturo;
H. : Kaladhatura;
Mal. : Kariyummatta, Madanam;
Mar. : Kaladhotra;
Oriya : Jambunodo;
Tam. : Ponnumattai, Turam;
Tel. : Dutturamu, Matulaputrakamu.

Distribution : Throughout India.

Description : Undershrubs. Leaves deeply toothed. Flowers white. Fruit spiny, globose.

Reproductive Cycle : June-September.

Habitat Ecology : Wastelands.

Material Examined : EBH-690, 11-08-2002.

Parts Used : Root. Leaves. Fruit.

Folk Uses : 3-5 ml extr. of flowers and fruits good against **phlegm.** (twice daily for 3-5 days). Ash of its burnt roots rubbed on chest to cure **pneumonia**, and poultice of fruits beneficial in **painful necks**. Leaf juice boiled in 'til oil' (*Sesamum indicum* L.) used as drops for relief from **earache.** Seeds considered **depressant** and also its paste useful against **goitre** and **lice.**

Active Constituents : Principle constituents are atropine, fastunine, fastudine, fastusidine, daturonolene and fastrosic acid (seeds), and scopolamine, hyoscine and hysocyamine (leaves).

Biological Activity : Scopolamine (*i.v.*) was ineffective against experimental cardiac arrhythmias induced in mice, rats, rabbits and guinea pigs.

Uses in Literature : Known earlier as anodyne, antidote for narcotic drugs, antispasmodic, aphrodisiac, emollient and suppurative boils abscess, magico-thereauptic (prevents abortion), mydriatic, narcotic (leaves), poison, psychoactive plant, toothbrush; and for amenorrhoea, asthma, bleeding wounds and burns (vet), cholera, chronic bronchitis, chronic rheumatism, cold (vet), cough (vet), diarrhoea (vet), dysentery (vet), eczema (vet), excessive secretion of milk, fainting, fever (vet), glandular enlargement, gonorrhoea, gout, hydrophobia, inducing permanent sterility in females, inflammation of breast, insanity, insect-bite (vet), intermittent fevers, intestinal colic, intestinal worms, keeping off evil spirits, madness (vet), malaria, making liquors more intoxicating, mumps, nose and chest troubles, painful swellings, rabies, relieving pain of patella and swelling in horses, rib-muscle pains, skin (vet), skin diseases, sprains, syphilis, tightening flabby breast, toothache, and giddiness, (**Aminuddin & Girach, 1991; Banerjee, 1996; Borthakur, 1993;**

Chauhan, 1999; Dagar & Dagar, 1996; Jain, 1968; Jain *et al.*, 1991, 1994; Khanna *et al.*, 1996a; Mahato & Mahato, 1996; Mandal & Basu, 1996; Masih, 2003; Mishra *et al.*, 1997; Mohanty & Padhy, 1996; Pandey & Verma, 2003; Rama Rao & Henry, 1996; Sahoo & Mudgal, 1995; Saini, 1996a; Sharma & Borthakur, 2003; Singh & Prakash, 1994; Siwakoti & Varma, 1996; Srivastava *et al.*, 2003; Sundaravelu *et al.*, 2003; Watt, 1972).

Major Ayurvedic Formulations : Alarex, Dhathuradi-pralep, Dhaturphal-bhasma, Dugdhavati, Kanak-sundara, Kankasava, Laghu-Vishgarbha-taila, Lakshimilavilasa-rasa, Maha-jvarankusa, Mahalakshmi-vilas-ras, Spasmolin, Suta-sekhara.

Trade : Dhaturaa-bija, Dhaturaa-patti.

Datura stramonium L.[+] (Pl. 22B).

Syn.: *D. stramonium* var. *tatula* (L.) Cl., *D. tatula* L., *D. ferox* Nees, *D. wallichii* Dunal, *D. inermis* Jacq., *Stramonium vulgatum* Gaertn.

Family : Solanaceae.

Vern.: Dhatura.

English, Sanskrit & Regional Names

Eng. : Apple of Peru, Devil's Apple, Devil's Trumphet, Dewtry, Jamestown Weed, Mad Apple, Stink Weed, Thorn Apple;

Sans. : Devika, Dhattura, Dhurta, Dhurtakrit, Dhustura, Dhultura, Ghantapushpa, Kalama, Kanaka, Kanakaohaya, Kantaphala, Madanaka, Mahamohi, Mahashatha, Matulaka, Savisha, Shivapriya, Shivasekhara, Shyama, Turi, Unmatta, Unmattaka;

B. : Sadadhutura;

G. : Dholo Dhaturo;

H. : Dhatura;

Kan. : Ummatta;

Mal. : Matulam, Ummam, Ummata;

Oriya : Dhutura, Sukladhutura;

P. : Dattura, Tattur;

Tam. : Emanamam, Simalyumattai, Turutturam, Umattai, Vellumattai;

Tel. : Dutturamu, Tellavummetta, Ummetta.

Distribution : Temperate Himalaya, Kashmir-Sikkim: 200-2200m.

Description : Shrubs. Leaves stalked, irregularly lobed and toothed. Flowers white. Capsule ovoid, covered with rigid, sharp prickles.

Reproductive Cycle : June-October.

Habitat Ecology : On wastegrounds, cultivated areas.

Material Examined : EBH-480, 12-10-2002.

Parts Used : Leaves. Flowers. Roots. Fruits. Seeds.

Folk Uses : Flowers considered **sacred** for **offering** to appease Lord Shiva. An amulet of its root worn on hand to cure **fever**, and inhalation of smoke of burning leaves believed to cure **asthma**. Fruits **intoxicating**, and 1g powdered seeds prescribed with cow's milk for 3 days to check **fever**.

Active Constituents : Various constituents are hyoscamine, hyoscine, scopolamine, fluorodaturative and homofluoro-daturatine (seeds), a new vitanolide-vitastra-monolide along with daturalactone and its keto derivative characterised as 5a,12a,27-trihydroxy-1-oxo-6a, 7a-epoxy-22R-vita 2,24-dienolide (leaves), hyoscine, 3a,6b-ditigloycoxytropane, 3a, 6b-ditigloyloxytropane-7b-OL, apoatropine, tigcoidine, tropine, meteliodine (Nigerian plant), 0.3 to 0.5% of alkaloids chiefly as hyoslyamine with smaller amount of hyoscine and atropine (dried leaves, flowering tops).

Uses in Literature : Recorded in India as anodyne, antirheumatic, antispasmodic, mydriatic, narcotic in treatment of bronchitis, psychoactive plant; and for asthma, boils, breast inflammation due to excessive formation of milk, cigarettes (leaves), controlling salivation and exparasite, dental caries, diarrhoea, dislocation of joints, dysmenorrhoea, enlarged testicles, epilepsy, haemorrhoids, headache, inflammatory pains, itch, jaundice, lactation (vet), milching (vet), nervous disorder (vet), nymphomania, painful diseases of rectum, paralysis, post encephalitic parkinsonianism, purpurea, rheumatism (vet), sciatica, skin diseases, stomach complaints, swelling of neck (vet), tonsilitis, toothache, travel sickness, ulcers, and vicious indulgence (**Chopra *et al.*, 1956; Jain, 1968, 1991; Jain *et al.*, 1994; Jha *et al.*, 1991; Manandhar, 1996; Singh, 1991; Upadhye *et al.*, 1994; Watt, 1972**).

Major Ayurvedic Formulations : Alarex, Dhathuradi-pralep, Dhaturphal-bhasma, Dugdhavati, Kanak-sundara, Kankasava, Laghu-Vishgarbha-taila, Lakshimilavilasa-rasa, Maha-jvarankusa, Mahalakshmi-vilas-ras, Spasmolin, Suta-sekhara.

Trade : National and Global. Traded as stramonium or 'Pahari dhatura'.

Datura suaveolens Humb. & Bonpl. *ex* Willd.* (Pl. 22C).

Syn: *Brugmansia suaveolens* (Humb. & Bonpl. *ex* Willd.) Bercht. & Presl.

Family: Solanaceae.

Vern.: Dhatura.

English Name

Eng. : Angel's Trumpet.

Distribution : Naturalised in the Himalayas upto 1700m.

Description : Large shrubs. Leaves large, ovate-oblong, entire. Flowers very large, white, pendulous funnel-shaped, sweet scented at night. Fruits oblong smooth.

Reproductive Cycle : May-January.

Habitat Ecology : Cultivated areas; sometimes as a hedge plant.

Material Examined : EBH-479, 11-10-2002.

Parts Used : Flowers. Seeds. Roots.

Folk Uses : Flowers offered to Lord Shiva. An amulet of its roots tied to wrist as **febrifuge**, and seeds **intoxicating.**

Active Constituents : Chief constituents isolated are 3α, diol, 3α-tigloylxytropan-6β, 7β-diol (metaloidine) and (-) and (±)3α-tigloylloxytropan-6β-ol hysocine, apohyoscine, norhyoscine, atropine and noratropine (aerial parts), and hysocine, tropine, atropine, metal iodine, littorine, 3α-acetoxytropane, 6β-(α-methyloxy)-3α-tigloyloxytopane, 3α, 6β-ditigloyloxytropane-7β-ol and cuscohygrine (roots).

Debregeasia hypoleuca Wedd.+ (Pl. 22D).

Syn.: *D. salicifolia* (Don) Rendle, *D. bicolor* (Roxb.) Wedd., *Boehmaria salicifolia* Don, *B. hypoleuca* Hochst., *Urtica bicolor* Roxb., *Missiessya hypopleuca* Wedd., *Morocarpus salicifolius* Blume

Family: Urticaceae.

Vern.: Siaru, Syarru.

Regional Names

P. : Pincho, Sansaru;

U.P. : Sansaru, Siaru, Tusarro, Tushiari.

Distribution : N. W. Himalaya: 1500-2400m.

Description : Sturdy shrubs. Leaves 3-nerved, long pointed with lower surface white tomentose. Flowers yellow, unisexual. Drupes yellow, small, aggregated in Heads.

Reproductive Cycle : May-August.

Habitat Ecology : Forests, shrubberies; common.

Material Examined : EBH-254, 03-07-2001.

Parts Used : Leaves. Wood.

Folk Uses : Wood used as **fuel**, and leaves considered as good **cattle feed.** 10-20 ml decoction of leaves said to be good against **hernia** and **epilepsy** (once daily for 15-20 days).

Uses in Literature : Used in India for cordage (bark), edible purposes, fishing lines, stomach pain, ulcers and killing maggots in wounds of cattle and goats (**Ambasta, 1986**; **Kapur, 1996**; **Rama Rao & Henry, 1996**; **Watt, 1972**).

Dendrocalamus hamiltonii Nees & Arn. *ex* Munro⁺ (Pl. 23A).

Syn.: *Bambusa monogyna* Griff., *B. maxima* Ham., *B. falconeri* Munro.

Family : Bambusaceae.

Vern.: Baingh, Maggar.

Regional Names

Ass. : Kokua;

Bo. : Pecha, Tama;

H. : Kagzi-Bans.

Distribution : S.E. Asia. & W. Himalaya.

Description : Stem sheath glabrous. Leaves broadly-lanceolate, cuspidate, spikelets ovoid, depressed, purple. Anthers black, hairy, twisted.

Reproductive Cycle : Flowering sporadically.

Habitat Ecology : As hedge; common around villages.

Material Examined : EBH-270, 12-11-2002.

Parts Used : Young shoots. Stem.

Folk Uses : Plant grown as **hedges**, and its stem extensively employed for making **baskets**, **house building**, and **as bier** for carrying corpse to funeral ground for performing last rites, and for checking **gastro-intestinal disorders** in animals. '**Bans lochan**' obtained from its stem used in preparation of '**sund**' which is given to ladies for overcoming **post delivery stress**.

Uses in Literature : Used so far as symbolic in religious ceremonies, vegetable (young culms), water conduits; and for bow, building material, headgear of bridegroom, and house building, (**Ambasta, 1986; Borthakur 1996; Chowdhery, 1996; Jain, 1991; Lalramnghinglova, 1996; Watt, 1972**) .

Desmodium multiflorum DC* (Pl. 23B).

Syn.: *D. floribundum* G. Don, *D. sambuense* (D. Don.) DC, *D. angulatum* DC, *D. dubium* Lindl., *Hedysarium floribundum* Don, *H. wallichainum* Spreng., *H. sambuense* D. Don.

Family : Fabaceae.
Vern.: Shaklathi, Shetikathi.

Distribution : Tropical & Temperate Himalayan regions: 600-2400m.

Description : Large erect shrubs. Leaves trifoliate, hairy on both sides. Flowers pink or purple. Pods sessile, hairy.

Reproductive Cycle : June-October.

Habitat Ecology : Shrubberies, open slopes, cultivated areas.

Material Examined : EBH-1038, 25-06-2003.

Parts Used : Leaves. Stem. Wood.

Folk Uses : Leaves used as **fodder** and stem for **scouring teeth**. Wood used as **fuel**.

Active Constituents : Active principles is due to hordenine.

Deutzia corymbosa R. Brown* (Pl. 23C).

Syn.: *D. compacta* Craib., *D. parviflora* Bunge, *Philadelphus (Leptospermum) corymbosus* Wall.

Family : Saxifragaceae.
Vern.: Nal.

Distribution : Temperate Himalaya. 2000-3350m.

Description : Shrubs. Flowers white, fragrant. Leaves nearly sessile, ovate-lanceolate, long-pointed.

Reproductive Cycle : May-October.

Habitat Ecology : Forests, shrubberies.

Material Examined : EBH-1037, 14-07-2002.

Parts Used : Leaves. Wood.

Folk Uses : Leaves used as **fodder** and wood for making **walking sticks**. This plant is usually grown near houses as **termite repellent**.

Active Constituents : The plant contains two new echinocystic acid glycosides-deutzicosides A and B.

Dioscorea bellophylla Voigt⁺ (Pl. 23D).

Syn.: *D. glabra* Roxb., *D. cliffortiana* Lam., *D. versicola* Herb., *D. pulchella* Roxb., *D. heterophylla* Roxb., *D. decemangularis* Herb., *D. thunga* Herb., *Helmia bulbifera* Kunth.

Family : Dioscoreaceae.
Vern.: Tardi.

Distribution : Sub-Tropical Himalaya.

Description : A climber with opposite, ovate-lanceolate leaves. Flowers in spikes. Capsule membranous. Seeds winged all round.

Reproductive Cycle : September-November.

Habitat Ecology : Scrub jungles, hedges.

Material Examined : EBH-1180, 25-11-2002.

Parts Used : Tuber. Leaves.

Folk Uses : Tubers consumed as **vegetable** and also one of the constituents of '**chywanprash**'. 5-6g powdered leaves given thrice daily for 10 days to cure **leucorrhoea**. Seed powder commonly used by magicians for performing water related **magics**.

Active Constituents : Tuber contains diosgenin.

Biological Activity : Antiinflammatory and antifertility activities found (+) ve.

Uses in Literature : Known in India for its edible tubers (**Ambasta, 1986; Jain, 1991; Lal *et al.*, 1996; Singh, 1996b; Sinha, 1996**).

Major Ayurvedic Formulations : 'Chyawanprash'.

Dioscorea bulbifera L.[+] (Pl. 24A).

Syn.: *D. sativa* auct. non Thunb., *D. pulchella* Roxb., *D. pulchella* Roxb., *D. pulchelia* Hoken., *D. tamnifolia* Salisb., *Helmia bulbifera* Kunth.

Family : Dioscoreaceae.

Vern.: Peela Khetha, Khitha.

English, Sanskrit & Regional Names

Eng. : Black Bearing Yam;

Sans. : Badar Kachna, Vishvaksenapriya;

Ass. : Kathalu, Mati-Alu;

B. : Banalu, Bandorechalu;

Bo.	:	Hadukaranda;
G.	:	Salvinavelya, Suariya;
H.	:	Bhirvolikanda, Karawakand;
Kan.	:	Heggenasu, Onthalaigasu, Tung-Genasu;
Mal.	:	Kattukachil;
Mar.	:	Dukarakanda;
Oriya	:	Pita-Alu;
P.	:	Zaminkand;
Tel.	:	Chedupaddudumpa, Malakakayapenda-Lamu, Pannukelangu.

Distribution : Himalaya upto 2100m.

Description : Climbers. Leaves ovate-cordate, alternate. Flowers yellowish-green but fading later to dull red. Seeds winged at the base.

Reproductive Cycle : June-August.

Habitat Ecology : Scrub jungles, roadsides, wasteplaces.

Material Examined : EBH-544, 02-08-2002.

Parts Used : Tuber.

Folk Uses : Tubers **edible**, and often employed for **sacrificial rites** by 'tantriks'. Also, a pinch of its powder added as one of constituents of local wine '**sur**'.

Active Constituents : Three new furanoids nortiterpenes-diosbulbins A, diosbulbin B and diosbulbin C isolated from root tubers.

Biological Activity : Antiinflammatory and antifertility activities (+) ve.

Uses in Literature : Recorded so far as contraceptives, psychoactive plant, vegetable (tubers), febrifuge, fish catcher; and for abdominal pain, boils, bone fracture, bulbils, dysentery, itching, jaundice, piles, rheumatism and gout, skin infections, syphilis, and ulcers (**Ambasta, 1986; Borthakur, 1996a; Brahmam *et al.*, 1996; Chowdhery, 1996; Girach *et al.*, 1986; Jain, 1991; Jain *et al.*, 1994; Lal *et al.*, 1996; Lalramnghinglova, 1996; Mandal & Basu, 1996; Rama Rao & Henry, 1996; Saini, 1996a; Singh & Prakash, 1996; Singh & Zaheer, 1996; Sinha, 1996; Siwakoti & Varma, 1996; Watt, 1972**).

Major Ayurvedic Formulations : 'Chyavanprash', Garbhadharini-vati, Mahamash-tel, Narsingh-tel.

Dioscorea deltoidea Wall. (Pl. 24B).

Family : Dioscoreaceae.

Vern.: Singli-Mingli.

Regional Names

Kash.	:	Kildri, Kins, Kithi, Krish;
P.	:	Kniss, Kriss, Tas, Kitra.

Distribution : Temperate Himalaya: 2000-3350m.

Description : Glabrous shrubs with usually alternate, long-pointed leaves and flowers in small, distant clusters. Wings broadly rounded.

Reproductive Cycle : May-August.

Habitat Ecology : Hill slopes, forests; not common.

Material Examined : EBH-1030, 23-07-2003.

Parts Used : Roots. Rhizome. The Whole plant.

Folk Uses : Rhizomatous tubers **edible** and also as **detergent**. 1-2g powdered roots taken with cow's milk twice a day for 10 days to check **conception**. Powdered plant (one teaspoonful, twice a daily for 3-5 days) given with honey as **febrifuge** and for **arthritis**.

Active Constituents : These are smilagenone, epismilagenin, deltonin and dioscin (rhizomes) and sapogeuin (plant).

Uses in Literature : Known to be used in India as detergent (tubers rich in saponin), febrifuge, idiopathic thrombocytopenic purpura, insecticide (vet), edible; and for birth control, dyeing, fishing-poison, gastric troubles, killing lice, ophthalmic disorders of allergic states, and rheumatic diseases (**Ambasta, 1986; Chauhan, 1999; Jain, 1991; Kapur *et al.*, 1996; Watt, 1972**).

Remarks : Diosgenin is used in the synthesis of modern drugs like cortico-steroids, which enter into preparation of sex hormones, contraceptive pills, and the anabolic agents.

Trade : Singli-mingli.

Diospyros kaki L.$^{+}$ (Pl. 24C).

Syn.: *D. chinensis* Blume, *D. schi-tse* Bunge, *D. roxburghii* Rev., *D. costata* Blume, *Embryopteris kaki* G. Don

Family: Ebenaceae.

Vern.: Japani-phal.

English & Regional Names

Eng. : Japanese Persimmon, Kaki Persimmon;

Ass. : Dieng-Iong, Soh-Tang-Jong;

H. : Halwa Tendu.

Distribution : Planted. Western & North India.

Description : A small tree with coriaceous leaves and globose fruit.

Reproductive Cycle : May-June.

Habitat Ecology : Planted.

Material Examined : EBH-500, 12-05-2002.

Parts Used : Fruit.

Folk Uses : Fruits **edible** for its sweetmeat and unripe ones given to check **mouth ulcers**. Wood used for **ornamental works**.

Active Constituents : Fruit yields delphinidin, cyanidin, lycopene, catechin-3-gallate, gallocatechin and gallocatechin-3-gallate.

Uses in Literature : So far recorded as an astringent, edible; and for boxes, dyeing, ornamental works, tannin (raw fruits) (**Ambasta, 1986; Jain, 1991; Watt, 1972**).

Dodonea viscosa Jacq.+ (Pl. 24D).

Syn.: *D. angustifolia* L., *D. dioca* Roxb., *D. burmanniana* DC, *D. pallida* Miq., *D. micropoda* DC, *D. wightiana* Blume, *D. petandra* Griff., *D. heterophylla* Hortul., *D. spathulata* Sm., *D. arabica* Hochst. & Steud., *Ptelea viscosa* L.

Family : Sapindaceae.

Vern.: Mehndu.

Regional Names

B. : Bandari, Zakhumi;
Bo. : Jakhmi, Lutchmi;
H. : Alliar, Sonalta, Walaytinahndi;
Kan. : Bandara;
Mal. : Unnataruvi, Virali;
Mar. : Lutchmi;
Oriya : Mohra;
P. : Banmehndu, Mendhar, Sonatta;
Tam. : Velari, Virali;
Tel. : Bandaru, Pullenu.

Distribution : Afghanistan-H.P. upto 1500m.

Description : Shrubs. Leaves alternate, viscid with yellow resinous dots. Flowers greenish-yellow. Capsule yellowish-brown, compressed.

Reproductive Cycle : February-April.

Habitat Ecology : Open slopes, often gregarious.

Material Examined : EBH-179, 12-02-2000.

Parts Used : Leaves. Wood.

Folk Uses : Leaves used as **cattle feed**, and its paste good for suppuration of **boils**, and also its decoction (10 ml) given once daily for 3-5 days to **regulate menstruation** and as **refrigerant**. Besides, poultice of its leaves along with those of *Vitex negundo*, and *Cuscuta reflexa* and a pinch of rock salt prescribed for **sprains** and **joint pains**. Wood used as **fuel**.

Active Constituents : Two novel prosapogenins-R1 barrigenol-21,22-diangelate and jegosapogenol-21-(2,3-dihydroxy-2-methylbutyroyl)22-angelate- isolated from stem-bark.

Biological Activity : Plant extr. exhibits cardio-inhibitory, and spasmolytic activity on smooth muscles and intestine. It also has sedative effect on virgin guinea pig uterus and hypotensive effect unaffected by atropine.

Uses in Literature : Used in India as aromatic, astringent, carminative, febrifuge, manure, purgative, substitute for hops (*Humulus lupulus* L.) in the preparation of beer and yeast, sudofic in gout and rheumatism; and for bone fracture, broken legs of cows, burns, cold, cuts, engraving, epilepsy, fermentations, firewood, fodder, intestinal worms, muscle pain, sprains, swellings, ticks (vet), tool-handles, turnery, walking sticks, worms (vet), and wound (**Ambasta, 1986**; **Geetha *et al.*, 1996b**; **Goud & Pullaiah, 1996b**; **Jain, 1991**; **Rama Rao & Henry, 1996**; **Sharma & Sood, 1997**; **Vishwanthan, 1997**; **Watt, 1972**).

Duranta repens L.[+] (Pl. 25A).

Syn.: *D. plumieri* Jacq.

Family : Verbenaceae.
Vern : Neel Kanta.

English Name

Eng.	:	Creeping Skin Flowers, Golden Dew Drop, Pigeon Berry;
Sans.	:	Vriti Kanta;
H.	:	Daranta.

Distribution : W. Himalaya: 200-1600m.

Description : A spiny shrub. Leaves opposite, broadly elliptic, entire. Spines stout. Flowers bright, lavender-blue with pale centres. Fruit golden-yellow, globular.

Reproductive Cycle : April-August.

Habitat Ecology : Very common as a hedge plant.

Material Examined : EBH-735, 11-04-2000.

Parts Used : Flowers. Leaves. Aerial Parts.

Folk Uses : The plant grown as **hedges**, and its decoction effective for **toothache** and **purification of blood**. Flowers constituent of local wine, **'sur'**. For **joint pains**, 10-20 ml decoction of its juvenile leaves (250g) in combination with black pepper (10g) and a pinch of black salt, given orally twice a day for 15-20 days.

Active Constituents : Leaves afford β-sitosterol, α-amyrin, scuttellarein, 4'methoxy-scutellarein, scutellarein-7-O-rhamnoside, pectolinarigenin and its 7-O-rutinoside.

Uses in Literature : Recorded earlier as insecticidal and for turnery **(Ambasta, 1986)**.

Ehretia acuminata R. Br. (Pl. 25B, C).

Syn: *E. serrata* Roxb., *E. pyrifolia* Don., *E. ovalifolia* Hassk., *Cordia thyrsiflora* Sieb. & Zucc.

Family : Cordiaceae.

Vern.: Punna, Punya.

Regional Names

- **Ass.** : Bual;
- **B.** : Kula-Aja;
- **H.** : Arjun, Koda, Punjlawai, Punyam;
- **P.** : Punna, Pursan, Sium;
- **U.P.** : Arjun, Kur-kuria, Narra, Shaursi.

Distribution : Sub-Tropical Himalaya.

Description : A tree with alternate, ovate-elleptic leaves. Flowers white. Drupe red, globose.

Reproductive Cycle : May-July.

Habitat Ecology : Common on dry slopes, roadsides.

Material Examined : EBH-1052, 11-06-2002.

Parts Used : Leaves. Wood. Fruit.

Folk Uses : Fruits **edible**. Leaves used as **fodder** and wood for **furniture** and **construction work**.

Uses in Literature : Reported to be used for agricultural implements, building and edible purposes, gun-stocks, pickling, scabbards, sword hits, tongue's sores and as fodder (**Ambasta, 1986**; **Jain, 1991**; **Mudgal & Pal, 1980**; **Negi & Gaur, 1991**; **Watt, 1972**).

Ehretia laevis Roxb. (Pl. 25D).

Syn.: *E. punctata* Roth., *E. affinis* Wall., *E. dichotoma* Rottl., *E. floribunda* Benth *ex* Royle, *Beurreria laevis* G. Don, *B. punctata* G. Don.

Family : Cordiaceae.

Vern.: Narala.

Regional Names

- **Ass.** : Hanbuok-Arong;
- **B.** : Tamboli;

H.	:	Chamror, Datrang, Darar, Koda;
Kan.	:	Adak, Begari, Kappura;
M.P.	:	Tamoiya, Chinor, Bhoiumbar;
Mal.	:	Chavandi;
Mar.	:	Datrang;
Oriya	:	Guachipo, Masonea;
P.	:	Sakar;
Tam.	:	Addula;
Tel.	:	Paldatam, Peddapulmera, Pogari.

Distribution : Throughout India; ascending to 1000m.

Description : Trees. Leaves ovate, entire, and leathery. Drupe depressed, globose, distinctly 4-ribbed when dry.

Reproductive Cycle : January-June.

Habitat Ecology : Drier areas, stony slopes.

Material Examined : EBH-697, 16-06-2002.

Parts Used : Leaves. Wood.

Folk Uses : Leaves used as **fodder** and wood as **fuel,** also for **building works** and **agricultural implements**.

Active Constituents : Stem yields bauerenol, bauerenol acetate and β-sitosterol.

Uses in Literature : Useful as cattle fodder; and for agricultural implements, brush-backs, building purposes, diarrhoea, diptheria, edible purposes (fruit and inner bark eaten in times of scarcity), match boxes, muscular and rheumatic pains, shoe lasts, splints, syphilis, and thatching (**Ambasta, 1986; Jain, 1991; Negi & Gaur, 1991; Rama Rao & Henry, 1996; Saini, 1996a; Singh & Maheswari, 1983; Watt, 1972**).

Elaeagnus umbellata Thunb. (Pl. 26A).

Syn.: *E. parviflora* Wall. *ex* Royle,
E. reflexa Dcne & Morren, *E. done* Ann

Family : Elaeagnaceae.

Vern.: Ghain.

Regional Name

P. : Bammerwa, Ghain, Kankoli.

Distribution : Afghanistan - S.W. China, from 900-3100m.

Description : A thorny shrub. Leaves silvery beneath. Flowers clustered, white, fragrant. Fruit red, ovoid with bony nut.

Reproductive Cycle : May-October.

Habitat Ecology : Shrubberies, cultivated areas.

Material Examined : EBH-889, 11-06-2002.

Parts Used : Fruits. Flowers. Wood.

Folk Uses : Plant used as **fuel wood**. Decoction of flowers (5 ml) given twice a day for 15-30 days for **cardiac ailments**. Ripe fruits **eaten**, and **pickled**.

Uses in Literature : Known to be used as an astringent, cardiac, diuretic, fuel, haemostatic, ornamental, stimulant, edible; and for cough, curries, dysentery, pickling, and pulmonary affections, (**Ambasta, 1986; Jain, 1991; Kapur & Nanda, 1996; Kapur & Singh, 1996; Negi & Gaur, 1991; Negi *et al.*, 1993; Parmar & Kaushal, 1981; Singh, 1996b; Singh *et al.*, 1996a; Sinha, 1996; Watt, 1972**).

Embellia tsjeriam-cottam A. DC[+] (Pl. 26B).

Syn: *E. reticulata* Wall., *E. villosa* Wall., *E. robusta* Roxb., *E. basaal* A. DC., *E. picta* A. DC, *Samara picta* Wall., *S. robusta* Kurz, *Ardisia basaal* Roem. & Sch., *A. tsjeriam-cottam* Roem. & Sch.

Family : Myrsinaceae.

Vern.: Baburangi.

Regional Names

B.	:	Dhadki, Jhanti;
Bo.	:	Ambai, Barbatti;
H.	:	Babrang, Baibrang, Bayabirang;
Mal.	:	Cheriyannattam;
Mar.	:	Ambati, Ambuti, Kokla, Waiwarung;
Oriya	:	Bayibidingo, Bidingo, Krymighno.

Distribution : Throughout India.

Description : A large rambling shrub. Leaves more or less pubescent beneath. Berry red.

Reproductive Cycle : July-September.

Habitat Ecology : Scarce in hills, common in tracts.

Material Examined : EBH-1063, 26-09-2001.

Parts Used : Fruits. Wood.

Folk Uses : Ripe fruit **edible**, and also prescribed as **carminative**, **laxative** and for **throat complaints**. Wood used as **fuel**.

Active Constituents : Dried berries yield 1.6% embelin, 0.15% potassium hydrogen oxalate and 7.5% fatty ingredients. However, vilangin isolated from ripe berries.

Biological Activity : Hypotensive activity confirmed.

Uses in Literature : Used as adulterant of black pepper, anthelmintic, carminative, laxative, tonic, vegetable; and for burns, carbuncle, cholera, diarrhoea, fever, itching, piles, pleuritis, pneumonia, pregnancy problems, sores, throat complaints, toothache, tonsilitis, and ulcers, (**Ambasta, 1986; Hembrom, 1991; Jain, 1968, 1991; Jain *et al.*, 1991; Jha & Varma, 1996; Joshi *et al.*, 1980; Tarafdar, 1986; Tribedi *et al.*, 1982; Watt, 1972**).

Trade : Embelia.

Emblica officinalis Gaertn.+ (Pl. 26C).

Syn: *Phyllanthus emblica* L., *Cicca emblica* Kurz., *Dichelastina nodicaulis* Hance.

Family : Euphorbiaceae.

Vern.: Ambla, Amla.

English, Sanskrit & Regional Names

Eng. : Emblic Myrobalan, Indian Goose Berry;
Sans. : Adiphala, Amalaka, Dhatri;
B. : Amla, Amlaki;
G. : Amla, Amlika, Aonla;
H. : Amla, Amlika, Aonla;
Kan. : Amalaka, Nelli Kayi;
Mal. : Nelli;
Tam. : Nelli;
Tel. : Amalakamu.

Distribution : Throughout India, often cultivated: 1400m.

Description : A deciduous tree with small, narrowly linear leaves resembling pinnate leaves and yellow axillary flowers. Fruit fleshy, globose and pale coloured.

Reproductive Cycle : April-September.

Habitat Ecology : Common along borders of agricultural fields, forests.

Material Examined : EBH-608, 23-07-2001.

Parts Used : Fruit. Stem. Wood.

Folk Uses : Fruits eaten **raw** or cooked as **vegetable** and also pickled as rich source of **vitamin-C**. 5-7g powdered seeds (twice a day, for 15-20 days) effective for treating **piles**, **constipation**, **skin diseases**, promoting **appetite**, **liver disorders**, **spermatorrhoea**, **epilepsy** and as **diuretic**, **refrigerant** and **stomachic**. Also fruits form an important constituent of **'chyawanprash'**. Wood used for performing **'lakdi-rites'** in the cremation of dead body.

Active Constituents : Fruits afford trigalloylglucose, terchebin, corilagin, ellagic acid, indole acetic acid and four other auxins-a1, a3, a4 & a5, two growth inhibitors R_1 & R_2, tannins and vitamin C.

Biological Activity : Water soluble fraction of methanol extr. of leaves exhibits anti-inflammatory activity in carrageenin and dextran-induced rat hind paw oedema.

Uses in Literature : Used so far as anticancerous, antiscorbutic, aperient, appetiser (vet), astringent, detergent, diuretic, edible (eaten raw or cooked), expectorant, hair dye, hair shampoos, important constituent of 'chayawanprash' and 'triphala', inferior quality furniture, laxative, local drinks, magico-therapeutic to treat toothache, psychoactive plant, refrigerant, rich source of vitamin C, symbolic abode of God (sacred), tonic; and for agricultural implements, allaying thirst, anaemia, anthrax (vet), asthma, bad colds, biliousness, bronchitis, burns, clearing muddy water, confusion of thoughts, conjunctiva, diabetes, diarrhoea, disturbed cerebral circulation attended by a burning sensation at the crown of head, dropsy (vet), dyeing, dysentery (vet), dyspepsia (vet), enlarged liver and ascites, epistaxis (vet), eye complaints, fever, fractured bones, gastropathies (vet), genito-urinary discharges, gonorrhoea, haemorrhage, headache, heart complaints, horses in dysentery, impaction (vet), indigestion (vet), inflammation (vet), insanity, insect bites, irritability of the bladder, jaundice, leucorrhoea, liver (vet), loss of hair, madness, making preserve, menorrhagia, piles, pulmonary tuberculosis, pus formation in ear and earache, reviving taste, rheumatism (vet), rinderpest (vet), scurvy, skin diseases, stomach complaints, stomach diseases (vet), syphilis, tanning skin and garments, ulcer, vaginal complaints, vomiting, washing head, well rings, and wound maggots, (**Ambasta, 1986; Aminuddin & Girach, 1991, 1993; Banerjee & Pal, 1996; Borthakur, 1996a; Borthakur & Gogoi, 1994; Chauhan, 1999; Ghosh *et al.*, 1996; Girach, 1992; Hosagoudar & Henry, 1996a; Jain, 1968, 1991; Jain *et al.*, 1994; Khanna *et al.*, 1996b; Kothari & Moorthy, 1996; Kulkarni & Kumbhjokar, 1996; Kumar & Pullaiah, 2003; Lalramnghinglova, 1996; Mandal & Basu, 1996; Mehrotra et al., 1995; Mishra *et al.*, 1997; Negi *et al.*, 1993; Pandey & Verma, 2003; Rama Rao & Henry, 1996; Rana *et al.*, 1996; Sahoo & Mudgal, 1995; Sharma & Borthakur, 2003; Sharma & Sood, 1997; Sikarwar & Kaushik, 1992; Singh 1996b; Singh & Pandey, 1996; Singh & Prakash, 1994; Sinha, 1996; Siwakoti & Varma, 1996; Srivastava *et al.*, 2003; Tirkey *et al.*, 2003a; Vihari, 1995; Vishwanathan & Singh, 1996; Watt, 1972**).

Major Ayurvedic Formulations : Agnitundi-vati, Ambalyadi-churna, Amlaki-Churn, Amla-kyadi-churna, Arogya-vardhini, Brahma-rasayan, Brahmi-amla-tail, Brahmya-rasayana, Dhatrilauha, Dhatryadi-ghrita, Fresh fruit pulp is the main ingredient of 'Chayavanprash', Ghrit, Phalatrikudi-kwath, Raktapittank-taila, Sanjivani-vati, SG-1, Switradilepa-medigasone, Triphala-Churn, Triphala-Karanjaadi Yog.

Trade : Aamla, Emblic Myrobalan.

Eriobotyrica japonica Lindl.+ (Pl. 26D).

Syn.: *Mesilpus japonica* Banks.

Family : Rosaceae.

Vern.: Lokat.

English & Regional Names

Eng. : Chinese Medlar, Japan Quince Loquat;
H. : Logat;
Kan. : Lakkote;
Tam. : Ilakota.

Distribution : Usually cultivated in many parts of India.

Description : Trees with thick, 12-15 nerved coriaceous leaves and crowded, fragrant flowers. Fruits yellow, 2-5cm long.

Reproductive Cycle : July-September.

Habitat Ecology : Borders of agricultural fields, roadsides; prominent.

Material Examined : EBH-600,19-07-2001.

Parts Used : Fruit.

Folk Uses : Fruits **edible** and made into **jams** and **jellies**, and as an **appetiser**.

Active Constituents : A new leucocyanin-loquatoside-isolated from fruit pulp together with fructose, glucose and sucrose.

Biological Activity : Ursolic acid exhibits potent anticarcinogenic activity against sarcoma 180 cells and significantly suppresses c-myc oncogene expression.

Uses in Literature : Reported earlier in India for rulers and drawing materials, jams, jellies, pies, preserves; and as desert, expectorant, and sedative (**Ambasta, 1986; Watt 1972**).

Erythrina suberosa Roxb.+ (Pl. 27A).

Syn.: *E. glabrescens* Prain., *E. alba* Roxb., *E. nahasuta* Ham., *E. reniformis* Ham., *Micropteryx suberosa* Walp.

Family: Fabaceae.
Vern.: Paraira.

English, Sanskrit & Regional Names

Eng. : Indian Coral Tree;
Sans. : Bahupushpa, Kantaki, Parijata;
Ass. : Mandal;
B. : Palita Madar;
Bo. : Pangara;
G. : Jagriyo Khakharo;
H. : Dauldhak, Mandara, Rungra;
Kan. : Kaduparivala, Mulluharivana;
Mal. : Kalyanamurrikku;
Mar. : Pangra;
Oriya : Bonopalpdhua, Piri;
P. : Gulnastar, Pariara;
Tam. : Mullumurukku, Munimoduga;
Tel. : Badisa, Modugu.

Distribution : Common throughout India, often planted: 900-1200m.

Description : A tree with prickly branches, three leaflets, red flowers, and glabrous, cylindric stalked pods.

Reproductive Cycle : March-June.

Habitat Ecology : Forests, cultivated areas.

Material Examined : EBH-663, 12-04-2001.

Plant/Parts Used : Twig. Stem. The Whole Plant. Bark.

Folk Uses : Plant considered **sacred** and **worshipped** as incarnation of **Lord Vishnu**. Its twig hung on front door of the house to drive away **evil spirits**. Paste of bark in cow's urine good for curing **boils** and **skin eruptions**. Stem juice helpful in checking **earache**.

Active Constituents : Main ingredients are erythraline, erysodine, erysotrine, hypaphorine (seeds), erysotrine (leaves), delphinidin-3,5-diglucoside, cyanidin-3,5-digluco-side, erysodine, erysopine, erythraline, 11â-methoxyerythraline and its N-oxide, erythraline, 8-oxoerythrinine, erthratidine (flowers), myristic (0.13), palmitic (12.7), stearic (5.25), oleic (54.5), linoleic (20.44), arachidic (4.37) and behenic (2.62%) acids (seed oil), β-sitosterol, campesterol and stigmasterol (seed coat), and a new prenylated flavanone characterised as 5,7,3′,4′-tetrahydroxy-2′-(3″-methylbutyl-2″-enyl) flavonone (roots).

Biological Activity : Lecithin from seeds shows anti-H specificity.

Uses in Literature : Useful as bottle stoppers, cordage, crown cork liners, drums, psychoactive plant, scabbards (wood); and for destroying maggots in ulcer, dyeing, dysentery, insulation boards, jars storing ghee, laddles, lacquered boxes, match boxes, packing cases, planking, sieve frames, splints, and water troughs (**Ambasta 1986; Jain 1991; Jain *et al.*, 1994; Watt, 1972**).

Eucalyptus umbellata Domin+ (Pl. 27B).

Syn.: *E. tereticornis* Sm.

Family : Myrtaceae.

Vern.: Safeda.

English & Regional Names

Eng. : Blue Gum, Eucalyptus;

H. : Safeda;

Tam. : Karupuramaram.

Distribution : Introduced in India upto an altitude of 1200m.

Description : A tree with bark peeling in long thin layers and umbellate flowers. Operculum 2-4 times longer than calyx tube.

Reproductive Cycle : March-April.

Habitat Ecology : Commonly planted along roadsides.

Material Examined : EBH-1251, 12-04-2001.

Parts Used : Leaves. Stem Wood.

Folk Uses : Poultice of leaves checks **headache**. 1-2g powdered leaves given with honey, twice a day for 5 days to treat **cough** and **cold**. Also, smoke of burning leaves good for **repelling mosquitoes**. Wood good for **construction works**.

Active Constituents : Active principles are oleanoic and maslinic acids (leaves), borneol, camphene, cineole, cironellal, citronellyl acetate, p-coumarin, limonene, linalool, phellandrene, α- and β-pinenes, piperitone, γ-terpinene, terpen-1-en-4-ol and α-terpineol (essential oil), gallic acid, 7β-D-glucosyloxy-5-hydroxy-2-methylchromone, β-sitosterol and (+) catechin (wood).

Uses in Literature : So far used in India for engineering construction, sleepers, ship building, and flooring (**Ambasta, 1986**).

Major Ayurvedic Formulations : Panchgun-taila.

Eugenia jambolana Lam.[+] (Pl. 27C).

Syn.: *E. frondosa* Wall., *E. moorei* f. Muell., *E. caryophyllifolia* Lam., *Syzygium cumini* (L.) Skeels, *S. jambolanum* DC, *Calyptranthes jambolana* Willd., *C. capitellata* Ham., *Myrtus cumini* Linn.

Family : Myrtaceae.

Vern.: Jamun.

English, Sanskrit & Regional Names

Eng. : Black Plum, Java Plum;

Sans. : Jambaram, Rajaphala, Shukapriya;

Ass. : Jamu, Kola-Jamu;

B. : Chotajam, Kalajam;

Bo. : Jhambul;

G. : Jambudi, Ravanam;

H. : Jamni Phalani, Jamun, Phalanda;
Kan. : Nerale;
Mal. : Naga, Perinnaral;
Mar. : Jaman, Rajjambula;
Oriya : Bhotajamo, Chuajamo;
P. : Jammu;
Tam. : Kottainagam, Takkolam;
Tel. : Jambuvu, Peddoneredu.

Distribution : Throughout India.

Description : Evergreen trees with opposite, coriaceous leaves; white, fragrant flowers and dark purple fruits, globose, succulent.

Reproductive Cycle : April-August.

Habitat Ecology : Planted, near water sources in fields.

Material Examined : EBH-1071, 11-08-2000.

Parts Used : Fruits. Seeds. Leaves. Bark.

Folk Uses : Fruits **edible** and as **astringent**, Wood used for **construction work** and **agricultural tools**. Bark constituent of local wine '**sur**'. 10-12g of its powder given as single dose in **snake bite**. 3-5g mixture of its powdered leaves and those of *Aegle marmelos* and 'neem' (*Melia azadirachta* L.) taken with luke warm water on empty stomach in morning for treating **arthritis**. One teaspoonful of powdered seeds and leaves taken in the morning before meals for 30-45 days to cure **diabetes**. For curing **spermatorrhoea** and **joint pains**, 1-2g dose of a mixture comprising 500g of its fruit pulp, 25g 'ashwagandha' (*Withania somnifera* Dunal), 25g 'jayphal' (*Myristica fragrans* Houtt.) 25g 'jwati', 'clove' (*Syzygium aromaticum* L.) Merrill & Perry), 25g 'dalchini' (*Cinnamomum zeylanicum* Breyn.), given twice daily with cow's milk for 10-15 days.

Active Constituents : These are betulinic acid, friedelin, friedelinol, kaempferol and its 3-O-glucoside, quercetin, sitosterol and its glucoside, sucrose (stem bark), heptacosane, nonacosane, triacontane, hentriacontane, octacosanol, triacosanol, dotriacosanol, betulinic acid and crotegolic (masilinic) acid (leaves), two new anthocyanins- delphinidin-3-gentiobioside and malvidin-3-laminariboside- along with petunidin-3-gentiobioside (fruits), gallic and ellagic acids, corilagin and related ellagitannins, 3,6-hexahydroxyphenoylglucose and its isomer 4,6-hexahydroxydiphenoyl-glucose, 1-galloylglucose, 3-galloylglucose and quercetin (seeds).

Biological Activity : Anorexigenic (plant extr.) and hypoglycaemic (seed extr.) activities found (+) ve.

Uses in Literature : Known to be used as antidote to Nux vomica poisoning, astringent, colouring material in Mithila paintings, constuction wood, fodder, foliage for silkworms, green manure, mouthwash, props for shafts and galleries in mines, sacred (temple sculptures and symbolic abode of gods), tonic, vegetable; and for agricultural implements, anaemia, asthma, blisters in mouth, boat building, branches (festivals and marriages), bronchitis, calculous affections, cancer, cart-wheels, cholera (vet), colic, colouring fish nets, diabetes, diarrhoea, digestive complaints, dyeing and tanning, dysentery, edible purposes (fruits), enlargement of spleen, expelling placenta after delivery, fermentation for rice beer, furniture, giddiness, gums, indigestion, intestinal worms, jams, jellies, leucorrhoea, liver disorders, natural shiny black hair, piles, pimples, preserves, purifying blood, redness of eyes, sleepers, squashes, stomach pains, stored grain pest, swelling (vet), thatching, tool handles, troughs, ulcers, vinegar (ripe fruits), well-curbs, and wound maggots (**Ambasta, 1986; Aminuddin & Girach, 1993; Borthakur & Sarma, 1996; Chowdhery, 1996; Chowdhury *et al.*, 1996; Dixit & Pal, 1996; Gogoi *et al.*, 2003; Henry *et al.*, 1996; Hosagoudar & Henry, 1996a; Jain, 1968, 1991; Jha & Basak, 1994; Khanna *et al.*, 1996b; Kulkarni & Kumhojkar, 1996;**

Lalramnghinglova, 1996; Mandal & Basu, 1996; Masih, 2003; Mishra *et al.*, 1996b; Radhakrishnan *et al.*, 1996; Rama Rao & Henry, 1996; Shome *et al.*, 1996; Sikarwar & Kaushik, 1992; Singh, 1996b, Sinha, 1996b; Siwakoti & Varma, 1996; Tirkey *et al.*, 2003a; Watt, 1972).

Major Ayurvedic Formulations : Jambarista, Jambuadi-taila, Panchapallara-yoga, Panchapallava-yoga, Pusyanuga-churna, Usirasava.

Trade : Jaamun-guthli, Jambul-stone.

Euonymus echinatus Wall. *ex* Roxb. (Pl. 27D).

Syn.: *E. scandens* Grahm.

Family : Celastraceae.

Vern.: Goprala.

Distribution : Pakistan-Sikkim: 1500-2700m.

Description : An evergreen climbing shrub with tufts of aerial rootlets; leathery leaves; white orbicular flowers; and globose, prickly capsule.

Reproductive Cycle : May-November.

Habitat Ecology : Forests, on rocky slopes.

Material Examined : EBH-1236, 21-07-2002.

Parts Used : Leaves.

Folk Uses : Leaves used as **cattle feed.**

Active Constituents : Bark affords triacontane, cetyl alcohal, β-amyrin acetate and betulin.

Uses in Literature : Useful as galactagogue (vet).

Euphorbia pulcherrima Willd.+ (Pl. 28A).

Syn.: *Poinsettia pulcherrima* (Willd. *ex* Klotzsch) Graham

Family : Euphorbiaceae.

Vern.: Lalpatta.

English & Regional Names

Eng.	:	Christmas Star, Poinsettia;
B.	:	Lalpatta;
H.	:	Lalpatta;
Mar.	:	Lalpala;
Oriya	:	Panchutia;
P.	:	Lalpatti;
Tam.	:	Magilkunni.

Distribution : Himalaya: ascending to 2000m.

Description : Deciduous woody shrubs with alternate, ovate to lanceolate, entire toothed leaves and yellow flower heads surrounded by large vermillion-red bracts.

Reproductive Cycle : November-January.

Habitat Ecology : Planted as a hedge plant.

Material Examined : EBH-944, 27-11-2001.

Parts Used : Leaves.

Folk Uses : Planted as **ornamental shrub**. Decoction of its leaves along with '**baraein**' (*Acorus calamus*) and a pinch of rock salt used to relieve **cough**.

Active Constituents : Leaves possess germanicol acetate (0.25), germanicol (0.025), brein (0.0054), β-sitosterol (0.042), β-sitosteryl-3-D-glucoside (0.01), kaempferol-3-O-rutinoside (0.11) and rutin (0.12%).

Uses in Literature: Known to be used in India as colouration material in Mithila paintings, laxative, ornamental shrub; and for burning micturination, increasing the secretion of milk in nursing mothers, skin diseases, and wounds (**Binoj & Balakrishnan, 1996; Girach *et al.*, 1996; Jain, 1991; Jha & Basak, 1994; Kurian, 1995; Upadhye *et al.*, 1994; Watt, 1972**).

Euphorbia royleana Boiss.+ (Pl. 28B).

Syn.: *E. pentagona* Royle

Family : Euphorbiaceae.

Vern.: Chhuii, Chhun.

Regional Names

H. : Shakarpitan, Suli, Thor;

P. : Dandathor.

Distribution : N.W. Himalaya: 1800m.

Description : Cactus like shrub having angled branches; sessile, spathulate leaves with thorny persistent stipules; yellow involucres;. brown glands; and glabrous capsule.

Reproductive Cycle : June-July.

Habitat Ecology : Rocky slopes. Dry river valleys. Cultivated areas.

Material Examined : EBH-672, 21-07-2001.

Parts Used : Latex. Stem.

Folk Uses : Stem (inner portion) cooked as **vegetable**. Latex used for filling dental carries to check **toothache, cuts** and **wounds** and as **fish poison** and also in preparation of **'kazal'** (eyeliner). For **stomach discomforts** in cattle, a ball made of wheat flour containing its latex and a pinch of rock salt is said to be beneficial.

Active Constituents : Latex yields triterpenoids, taraxerol, epitaraxerol, euphol, ellagic acid, euphol, taraxerol, sitosterol m-hydroxy benzoic acid, 7-hydroxy-5,4-benzcoumarin, 7-methoxy-3,4-benzcoumarin, 2′,7-dihydroxy-3,4-benzcoumarin, α-amyrin, octa-cosanol, hexacosanol, tetracosanol, β-sitosterol, stigmasterol, and campesterol.

Biological Activity : Antiviral activity found to be (+) ve.

Uses in Literature : Used in India as anthelmintic, cathartic (latex), fuel wood, vegetable and 'Rayata'; and for boils, burns and cuts, diarrhoea and dysentery, ear complaints, fish poison, leprosy, skin diseases, sprains, and wounds (**Ambasta, 1986**; **Binoj & Balakrishnan, 1996**; **Jain, 1984, 1991**; **Jain *et al.*, 1991**; **Kapur & Nanda, 1996**; **Manandhar, 1987a**; **Shah & Joshi, 1971**; **Singh 1996b**; **Watt, 1972**).

Ficus bengalensis L.⁺ (Pl. 28C).

Syn.: *F. indica* L., *Urostigma benghalense* Gasp.

Family : Moraceae.

Vern.: Bad, Bare.

English, Sanskrit & Regional Names

Eng. : Banyan;

Sans. : Avaroha, Bahupada, Bhandira, Bhringi, Chira, Dhruva, Jatala, Jatiljati, Karmaja, Kshiri, Mahachhaya, Mana, Nandi, Nila, Nyagrodha, Padarohona, Raktaphala, Rohina, Skandhaja, Vaishravanodaya, Yakshaves, Yamapriya;

Ass. : Bot;

B. : Bar, Bot, But;

G. : Vad, Vadlo, Vor;

H. : Bar, Bargad, Bargat, Ber, Bor;

Kan. : Ala, Alada, Mara, Vata;

Mal. : Ala, Nyagrodham, Peral, Peralim, Peralu, Vatam;

Mar. : Vada, Wad, War;

Oriya : Boro, Boto;

P. : Bargad, Bera, Bohar, Bohir, Bor;

Tam. : Al, Ala, Alam, Eguvasam, Kadavam, Kagavaugam, Pali, Palu, Peral, Pudam, Pudavirujam, Vadavirukkam, Vekkurodam,;

Tel. : Bhandiramu, Mari, Matti, Pedamatti, Rohi, Vrikshadanamu.

Distribution : Throughout India upto 1400m.

Description : A large-sized spreading tree with numerous aerial roots; leathery, ovate, leaves; and pubescent, sessile, axillary, red figs.

Reproductive Cycle : February-June.

Habitat Ecology : Commonly planted as a shade tree at roadsides.

Material Examined : EBH-1089, 21-06-2002.

Parts Used : Roots. Stem. Fruits. Latex.

Folk Uses : Tree considered **sacred** and **worshipped**. Fruits edible and stimulate digestion. Latex applied on **cuts** and **wounds**, and also for suppuration of **boils**. Decoction of its roots checks **vomiting**. A dose (10g) comprising mixture of its powdered young stems and bark of *Ficus glomerata* (gular) in equal parts prescribed with cow's milk for **body tone** and **sexual vigour**.

Active Constituents : Chief active principles of stem-bark are methyl ethers of leucoanthocyanins-delphinidin-3-0-a-L-rhamnoside (I), pelar gonidin-3-0-a-L-rhamnoside (II) and leucocyanadin-3-O-b-D-galactosylcello-bioside (III) along with a methyl ether of leucoanthocyanidin, 20-tetra triacontene-2-one, pentatriacontan-5-one and 6-heptatriaconten-10-one, b-sitosterol-a-D-glucoside and mesoinositol.

Biological Activity : Bark exhibits hypoglycaemic activity in both normal and alloxan diabetic rats.

Uses in Literature : Reported so far as an antidote, aphrodisiac, astringent, diaphoretic, edible, fodder, plates, spermator, tonic, tooth-brush, vegetable (leaf buds); and for atrophy, bangy-poles, bird lime, bleeding and swelling of gum, boils, bone fracture, brass and bronze ornaments, burning sensation of burnt part of body, cart-shafts, cart-yokes, children in dysentery, cholera, coarse ropes, conception, cough, cracked or inflammed heel and toes, cuts, diabetes, diarrhoea, dysentery, eye complaints, fever, fibre, furniture, gastric (vet), gonorrhoea, haemoptysis, insanity, leucorrhoea, long and luxuriant growth of hair, lumbago, menorrhagia, neck injuries (vet), obstinate vomiting, oxidation of copper, pains and bruishes, paper-pulp, piles, pregnancy, rheumatism, scabies, sexual diseases, skin diseases, snake bite, sores in mouth, spermatorrhoea, stipules, syphilis, toothache, urinary complaints, well-curbs, wood in tent-poles, yokes, and in religious ceremonies (**Ambasta, 1986; Datt, 1996; Girach, 1992; Huidrom, 1996; Jain, 1991; Jain & Sharma, 1996; Kulkarni & Kumbhojkar, 1996; Manandhar, 1996; Masih, 2003; Mohanty & Padhy, 1996; Mukherjee & Namhata, 1990; Rama Rao & Henry, 1996; Sahoo & Mudgal, 1995; Saini, 1996a; Singh & Pandey, 1996; Singh & Prakash, 1994; Singh & Rao, 2003; Sinha 1996; Siwakoti & Varma, 1996; Tirkey** ***et al.*****, 2003a; Upadhye** ***et al.*****, 1994; Vihari, 1995; Watt, 1972**).

Major Ayurvedic Formulations : Nyagrodhadi-kvetha-churna, Nyagrodhadi-churna, Vatajatadi-taila.

Ficus clavata Wall. (Pl. 28D).

Syn.: *F. trachycarpa* Miq., *F. candata* Wall., *F. chincha* Roxb., *F. subincisa* Buch.-Ham. *ex* Sm.

Family : Moraceae.
Vern.: Ban Feguda.

Distribution : Lower hills of Himalaya: ascending to 1500m.

Description : An erect shrub having leaves with a narrow long pointed, axillary, shortly-stalked figs which are pale yellow on ripening.

Reproductive Cycle : June-September.

Habitat Ecology : Shady ravines, forests.

Material Examined : EBH-969, 23-07-2001.

Parts Used : Fruits. Leaves.

Folk Uses : Fruits **edible**, and leaves used as ocassional **cattle feed**.

Uses in Literature : Known to be used as fodder and vegetable (**Ambasta, 1986; Jain, 1991**).

Ficus foveolata Wall. (Pl. 29A).

Syn.: *F. sarmentosa* Ham., *F. erecta* Miq., *F. nipponica* Franch. & Savat., *F. thunbergii* Maxim., *F. impressa* Benth., *F. wrightii* Benth., *F. luducca* Roxb., *F. pubigera* Wall., *F. ludens* Wall., *Pogonotrophe pubigera* Miq., *P. verrucosa* Miq., *P. reticulata* Miq.

Family : Moraceae.
Vern.: Phaegdi.
Regional Name
P. : Garelu, Karmbal, Phogri.

Distribution : Outer Himalaya: 700-2400m.

Description : A shrub creeping over rocks and forming mats or climbing over trees. Leaves pubescent or glabrous. Figs axillary, solitary. Achenes oblong-reniform.

Reproductive Cycle : March-July.

Habitat Ecology : Steep banks.

Material Examined : EBH-297, 17-07-2001.

Parts Used : Leaves. Fruits.

Folk Uses : Leaves used as **fodder** and for **edible** fruits.

Uses in Literature : Reported earlier for edible purposes and fodder (**Ambasta, 1986; Jain, 1991; Watt, 1972**) .

Ficus glomerata Roxb.+ (Pl. 29B).

Syn.: *F. chitagonga* Miq., *F. racemosa* Wall., *F. mollis* Miq., *F. goolereca* Roxb., *Covellia glomerata* Miq., *C. mollis* Miq.

Family : Moraceae.

Vern.: Gular, Umara.

English, Sanskrit & Regional Names

Eng. : Cluster Fig;
Sans. : Jantuphala, Phalasambendha, Suprastish-thida;
Ass. : Jangedumuru, Yagyadimru;
B. : Jagyadumar;
Bo. : Umbar;
G. : Umbro;
H. : Gular, Umarai;
Kan. : Athi;
Kash. : Rumbal;
Mal. : Rumadi, Umbaram;
Oriya : Dimri, Odumboro;
P. : Batbar, Kakallam, Kathagular;
Tam. : Adavu, Anjiram, Malaiyinmmunivam, Tumbaram;
Tel. : Atti, Medi.

Distribution : Occurs in the Sub-Himalayan tract; outer hill ranges upto 1000m.

Description : Trees having tapering leaves which are pubescent above and glabrous beneath; reddish receptacles showing depressed umbilicous.

Reproductive Cycle : April-July.

Habitat Ecology : Commonly found growing in moist localities and bordering streams.

Material Examined : EBH-

Parts Used : Leaves. Flowers. Root-bark

Folk Uses : Plant considered **sacred** and **worshipped**. Wood used for performing '**havans**' during religious ceremonies. Paste of leaves considered antiseptic for healing open **wounds**. Decoction of roots either alone or in combination with 'kala zeera' (*Carum carvi*) and 'gur' (crude sugar) effective for treating **piles**, **diarrhoea**, and **gonorrhoea**. 2-4g mixture of its powdered bark (500g), 'saunf' (50g) (*Foeniculum vulgare*) and 50g sugar, given twice a day for 15 days to cure **liver disorders**.

Active Constituents : Active properties are due to lupeol, friedelin, b-sitosterol and stigmasterol, ceryl behenate, lupeol, its acetate, a-amyrin acetate (trunk bark), a tetracyclic triterpene, glucanol acetate (leaves) and leucopelarogonidin-3-0-a-rhamnopyrano-side, b-sitosterol and lupeol (root-bark).

Biological Activity : A glycoside rich fraction from leaves reported to have hypotensive, vasodilator and cardiac depressive effect. Also found to be antiprotozoal, hypoglycaemic and anticancerous.

Uses in Literature : Useful in India as an astringent, breakfast food, carminative, febrifuge, fodder, edible (fruits), galactagogue, powerful tonic, sacred, spermator, stomachic, toothbrush, vegetable, vulnerary, wood for shoring wells; and for asthma, bilious affections, blisters, blood dysentery, blood dysentery (vet), blood tissues, boils, bronchitis, cancer, curing piles, dermatitis, diabetes, diarrhoea, diseases of kidney and spleen, dislocation of joints, door, dry cough, dyeing, fuse-box, glandular enlargements, gonorrhoea, gum, habitual abortion, haemoptysis, healthy foetus (vet), hydrocele, lactating mothers to increase the secretion of milk, laryngitis, leprosy, leucorrhoea, match boxes, menorrhagia or excessive bleeding during mensturation, mouth sores, mumps, muscular pain, nose bleeding and for expelling intestinal worms, pain in throat, piles, plague (vet), planks and shutters, pneumonia, promoting fertility, pulmonary disease in children, rinderpest (vet), scabies, scorpion sting, skin crack on heels and lips, small pox, tone and vitality of the body, urinary and venereal complaints, washing wounds, well frames; and also to make effigy if corpse is lost or not available (**Ambasta, 1986; Hosagoudar & Henry, 1996a; Jain, 1991; Jain *et al.*, 1991; Jha *et al.*, 1996; Khanna *et al.*, 1996b; Kumar, 1996; Kurian, 1995; Malhotra & Moorthy, 1973; Mishra *et al.*, 1996a; Painuli & Maheswari, 1996; Rajendran *et al.*, 1996; Rama Rao & Henry, 1996; Saini, 1996a; Sharma, 1996; Singh & Pandey, 1996; Sinha, 1996; Watt, 1972**).

Major Ayurvedic Formulations : Mutrasamgrahaniya-kasaya-churna, Nyagrodhi-kvatha-churna, Udumbara-kavaleha, Udumba-ramrta, Udumbara-sara.

Ficus hispida L.[+] (Pl. 29C).

Syn.: *F. oppositifolia* Willd., *F. daemonum* Koem., *F. prominens* Wall., *F. mollis* Willd.,*F. scabra* Jacq., *Covellia daemonum* Miq., *C. oppositifolia* Gasp. Rich., *C. setulosa* Miq., *C. courtallensis* Miq., *C. wightiana* Miq., *C. assamica* Miq., *C. dasycarpa* Miq., *Sycomorpha roxburghii* Miq.

Family : Moraceae.

Vern.: Daebri.

Sanskrit & Regional Names

Sans.	:	Bhadrodumbarika, Kushthaghni, Phalguni;
Ass.	:	Khoskadumar;
B.	:	Dumoor, Kakdumar;
Bo.	:	Dhedu, Mira;
G.	:	Dhedaumuro, Janglianjir;
H.	:	Gobla, Katugularia, Totmila;
Kan.	:	Adavi atti, Kadatti;
Mal.	:	Erumanakku, Peyalti;
Mar.	:	Bhokada, Dherumbar,Kharoti;

Oriya : Bhaidimiri, Korotosamo;
P. : Daduri, Rumbal;
Tam. : Ottanalam, Sonatti;
Tel. : Boddamatti, Kukkabodda;
U.P. : Ban-gular.

Distribution : H.P. - Bhutan: 450-1100m.

Description : A small tree with broadly ovate opposite leaves; and shortly-stalked, globose, pale yellow figs.

Reproductive Cycle : June-August.

Habitat Ecology : Rocky slopes, occasionally along borders of agricultural fields.

Material Examined : EBH-1075, 14-08-2000.

Parts Used : Fruits. Latex. Stem Wood.

Folk Uses : Ripe fruits **edible** and **pickled**; and unripe ones and tender leaves cooked as **vegetable** for checking **flatulence**. 3-4g powdered bark in combination with 'soa' (*Anethum sowa* Roxb.) given with cow's milk for **spermatorrhoea**. Latex applied in **dental caries** as a relief from **toothache**. Wood used for making a roller in '**ghrat**' (countryside flour grinder).

Active Constituents : Therapeutic properties of bark are due to bergapten, psoralen, β-amyrin and β-sitosterol, hispidine phenanthraindolizidines, n-triacontanyl, and gluanol acetate.

Biological Activity : Stem bark antibacterial, antiprotozoal, anthelmintic and antiviral whereas root-bark found to be CNS stimulant and hypoglycaemic.

Uses in Literature : So far recorded to be used as an antiperiodic, astringent, cattle fodder, edible, emetic, fibre, galactagogue (vet), purgative, tonic, vegetable (leaves sometimes cooked with pork); and for blood dysentery, bubo (vet), curries, cuts, delivery (vet), diarrhoea (vet), earache, easy shedding of placenta (vet), giddiness, gonorrhoea, jams, headache, milching cattles to dry up their milk, preserving foetus in the womb, removal of placenta (vet), stomach ulcers, ulcers in mouth, warts, and wounds (**Ambasta, 1986; Geetha *et al.*, 1996b; Girach *et al.*, 1996, 1997; Jain, 1991; Mohanty & Padhy, 1996; Rama Rao & Henry, 1996; Saini, 1996a; Singh *et al.* 1996b; Sinha, 1986, 1987; Siwakoti & Varma, 1996; Watt, 1972**).

Ficus palmata Forssk.+ (Pl. 29D).

Syn.: *F. caricoides* Roxb., *F. pseudo-sycamorus* Decne., *F. virgata* Roxb.

Family : Moraceae.

Vern.: Dhuda, Fegura, Khasra, Phegura.

Regional Names

G.	:	Pepri;
H.	:	Anjiri, Khabara;
P.	:	Daholia, Jamir, Phaguro;
Tel.	:	Manjimedi;
U.P.	:	Ban-Gular.

Distribution	:	Himalaya: upto 2300m.
Description	:	Small trees. Leaves broadly ovate, toothed towards the top. Figs axillary, stalked, solitary, pear-shaped and purple colured.
Reproductive Cycle	:	April-September.
Habitat Ecology	:	Rocky slopes; along agricultural fields.
Material Examined	:	EBH-734, 15-09-2002.
Parts Used	:	Twigs. Fruit. Latex.

Folk Uses : Ripe fruits **edible** for curing **diabetes** and **lung affections** and unripe ones cooked and considered good as **laxative** and **appetiser**. Also, 1-2g mixture of its powdered fruits (500g) in combination with 25g black pepper and 25g rock salt, prescribed twice daily for 3-5 days with cow's milk to aid in **gastro-intestinal** disorders. Latex applied as an emollient for **ringworm**, and also to draw out **thorns** from skin.

Active Constituents : Leaves yield β-sitosterol and bergapten.

Uses in Literature : Recorded in India for its use as cattle fodder, demulcent, edible (fruits), laxative; and for bladder, building purposes, diseases of lungs, dysentery, high blood pressure, indigestion, jam, squash, pimples; and also to draw out thorns (**Ambasta, 1986; Ansari, 1991; Aswal, 1996; Jain, 1991; Kapur & Nanda, 1996; Lal *et al.*, 1996; Negi *et al.*, 1993; Parmar & Kaushal, 1982; Rana *et al.*, 1996; Saini, 1996a; Singh, 1996b; Sinha, 1996; Watt, 1972**).

Ficus religiosa L.⁺ (Pl. 30A, B).

Syn.: *F. affinior* Griff., *Urostigma religiosum* Gasp., *U. affine* Miq.

Family : Moraceae.

Vern.: Peepal.

English, Sanskrit & Regional Names

Eng. : Peepul Tree, Pipul Tree;

Sans. : Chaityavriksha, Pavitraka, Vriksharaja, Yajnika;

Ass. : Ahot;

B. : Ashathwa;

Bo. : Arli, Jari;

G. : Pipar, Pipro;

H. : Pipal, Pipli;

Kan. : Arali, Ashwuattha;

Kash. : Bad;

Mal. : Arayal, Chalodalam;

Mar. : Ashvatha, Pimpla;

Oriya	:	Oshwottho, Pippolo;
P.	:	Bhor, Pipal;
Tam.	:	Achuvattam, Podoroganam, Vanagardi;
Tel.	:	Ashvaddhamu, Bogaha.

Distribution : Widely cultivated in the Himalayas upto 1400m.

Description : A large glabrous tree with long-petioled, coriaceous leaves, which are shining above and dry beneath; and dark purple receptacle.

Reproductive Cycle : March- October.

Habitat Ecology : Usually planted alongwith *Ficus benghalensis.*

Material Examined : EBH-1197, 12-08-2000.

Parts Used : Stem. Leaves. Bark. Roots. Latex.

Folk Uses : Plant considered **sacred** and **worshipped**, and also its branches used in '**havans**' for performing religious rites. Ash of its burnt branches and branchlets applied on tongue to heal **tongue sores**. Poultice of leaves applied on tongue on boils for quick **discharge of pus**, and its powder (3-5g) given with cow's milk (twice a day for 5-15 days) to **regulate menstruation** and for **jaundice**, and **mental debelity**. Decoction of bark beneficial in **jaundice** and **spermatorrhoea**. 3-4g powdered fruits given thrice a day with luke warm water for curing **asthma** and **gastro-intestinal disorders**. Seeds given to women after third day of menstrual cycle to ensure **conception**.

Active Constituents : Bark contains tannins, b-sitosterol and D-glucoside.

Biological Activity : The stem-bark exhibits antiprotozoal activity against *E. histolytica*, anthelmintic activity against *Ascardia galli* and antiviral against Ranikhet disease virus.

Uses in Literature : So far reported for its use as an abortifacient, absorbent in inflammatory swelling, alterative, antidote to snake bite, astringent, colouration material and fixative, contraceptive, edible (fruits and tender leaf buds eaten in times of scarcity), fodder, laxative, psychoactive plant, sacred tree (worshipped); cutting tree more sinful than killing a saint, tonic, tree of fertility; and for after-pain (postnatal), aphthae, asthma, bird-lime, blisters, boils, bowels, bronchitis (vet), burns, carbuncle, cholera, chronic liver disorders, constipation (vet), cough (vet), dyeing, dysentery, eczema, enhancing memory, fever, fibre, fistula, foot and mouth disease (vet), gonorrhoea, gum, heamaturia, match-boxes, obstinate hiccups, paintings (Mithila), over intoxication by opium, packing cases, paralysis, rheumatism, scabies, skin affections, small pox, sores in mouth, spermatorrhoea, spoons, tanning, toothache, ulcer, urinary complaint, vermifuge (vet), whooping cough, and wounds (vet) (**Ambasta, 1986; Borthakur, 1996a; Gogoi & Borthakur, 1991; Gogoi *et al.*, 2003; Hosagoudar & Henry, 1996a; Jain, 1991; Jain *et al.*, 1991, 1994; Jha & Basak, 1994; Khanna *et al.*, 1996a, b; Masih, 2003; Mishra *et al.*, 1996 a; Molla & Roy, 1985; Rama Rao & Henry, 1996; Saini, 1996b; Sathapathy & Brahmam, 1996; Sharma, 1996; Sikarwar & Kaushik, 1992; Sikarwar, 1996; Singh & Prakash, 1996; Singh & Rao, 2003; Singh & Zaheer, 1996, Sinha, 1996; Siwakoti & Varma, 1996; Srivastava, 2003; Virjee *et al.*, 1984; Watt, 1972**).

Major Ayurvedic Formulations : Nyagrodhadi-churna, Nyagrodhadi-kvatha-churna.

Ficus roxburghii Wall. *ex* Miq.+ (Pl. 30C).

Syn.: *F. auriculata* Lour., *F. macrophylla* Griff., *F. regia* Miq., *F. scleroptera* Griff., *Covellia macrophylla* Miq.

Family : Moraceae.

Vern.: Traimbal.

Regional Names

B. : Demur, Doomoor;

H. : Timla, Tirmal;
P. : Dadari, Tirnal, Trimal, Trimbal, Urmul.

Distribution : Outer Himalaya: ascending to 1600m.

Description : A spreading tree. Leaves stalked, broadly ovate. Figs stalked, clustered on short thick, leafless branches, red or dull purple when ripe.

Reproductive Cycle : April-August.

Habitat Ecology : Wastelands; Rocky Slopes; Forests.

Material Examined : EBH-

Parts Used : Leaves. Fruit. Bark. Gum.

Folk Uses : Plant used as **fodder** for feeding **livestock.** Fruits **edible**, and its powder (250g) along with 'shilajeet' (25g) and 'saunf' (*Foeniculum vulgare*) (25g) beneficial in **spermatorrhoea** (twice daily for 5-7 days). Paste of its bark applied on **boils** for quick discharge of **pus**. Also, **ropes** are made from its bark.

Active Constituents : Constituents isolated from the plant are â-sitosterol, fridelin and epifriedelanol.

Uses in Literature : Used in India as edible (fruit used in curries), fodder, galactagogue (vet), vegetable; and for breast ulcers, delivery (vet), leucoderma, meal plates for feasts in villages, ring worm and ropes (bark) (**Jain, 1991; Parmar & Kaushal, 1982; Rama Rao & Henry, 1996; Watt, 1972**).

Ficus rumphii Bl.$^{+}$ (Pl. 30D).

Syn.: *F. cordifolia* Roxb., *Urostigma cordifolium* Miq., *U. rumphii* Miq.

Family : Moraceae.
Vern.: Badarein.

Regional Names

Ass.	:	Jakri, Pakri;
B.	:	Gelasvattna;
H.	:	Gagjaira, Pikan;
Kan.	:	Betaarali;
Mar.	:	Ashta, Payal;
P.	:	Badha, Pulakh, Rumbal.

Distribution : Throughout India: ascending to 1650m.

Description : Trees with long-petioled, broadly-ovate leaves and axillary black figs.

Reproductive Cycle : April-July.

Habitat Ecology : Shrubberies, open slopes.

Material Examined : EBH-723, 17-07-2001.

Parts Used : Wood. Leaves. Fruits.

Folk Uses : Wood used for performing **'havan'** and leaves lopped for **fodder**. Ripe fruits kept in storage containers for **repelling insects** and **borers**.

Active Constituents : Bark yields β-sitosterol and 3-hydroxy-3'-methoxy-flavone glucoside.

Uses in Literature : Known to be used as anthelmintic (fruit juice), edible, fodder (leaves and twigs), sacred; and for asthma, charcoal, fibre, fuel, gum, lice (vet), rearing lac insect, vomiting, wounds (vet) (**Ambasta, 1986; Jain, 1991; Watt 1972**).

Ficus semicordata Buch.-Ham. *ex* Smith⁺ (Pl. 31A).

Syn.: *F. cunia* Buch.-Ham. *ex* Roxb., *F. conglomerata* Roxb.

Family : Moraceae.
Vern.: Plakharein.

Distribution : India. S. China. Burma. S.E. Asia: 2000m.

Description : A small spreading tree with rough leaves, which are unequally lobed at the base, the lower lobe more rounded and much longer and reddish brown fruit.

Reproductive Cycle : May-September.

Habitat Ecology : Forests; common.

Material Examined : EBH-1191, 14-08-2001.

Parts Used : Leaves. Latex.

Folk Uses : Leaves used as **fodder**, and also used as **sand paper** while painting and varnishing. Its latex good against **dehydration**.

Uses in Literature : Recorded for its use as edible, fodder, sand paper (leaves), vegetable; and for bladder complaints, choking in throat (vet), dysentery, fibre, lac, leprosy, and viscereal obstructions (**Jain, 1991**; **Jain & Saklani, 1992**; **Lalramnghinglova, 1996**; **Negi & Gaur, 1991**; **Sahu, 1996**; **Siwakoti & Varma, 1996**; **Watt, 1972**).

Flacourtia ramontchi L' Herit+ (Pl. 31B).

Syn.: *F. indica* (Burm. f.) Merr., *F. sapida* Roxb., *F. sepiara* Roxb., *F. latifolia* Cooke, *F. occidentalis* Blatter, *Gmelina indica* Burm. f.

Family : Flacourtiaceae.

Vern.: Kaangu.

English, Sanskrit and Regional Names

Eng. : Governor's Plum, Madagascar Plum;

Sans. : Swadukantaka;

B. : Binja, Tambat;

Bo. : Kakad, Swadu;

G. : Kankod;

H. : Bhanber, Kakein, Katar;

Kan. : Hattarimullu, Humunki;
M.P. : Kankar;
Mar. : Kaker, Paker;
Oriya : Bonisha, Potnabonisho;
P. : Kakao, Kukai;
Tam. : Mullikkarai, Sottaikala;
Tel. : Bontakandregu, Nakkaneredu;
U.P. : Katai.

Distribution : Throughout India: upto 1000m.

Description : Trees. Armed with axillary thorns. Leaves variable, elleptic oblong, ovate. Flowers greenish-yellow. Fruit dark brown to red.

Reproductive Cycle : March-April.

Habitat Ecology : Edges of agricultural fields, wasteplaces.

Material Examined : EBH-695, 17-04-2000.

Parts Used : Leaves. Fruits. Stem Wood.

Folk Uses : Wood used for **turnery** and **agricultural implements**, and also as **fuel**. Ripe fruits **eaten** and **pickled**. Leaves lopped for **fodder**, and its paste good against **skin disorders.**

Active Constituents : These are tannin and flacourtin, a phenolic glucoside ester (bark), new butyrolactone lignan disaccharide-romontoside diphyllin-4-O-[β-D-glucosyl (1>4)]-β-2,3-di-O-methyl-D-methyl-D-xyloside (heartwood).

Biological Activity : Hypotensive activity and effect on CNS found positive.

Uses in Literature : Recorded for its use as an antidote to snake bite, appetiser, astringent, cattle fodder, diuretic, tarts, tonic after delivery; and for agricultural implements, backache in ladies, bite of mad dog, burning sensation in chest, carbuncle, chest pain, cholera, dysentery, eczema, enlarged spleen, excess body heat, facilitating child birth, fodder, foot and mouth disease (vet), gout, intermittent fever, leucorrhoea, liver disorders, pickling, rheumatic pains after delivery in women, skin diseases, small-pox, sores, syphilis, tanning, tonsilitis, turnery, wounds; and also to reshape abdomen after delivery (**Ambasta, 1986; Aminuddin & Girach, 1996; Brahmam & Saxena, 1990; Jain, 1991; Jain *et al.*, 1991; Katewa & Arora, 1997; Maheswari *et al.*, 1986; Nazarudeen *et al.*, 1996; Negi & Gaur, 1991; Painuli & Maheswari, 1996; Parmar & Kaushal, 1982; Rama Rao & Henry, 1996; Ramachandran & Nair, 1981; Saini, 1996a; Shah & Singh, 1990; Sikarwar & Kaushik, 1992; Singh, 1996b**).

Flemingia macrophylla Blume *ex* Miq.[+] (Pl. 31C).

Syn.: *F. congesta* Roxb., *F. prostrata* Roxb., *F. angustifolia* Roxb., *F. wallichii* W. & A., *F. semialata* Roxb., *F. stricta* Roxb., *Crotolaria macrophylla* (Willd.) Alston, *Maughania macrophylla* (Willd.) O. Kuntze, *M. prostrata* (Roxb.) Mukherjee, *M. semialata* (Roxb.) Mukherjee, *M. stricta* (Roxb.) O. Kuntze, *Rhynchosia crotalariodes* DC

Family : Fabaceae.

Vern.: Bada Salpan, Jangli Chana.

Regional Names

B.	:	Barasalpan, Bhalia;
Bo.	:	Dowdowla;
H.	:	Barasalpan, Bhalia, Kusunt, Supta;
Mal.	:	Kamatteri;
Mar.	:	Dowdowla;
Oriya	:	Bonokandulo.

Distribution : N.W. & C. Himalaya.

Description : An undershrub with trifoliate leaves and racemes about as long as petioles. Flowers white.

Reproductive Cycle : June-September.

Habitat Ecology : Sal forests, Forest clearings, Secondary forests.

Material Examined : EBH-603, 28-07-2001.

Parts Used : Twigs. Roots.

Folk Uses : Plant used as **green feed** for live-stock. Poultice of its roots effectively cures **boils** and for **neck swelling**.

Active Constituents : Chief principles are flemichins A, B, C, and D, flemiflavanones A, B and C (roots), vicenin-2, reynoutrin, rutin, myricitrin and robinin (leaves) and flemiphyllin (stems).

Biological Activity : Flemiflavanone D exhibits antimicrobial activity *in vitro* against *Staphylococcus aurens* and *Mycobacterium smegmatis*.

Uses in Literature : Used in India as an anthelmintic, edible (pods); and for blindness, cholera, cold, dyeing, dysentery, fever, fistula, pain in body muscles, skin diseases, small-pox, spleen complaints, stomachache, swellings, and ulcers (**Ambasta, 1986; Brahmam & Saxena, 1990; Jain, 1991; Watt, 1972**).

Flemingia strobilifera R. Br.+ (Pl. 31D).

Syn.: *F. abrupta* Wall., *Moghania strobilifera* Baker, *M. fruticulosa* Wall. *ex* Benth, *Hedysarum strobiliferum* L., *Zornaria Strobilifera* Pers.

Family : Fabaceae.

Vern. : Sophang.

Regional Names

Bo. : Kanphutti, Nundar;

H. : Kassunt;

Kan. : Kumala;

Mal. : Kumbilteri;
Mar. : Bundar, Kanphute;
Tel. : Nallbaddu.

Distribution : Himachal Pradesh-Bhutan. 300-2300m.

Description : Trailing shrubs with simple leaves, having rounded base; pink flowers; and oblong, turgid pod.

Reproductive Cycle : August-September.

Habitat Ecology : Sal forests; shrubberies.

Material Examined : EBH-747, 14-09-2001.

Parts Used : Leaves. Roots.

Folk Uses : Leaves as **fodder**, and also chewed for **expelling worms** from the body. Decoction (1 glass) of roots with buttermilk and a pinch of black pepper given as an **antidote** to **snake bite**.

Uses in Literature : So far used in India as an analgesic, antidote to insect sting, fodder, insecticide, psychoactive plant, sedative, tonic, tonic (vet), vermifuge; and for bodyache, epilepsy, fever, gastric complaints, hysteria, inducing sleep, leucorrhoea, relieving pain, skin (vet), skin diseases (**Ambasta, 1986; Bhandary *et al.*, 1996; Hosagoudar & Henry, 1996b; Jain, 1991; Jain *et al.*, 1994; Tosh, 1996; Watt, 1972**).

Fuchsia splendens (Plum.) L. (Pl. 32A).

Family : Onagraceae.
Vern. : Ghanti-phool.

Distribution : Planted.

Description : Densely branched shrub attaining 2m or more with light green pointed heart-shaped leaves and drooping flowers. Petals scarlet coloured and with pale green tips.

Reproductive Cycle : May-September.

Habitat Ecology : Cultivated.

Material Examined : EBH-996, 12-08-2002.

Parts Used : Flowers.

Folk Uses : Planted as an **ornamental.**

Gardenia angusta (L.) Merr. (Pl. 32B).

Syn.: *G. jasminoides* Ellis, *G. florida* (L.) Merr., *Verneria angusta* L.

Family : Rubiaceae.
Vern. : Taggar.
English, Sanskrit & Regional Names

Eng.	:	The Cape Jasmine;
Sans.	:	Gondharaja;
B.	:	Gandharaja;
G.	:	Gondhorojo;
H.	:	Gandharaja;
Mar.	:	Karinga;
Tam.	:	Karinga.

Distribution : Cultivated.

Description : A glabrous unarmed shrub with leaves narrowed at both ends; whitish-yellow fragrant flowers and orange, fleshy fruits.

Reproductive Cycle : July-October.

Habitat Ecology : Planted near houses.

Material Examined : EBH-130, 14-08-2000.

Parts Used : The Whole Plant. Leaves. Fruits.

Folk Uses : Grown as an **ornamental** plant. Leaves used for **perfuming tea**. 1-3g powdered fruits prescribed with cow's milk, twice a day for 8-10 days as **nervine tonic** and for **liver disorders**.

Active Constituents : Various constituents isolated from the plant are genipin-1-b-D-gentiobioside, geniposide, gardoside and scandoside methyl ester (fruits), genepin-b-D-glucoside (seeds), geniposidic acid agylcone, acetyl oleanolic acid, mannitol and stigmasterol (stems and roots), triterpenoid-gardenolic acid A, two new cycloartane triterpenoids–gardenic acid and gardenolic acid (flowers).

Biological Activity : Genipin and geniposide exhibit hypolipadaemic activity in hyperlipadaemic rats.

Uses in Literature : Known for its use as diuretic, emetic, stimulant; and for dyeing fabrics, dyspepsia, headache during pregnancy, hysteria, jaundice, nervous disorders, perfuming tea, pulmonary, and renal troubles (**Ambasta, 1986; Watt, 1972**).

Glochiodion velutinum Wight[+] (Pl. 32C, D).

Syn. : *Bradleia ovata* Wall., *Phyllanthus velutinus* Muell., *P. nepalensis* Muell.

Family : Euphorbiaceae.

Vern. : Shama, Shamba.

Regional Names

Ass.	:	Dolpoduli, Uding-That;
H.	:	Anwin, Chamari, Katu-manwa, Katumana, Koria;
M.P.	:	Koria;
Mal.	:	Kayara;
Mar.	:	Paritza, Showra;

P.	:	Gol-Kamila, Pundna, Sama;
Tam.	:	Panniccavu;
U.P.	:	Chamari, Katu, Manwa.

Distribution : Outer Himalaya upto 1650m.

Description : Trees with leaves in two rows; small, green yellow flowers in thick erect spikes and depressed globose capsule.

Reproductive Cycle : May-August.

Habitat Ecology : Grasslands, wasteplaces.

Material Examined : EBH-585, 12-08-2000.

Parts Used : Bark. Wood. Stem. Leaves.

Folk Uses : Wood used for **construction works** and as **fuel**, and its bark for **tanning** leather. Leaves and stem good for **scouring teeth.**

Active Constituents : Leaves afford glocidone.

Uses in Literature : So far recorded in India as fuel, leech repellent; and for tanning (**Ambasta, 1986; Hosagoudar & Henry, 1996a; Watt, 1972**).

Gmelina arborea Roxb.+ (Pl. 33A).

Syn. : *G. rheedii Hook., Premna arborea Roth, P. tomentosa Miq.*

Family : Verbenaceae.

Vern. : Gumbhari.

English, Sanskrit & Regional Names

Eng.	:	Canandhar Tree, Cashmere Tree, Coomb Teak, White Teak;
Sans.	:	Ashveta, Krishnavantika, Sarvatobhadrika;
Ass.	:	Gomari;

B. : Gamari, Gamargachha;
Bo. : Kasmari;
G. : Shivan;
H. : Gamari, Kambhar, Shiwan;
Kan. : Shivani, Kasmiri-Mara;
Mal. : Kumbili;
Mar. : Gamar, Shewan;
Oriya : Bhodroparni;
P. : Khambari, Kumhar;
Tam. : Kattanam, Pulirgumil;
Tel. : Adavigummudu, Peddagumudutekhu.

Distribution : Lower ranges of Himalaya upto 1000m.

Description : An unarmed tree with more or less acuminate leaves; corolla brown-yellow; and ovoid drupe.

Reproductive Cycle : March-June.

Habitat Ecology : Fairly common.

Material Examined : EBH-800, 12-05-2002.

Parts Used : Root. Bark. Wood. Flowers.

Folk Uses : Ripe fruits **edible** and unripe ones cooked as **vegetable**. Wood used for **ornamental** and **panelling** work, and bark constituent of '**chyawanprash**'. For proper secretion of milk during breast-feeding, its powdered roots (2-3g) given with 'mulathi' (*Glycyrrhiza glabra* L.) and honey, thrice a day for seven days. Decoction of leaves (5-10 ml) useful as **diuretic** (twice a day for five days).

Active Constituents : These are b-sitosterol, stigmasterol, stigmastanol, campesterol a-sitosterol and betulinol, oleic and linoleic acids (wood), hentriacontan-1-ol, ceryl alcohol, b-sitosterol, octacosanol and gmelinol (roots), and ketolignans-arborone, 7-oxodihydrogmelinol, paulownin acetate, epieudesmin, methyl trans-p-methoxylinnaniate and trans-p-hydroxy cinnamate acid (heartwood).

Uses in Literature : Known to be used in India as an anthelmintic, antidote to poisons, blood purifier, carminative, coagulant, demulcent, edible, fodder, psychoactive plant, soothing medicine, stomachic, tonic; and for agricultural implements, artificial limbs, asthma, axles, billious affections, boat decks, bone fracture, bridges, bronchial problem (vet), bronchitis, carriages, catarrh of bladder, cholera, colic, combs, constipation, cough, diarrhoea, dropsy, drums, dyeing, dysentery, epilepsy, fever, furniture, gonorrhoea, headache, house building, increasing secretion of milk, indigestion, lacquered boxed, malaria, musical instruments, ornamental panelling, phthisis, picture frames, planking, rearing silkworms, rheumatism, religious sculptures, rutting (vet), shafts, ship building, small-pox, sore (vet), sore of nails and soles (vet), spleen complaints, stomach disorders, swelling of throat (vet), syphilis, tea chest-plywood, ulcers, urticaria, venetian blinds, washing vaginal ostium, whitlow, wound (vet), and wounds (**Ambasta, 1986; Borthakur, 1996a; Chowdhery, 1996; Chhetri, 1994; Gogoi & Borthakur, 1991; Jain, 1991; Jain *et al.*, 1994; Lalramnghinglova, 1996; Mohanty & Padhy, 1996; Negi & Gaur, 1991; Rama Rao & Henry, 1996; Ranjan, 1996; Watt, 1972**)

Major Ayurvedic Formulations : Shriparnaadi-kwatha, Shriparni-taila, Brihat-panch-mulaadi-kwatha, Arvindasava, Draksadi-kvatha-churna, Dasamularishta, Dasamulaharita, Dasa-mula-satpalaka-ghrita, Amritarista, Brhatpanca-mula-kvatha.

Gossypium arboreum L.[+] (Pl. 33B).

Syn. : *G. indicum* Tod., *G. neglectum* Tod., *G. sanguineum* Hassk., *G. intermedium* Tod., *G. obustifolium* Roxb., *G. nanking* Meyen, *G. arboreum* L.

Family : Malvaceae.

Vern. : Kapah.

English, Sanskrit & Regional Names

Eng. : Tree Cotton;
Sans. : Karpasamu;
Bo. : Deokapas;
H. : Deokapas, Nurma;
Kan. : Hathi;
Mal. : Chemparutti;
Mar. : Devakapusa;
Oriya : Karposa, Kopa;
P. : Kapas;
Tam. : Sembarutti;
Tel. : Patti, Karpasamu.

Distribution : In the plains of India upto 1800m.

Description : A shrub. Leaves palmately 5-7 lobed. Flowers purple or yellow with a crimson centre.

Reproductive Cycle : July-September.

Habitat Ecology : Planted, near houses, temples, gardens.

Material Examined : EBH-985, 17-07-2001.

Parts Used : Seed-coat.

Folk Uses : Cotton fibres used for making **sacred thread** 'janau', and also used for making **cloth**.

Active Constituents : Active properties are due to (-) gossypol isolated from seeds.

Biological Activity : Gossypium-8-O-rhamnoside along with quercitin-3-7- and 3′-O-glucoside of quercitin exhibits antibacterial activity against *Pseudomonas maltophila* and *Enterbacter cloacae*.

Uses in Literature : Known earlier as soothing agent, symbolic for making sacred thread; and for conjunctivitis of infants, earache, fever, fibre, gonorrhoea, insanity, labour complaints, skin eruptions, strangulary, and patellar dislocation (**Khanna *et al.*, 1996b; Sharma & Borthakur, 2003; Watt, 1972**).

Grevillea robusta A. Cunn. *ex* R. Br.⁺ (Pl. 33C).

Family : Proteaceae

Vern. : Silver Oak.

English & Regional Name

Eng.	:	Silky Oak, Silver Oak;
Tam.	:	Savukkumaram.

Distribution : Planted in gardens and avenues as a shade tree.

Description : An evergreen tree with pinnate leaves; leaflets deeply pinnate. Flowers orange in clusters. Fruits obliquely-oblong, compressed. Fruit brown.

Reproductive Cycle : March-July.

Habitat Ecology : Often planted along roadsides in Himalaya.

Material Examined : EBH-451, 18-05-2000.

Parts Used : Wood.

Folk Uses : Plant considered excellent **source of nectar** for bee keeping. Wood extensively employed for **ornamental work.**

Active Constituents : Wood yields grevillo, sitosterol, epilyeniresinol, its 2-O-glucoside, rhamnocitrin-3-O-rutinoside and kaempferol-3-O-rutinoside.

Uses in : So far known as edible, green manure; and for floors,

Literature furniture, gum, ornamental panelling, parquet, plywood, tannin, toys, and veneering (**Ambasta, 1986; Sinha, 1996**).

Grewia oppositifolia Roxb. (Pl. 33D).

Syn. : *G. optiva* Drumm.

Family : Tiliaceae.

Vern. : Beul.

Regional Names

H. : Biul, Biung, Bhimal;

Kan. : Thidsal;

P. : Behal, Dhaman, Pharwa.

Distribution : N. W. Himalaya upto 1800m.

Description : Trees having 3-nerved, toothed leaves. Petals yellow. Drupe olive-green, black when ripe.

Reproductive Cycle : April-June.

Habitat Ecology : Edges of agricultural fields, wasteplaces.

Material Examined : EBH-48, 15-04-1999.

Parts Used : Leaves. Branches. Wood. Bark.

Folk Uses : Leaves and branches lopped for **fodder** and fruits **edible**. Fibres from branches used for making **ropes**. Branches used as **torch** in religious festivals. Ash of burnt wood and a pinch of rock salt applied as a poultice to ensure relief from **internal injuries**. Extr. of roots used as **shampoo**. Paste of its bark applied for killing lice in humans as well as in animals, and its poultice in cow's urine effective for healing **fractured bones** and **rheumatic pains**.

Uses in Literature : Used in India as detergent, edible (fruit), fuel, galactagogue (vet), lubricant during child birth, nutritive fodder, shampoo, symbolic torch in religious festivals; and for agricultural implements, axe-handles, banghy-poles, basketry, boils, blisters, bone fracture (vet), bows, clothing, cordage, cot frames, cough, dyeing, eruptions, fever, gastric problem (vet), increasing milk-giving powers, indigestion (vet), material requirement, oar-shafts, paddles, ropes, tool and axe handles (**Ambasta, 1986; Hosagoudar & Henry, 1996a; Jain, 1991; Kapur, 1996; Lata, 1996; Rana *et al.*, 1996; Singh, 1996b; Watt, 1972**).

Hedera helix auct (non L.) Clarke⁺ (Pl. 34A).

Syn : *H. nepalensis* K. Koch, *H. himalaica* Tobler, *H. rhombea* Sieb. & Zucc.

Family : Araliaceae.

Vern. : Dakari.

English & Regional Names

Eng.	:	Barren Ivy, Beene Tree, Bent Wood, Black Ivy, Creeping Ivy, Nepal Ivy, Wood Bind;
Ass.	:	Mej-Peosree;
H.	:	Lab-Lab;
Mal.	:	Maravala;
P.	:	Banbakri, Banda, Kadloli, Karbaru;
Tam.	:	Maravalai.

Distribution : N.W. Himalaya: 1800-3000m.

Description : A woody climber, climbing by aerial rootlets. Leaves thick, 3-5 lobed, shining ovate. Flowers yellow-green. Fruits black.

Reproductive Cycle : June-November.

Habitat Ecology : Forests.

Material Examined : EBH-69, 11-08-1999.

Parts Used : Young Branches. Stem. Leaves. Bark.

Folk Uses : Berries **edible**, its branches and leaves lopped for **fodder**. Leaves and bark constituent of local wine, '**sur**'. One teaspoonful of powdered climber prescribed with cow's milk, twice a day for **gastric disorders**. 10 ml decoction of bark prescribed twice daily, for 3-5 days as **stomachic**. Also, massage of paste of its stem along with seeds of *Pinus roxburghi* and 'alsi oil' (*Linum usitatissimum* L.) effective against **arthritis**.

Active Constituents : These contain rutin, kaempfecol-3-glucoside, coumaric acid, ferulic acid, chlorogenic acid, scopoline (bark), a-hederin, hederasaponin C, a-hederin and two saponin complexes- CS60 and CS90 (leaves), eleven triterpene glycosides - kizuta saponine K2, K4, K5, K7, K7A, K7B, K7C, K8, K9, K11 and K13 (stem-bark).

Biological Activity : Hederagenin exhibits highly significant anti-inflammatory activity against carrageenin-induced oedema and formaldehyde induced arthritis in cat.

Uses in Literature : Recorded in India as cathartic, diaphoretic, purgative, stimulant; and for fodder, mumps, and sores (**Ambasta, 1986**; **Chhetri, 1994**; **Watt, 1972**).

Hibiscus rosa sinensis L.+ (Pl. 34B).

Family : Malvaceae.

Vern. : China Rose, Shoe Flower.

English, Sanskrit & Regional Names

Eng.	:	Chinese Hibiscus, Shoe Flower;
Sans.	:	Hariballabha, Odhrapushpa;
Ass.	:	Joba;
B.	:	Jiwa, Oru;
Bo.	:	Jasavanda;
G.	:	Jasuva;
H.	:	Gulhar; Jasum, Jasut;
Kan.	:	Dasavala;
Mal.	:	Ayamparutti, Jampa;
Mar.	:	Dasindachapula, Jasavanda;

Oriya	:	Mondaro, Ongribollika;
P.	:	Jasum;
Tam.	:	Irattaichegappuchembrattam.

Distribution : Cultivated Throughout India.

Description : Shrubs. Leaves coarsely toothed at apex and entire at base. Corolla red, capsule many seeded.

Reproductive Cycle : February-October.

Habitat Ecology : Cultivated in gardens.

Material Examined : EBH-1190, 28-01-2002.

Parts Used : Flowers.

Folk Uses : Flowers **sacred** and offered to deities in **religious worshipping**, and its paste kept overnight in bronze vessel, given on empty stomach in morning for the cure of **piles**.

Active Constituents : Active properties are due to quercetin-3-diglucoside 3,7-diglucoside, cyanidin-3,5-diglucoside and cyanidin-3-sophoroside-5-glucoside (deep yellow flowers), and kaempferol-3-xylosylglucoside (ivory white flowers).

Biological Activity : Hypotensive and antispasmodic activities found (+) ve.

Uses in Literature: Known to be used in India as abortifacient, colouration material (Mithila paintings), anodyne, demulcent, dye, emollient, laxative, refrigerant drink in fevers, symbolic (offering in chronic diarrhoea), vegetable (flowers eaten raw or pickled); and for antifertility, appendicitis, asthma, bleeding piles, boils, bronchitis, cancerous swellings, child birth, colouring syrups, consumption, cough, developing foetus (vet), diarrhoea, dye, flu, gonorrhoea, hair dye, headache, impotency, inflammation, irritable conditions of genito-urinary tract, kidney diseases, marriage ceremonies (flowers), menstrual complaints, mumps, offerings and adornments, postnatal haemorrhage, scabies (vet), skin diseases, soothing internal and external wounds, sores, and syphilis (**Ambasta, 1986; Banerjee, 1996; Banerjee & Pal, 1996; Chowdhury** ***et al.*****, 1996; Datt, 1996; Gill** ***et al.*****, 1993; Girach, 1992; Girach** ***et al.*****, 1996; Huidrom, 1996; Jain, 1991; Jha & Basak, 1994; Jha** ***et al.*****, 1996; Kurian, 1995; Prasad** ***et al.*****, 1996; Singh & Pandey, 1996; Vihari, 1995; Watt, 1972**).

Major Ayurvedic Formulations : Japa-taila.

Holarrhena antidysentrica Wall. (Pl. 34C).

Syn: H. pubescens Wall., H. codaga G. Don, H. malaccensis Wight, *Echites antidysentrica* Roxb., *E. pubescens* Buch.-Ham., *Wrightia antidysentrica* Grah.

Family : Apocynaceae.

Vern. : Keor, Kurchi.

English, Sanskrit & Regional Names

Eng.	:	Antidysentric Holarrhena, Conessi Bark, Kurchi;
Sans.	:	Girimallika,Mallikapushpa, Shakraparyaya;
Ass.	:	Dudcory, Dhutkhuri;
B.	:	Kurchi, Titaindaraju;
Bo.	:	Dolakura, Pandhrakura;
G.	:	Dhowda, Indrajwannujad;
H.	:	Dhudi, Karvaidrajau;
Kan.	:	Beppak, Koodsaloo, Korchei;
Kash.	:	Keor, Kewar;
M.P.	:	Kudo, Kurato;
Mal.	:	Kaippakkotakappala, Venpala;
Mar.	:	Bedaki, Pandharakuda;
Oriya	:	Kherwa, Patrukurwan, Potrakaya;
P.	:	Kewar, Koeva, Kogur;
Tam.	:	Erukkalaipalai, Vattagam;
Tel.	:	Istrarakupala, Peddankuda, Vistarakuphala.

Distribution : Throughout India; ascending to 1200m.

Description : A small deciduous tree. Leaves with rounded base, arched. Flowers white, inodorous. Seeds glabrous, crowned with a tuft of long silky hairs.

Reproductive Cycle : June-February.

Habitat Ecology : Often gregarious in deciduous forests and open slopes.

Material Examined : EBH-705, 08-02-2002.

Parts Used : Root. Seeds. Bark.

Folk Uses : Powdered roots used as **snuff** to check **nose bleeding**. 5-10 ml decoction of its seeds and/or bark useful against **dysentery** and **gastric disorders** (thrice daily for 3-5 days).

Active Constituents : The principle alkaloids isolated from bark are conessine, kurchicine, kurchine, conamine, conkurchine, holarrhenine, holarrhinine, holarrhine and lettocine.

Biological Activity : Antiprotozoal, anticancerous, hypoglycaemic and spasmolytic activities found positive. (**Jain *et al.*, 1991**).

Uses in Literature : So far known in India as an anthelmintic, antidote to snake bite, antidysenteric, antiperiodic, antipyretic, antitetanic, aphrodisiac, appetiser (vet), astringent, bitter carminative, conceptive, diuretic, expectorant, febrifugal, fodder, lithnotropic, psychoactive plant, spermator, tonic; and for abnormal pain, anaemia, anthrax (vet), asthma, biliousness, bleeding from nose, blisters (vet), bobbins, boils, bowels, bronchitis, bronchopneumonia, carved boxes, chest complaints, cold, colic, combs, constipation, cough, curdling milk, cuts, dandruff, diarrhoea, diarrhoea (vet), dropsy, dyeing, dyspepsia, easing delivery, eczema, epilepsy, fatigue, flatulence, furniture, gastric disorder, gout, gums, haematuria, haemorrhages after child birth, hallucinations, headache, increases secretion of milk, inflammation of uterus, influenza, killing lice, lessening inflammation to strengthen and giving tone to the stomach, leucoderma, liver ailments, lumbago or pain in the lumbar region, mathematical instruments, menorrhagia, neck sore (vet), ornamental beads worn around the neck as charm, paralysis, pen holders, picture frames, piles, plague, reviving taste, rheumatic pain, skin affections, spleen complaint, spoons, stomach ulcers, stomachache, toys, malarial fever, ulcers, vomiting and nausea, worms, and wounds (vet) (**Ambasta, 1986; Aminuddin & Girach, 1991, 1996; Baruah & Sarma, 1984; Borthakur & Sarma, 1996; Brahmam & Saxena, 1990; Chauhan, 1999; Datt, 1996; Girach, 1992; Girach & Aminuddin, 1995; Gupta, 1981; Jain, 1968; Jain *et al.*, 1991, 1994, Kothari & Moorthy, 1996; Kumar, 1996; Kurian 1995; Lalramnghinglova, 1996; Masih, 2003; Rama Rao & Henry 1996; Shah *et al.*, 1983; Sharma & Sood, 1997; Singh & Rao, 2003; Singh 1996a; Siwakoti & Varma, 1996; Srivastava *et al.*, 2003; Tirkey *et al.*, 2003a; Upadhye *et al.*, 1994; Varma, 1997; Watt, 1972**).

Major Ayurvedic Formulations : Amiritastak-kwatha, Amritarishta, Bhuinimbadi-churna, Brihat-sudarshan-churna, Diarex, Karpur-ras, Kotari-rasa, Krimikuthar-ras, Kuja-rishta, Kutajaghana-vati, Kutajapartpati, Kutajarishta, Kutajastakakvatha, Kutajavaleha, Kutajghan-vati, Punarnava-mandur, Sarvajawarhava-laghuganga-dhar-churna, Sudarshan-churna, Yograja-gugglu

Trade : Indrajau karwaa.

Hydrangea grandiflora Sieb.

Syn : *H. paniculata* Sieb. var. *grandiflora*

Family : Hydrangeaceae.

Vern. : Hansraju.

Distribution : Planted.

Description : A hardy shrub with broad oval, prominent veined leaves and large, flattened or domed heads of bloom in shades of pink, purpkle and white.

Reproductive Cycle : April-September.

Habitat Ecology : Commonly met as an escape.

Material Examined : EBH-477, 18-05-2000.

Parts Used : Inflorescence.

Folk Uses : Grown for its **beautiful inflorescence**.

Uses in Literature : Blossoms contain rutin used in capillary fragility (**Ambasta, 1986**).

Hypericum cernuum Roxb.⁺ (Pl. 35A).

Syn.: *H. oblogifolium* Choisy, *H. speciosum* Wall., *H. govanianum* Wall.

Family : Hypericaceae.
Vern. : Pinli.

Distribution : W. Himalaya: 800-2100m.

Description : Glabrous shrubs. Leaves ovate, each pair of leaves placed at right angles to the next pair, above or below it. Flowers yellow. Capsule conical.

Reproductive Cycle : April-October.

Habitat Ecology : Shrubberies, Open slopes; common.

Material Examined : EBH-379, 14-08-1999.

Parts Used : Twigs. Seeds.

Folk Uses : Twigs used as **toothbrush** and seeds as a condiment for **flavouring curries**.

Uses in Literature : Known earlier as an aromatic, poison (vet), stimulant; and for old sores, eczema, and toothache (**Kapur, 1996**).

Hypericum patulum Thunb.[+] (Pl. 35B).

Syn.: *H. uralum* Ham., *H. oblogifolium* Wall.

Family : Hypericaceae.
Vern. : Kuradu.
Regional Name
Eng. : St. John's Wort;
Ass. : La-syn-rit.

Distribution : Temperate Himalaya: 1000-2350m.

Description : A shrub with oblong-lanceolate leaves. Each pair placed at right angles to the next.

Reproductive Cycle : August-September.

Habitat Ecology : Hilly slopes, moist places.

Material Examined : EBH-1208, 24-08-2002.

Parts Used : Flowers. Root.

Folk Uses : Flowers **poisnous** to eat. Poultice of crushed roots useful against **boils**.

Uses in Literature : Used in India as an aromatic (seeds), and stimulant (**Ambasta, 1986; Watt, 1972**).

Ilex dipyrena Wall.[+] (Pl. 35C).

Family : Aquifoliaceae.
Vern. : Charkha, Kalucha.

English Name
Eng. : Himlayan Holly.

Distribution : N. W. Himalaya: 1500-3000m.

Description : Trees. Leaves ovate-lanceolate with strong thorny teeth. Flowers white. Drupe scarlet, globose.

Reproductive Cycle : April-October.

Habitat Ecology : Forest, damp ravines, common in Oak forests.

Material Examined : EBH-894, 14-04-2001.

Parts Used : Leaves. Wood.

Folk Uses : Wood used for low grade **furniture** and leaves as **fodder**, and also its paste considered good for **nail disorders**.

Uses in Literature : Reported earlier for its use as diuretic, ocassional fodder, purgative, fodder, inferior quality timber (**Ambasta, 1986; Kapur, 1996; Watt, 1972**).

Ilex excelsa (Wall.) Hook. f.+ (Pl. 35D).

Syn : *I. doniana* DC, *Cassine excelsa* Wall.

Family : Aquifoliaceae.
Vern. Name : Mathedgi.

Distribution : Sub-Tropical Himalaya. H.P.-Sikkim.

Description : Trees. Leaves fleshy. Flowers greenish-white. Drupes with trigonous stones, bright red.

Reproductive Cycle : May-October.

Habitat Ecology : Forests, ravines, hilly slopes.

Material Examined : EBH-1104, 26-07-2002.

Parts Used : Leaves.

Folk Uses : Leaves lopped for **fodder** and considered excellent for **milching cattles**.

Uses in Literature : Used earlier as inferior quality timber (**Ambasta, 1986**).

Indigofera gerardiana Wall.+ (Pl. 35A).

Syn: *I. rubroviolacea* Dunn, *D. heterantha* Wall. *ex* Brandis, *I. dosua* Wall., *I. virgata* Roxb., *I. polyphylla* DC

Family : Fabaceae.

Vern. : Kali-kathi.

English and Regional Name

Eng. : Himalayan Indigo;

P. : Kati, Katsu, Khenti, Kutz, Mathu, Shagali.

Distribution : Temperate Himalaya: 1500-3000m.

Description : Shrubs with silvery pubescent branches. Leaflets 9-17. Flowers in pedunculate receme, pale-red.

Reproductive Cycle : May-July.

Habitat Ecology : Shrubberies; W.Himalaya.

Material Examined : EBH-1237, 21-07-2002.

Parts Used : Stem. Bark. Roots. Wood.

Folk Uses : Stem used for **scouring teeth**. Powdered stem-bark useful against **toothache** (2-3g twice daily for 5 days). For healing **internal injuries**, its powdered root along with 'haldi' (*Curcuma longa* L.), in equal parts given with milk, twice a day for 10-15 days. Wood used as **fuel**.

Active Constituents : The plant yields apigenin, acacetin, isoquercitin, ursolic acid and β-sitosterol.

Biological Activity : Aerial parts anticancerous.

Uses in Literature : Known to be used as vegetable; and for basket work, rope bridges, toothache, urine complaints (vet) (**Ambasta, 1986**; **Jain, 1991**; **Kapur & Singh, 1996**; **Watt, 1972**).

Indigofera pulchella Roxb.+ (Pl. 36B).

Syn: *I. cassoides* Rottl. *ex* DC, *I. purpurascens* Roxb., *I. glaucescens* Grah., *I. elliptica* Roxb., *I. violacea* Roxb., *I. arborea* Roxb., *I. jirahulia* Hamilt, *I. gibsonii* Grah.

Family : Fabaceae.

Vern. : Kathi.

Regional Names

B.	:	Nil;
G.	:	Gali, Gari, Nil;
H.	:	Hakna, Nil;
Kan.	:	Gogge;
Mal.	:	Manali, Nilam;
Mar.	:	Baroli;
Oriya	:	Girili;
P.	:	Kainthi, Kathi;
Tam.	:	Nairnji;
Tel.	:	Siralli, Vuyye.

Distribution : Kashmir-Bhutan. 300-1500m.

Description : Shrubs. Leaflets 11-21, opposite, oblong-ovate. Flowers bright pink. Pod straight, turgid, glabrous.

Reproductive Cycle : July-November.

Habitat Ecology : Forests, open slopes, grows common in forests having *Pinus roxburghii* and *Shorea robusta*.

Material Examined : EBH-318, 17-10-2000.

Parts Used : Flowers. Leaves.

Folk Uses : Extr. of flowers useful against **bleeding from nose**. 3-4g mixture of its powdered leaves (500g) with those of 100g 'bahera' (*Terminalia bellerica*), 100g 'amla' (*Emblica officinalis*), 25g 'black pepper' (*Piper nigrum*), given with milk for **kidney affections, liver disorders** and **discomforts in stomach**. Also, powdered roots (1-2g) prescribed with honey for **cough** and **chest pain**.

Biological Activity : EtOH (50%) extr. of root exhibits antiviral and spasmolytic activities.

Uses in Literature : Recorded for its use as vegetable (flower buds and young flowers), blennorrhagia, conception, cough, epilepsy, fever, green manure, hepatitis, labour, menorrhagia, mucous in urine, nervous disorders, old ulcers, piles, sores, swelling of face in labour, urinary complaints (**Ambasta, 1986; Dwarakan & Ansari, 1996; Jain, 1991; Rama Rao & Henry, 1996; Watt, 1972**).

Inula cappa DC⁺ (Pl. 36C).

Syn: *I. pseudo-cappa* DC., *I. eriophora* DC., *I. saliodara* Schultz.-Bip, *Conyza cappa* Ham., *C. lanuginosa* Wall., *C. eriophora* Wall., *C. argentea* Wall., *Duhaldea chinensis* DC.

Family : Asteraceae.
Vern. : Basanti.

Distribution : H.P. - S.W. China: 1000-2400m.

Description : Aromatic shrubs. Leaves oblong-lanceolate. Flowers yellow, numerous. Pappus thick, leathery.

Reproductive Cycle : May-October.

Habitat Ecology : Shrubberies, open slopes, often gregarious.

Material Examined : EBH-646, 14-08-2001.

Parts Used : Buds. Flowers.

Folk Uses : Dried buds and flowers used for **decorative** purposes.

Active Constituents : 3,5,7,2′-tetrahydroxy-5′-methoxyflavanone (I) 5,7,2′,5′-tetrahydroxyflavanone (II) and 3,5,2′-tri-hydroxy-7,5′-dimethoxyflavone (III), four inositol tetraangelayes isolated from aerial parts.

Uses in Literature : Known to be used for fever and indigestion (**Manandhar, 1980b**).

Ipomoea cairica (L.) Sweet⁺ (Pl. 36D).

Syn: *I. palmata* Forssk., *I. senegalensis* Lamk., *I. stipulaceae* Jacq., *I. pendula* Br *I. pulchella* Roth, *I. tuberculata* Roem. & Sch., *Convulvulus cairicus* L., *C. tuberculatus* Lamk., *C. bellus* Spreng., *C. digitalis* Roxb., *C. heptaphyllus* Rottl., *C. lupulifoilus* Griff.

Family : Convolvulaceae.

Vern. : Railway Creeper.

English Name

Eng. : Railway Creeper.

Distribution : Deccan peninsula, Nepal, H. P.

Description : Twinner. Leaves palmate, 5-7 partite. Flowers white-light pink.

Reproductive Cycle : June-November.

Habitat Ecology : Roadsides, forests edges, hedges, wasteplaces; common.

Material Examined : EBH-40, 12-08-1999.

Parts Used : Leaves. Flowers.

Folk Uses : Leaves used as **fodder**. Decoction of its leaves and flowers useful against **constipation**.

Active Constituents : Leaves possess ergosnine, ergocornine and ergocistrine.

Uses in Literature : So far known as cyanogenetic, food, purgative; and for coughs, and pain (**Ambasta, 1986**; **Jain, 1991**).

Ipomoea carnea Jacq.$^{+}$(Pl. 37A).

Syn: *I. fistulosa* Mart. *ex* Choisy, *I. crassicaulis* (Benth.) Robinson, *Balatas crassicaulis* Benth.

Family : Convolvulaceae.
Vern. : Bisharm Vuti, Nakti Buti, Vilayati Ak.

Distribution : W. Himalaya & Nepal: 1400m.

Description : A shrub with hollow stem. Leaves long-stalked, ovate-triangular. Flowers light pink. Seeds brown, villous.

Reproductive Cycle : July-September.

Habitat Ecology : Wastelands, hedges, ditches; often gregarious.

Material Examined : EBH-682, 21-07-2001.

Plant/Parts Used : Leaves. The Whole Plant.

Folk Uses : Commonly planted as a **hedge plant** to check **soil erosion**. A gargle of leaf extr. beneficial in healing **mouth sores**, and **pyorrhoea**.

Active Constituents : Principle constituents of leaves are b-sitosterol, triacontane and saponin.

Biological Activity : Ethanol extr. of the plant CNS depressant.

Uses in Literature : Known to be used in India as poison (vet), purgative, vegetable; and arthritis, cuts, sprain, ulcer, and wounds **(Ambasta, 1986; Jain, 1991; Singh & Prakash, 1996).**

Ipomoea eriocarpa R. Br. (Pl. 37B).

Syn : *I. hispida* (Vahl.) Roem. & Sch., *I. sessiliflora* Roth., *I. ligulata* Bojer, *I .sphaerocephala* Sweet, *I. horsefieldiana* Miq., *Convolvulus eriocarpus* Spreng., *C. sessiliflorus* Spreng., *C. rampiana* Wall., *C. hispidus* Vahl.

Family : Convolvulaceae.

Vern. : Safed Ghordan.

Sanskrit & Regional Names

Sans. : Nakhari;
Ass. : Kalman;
Mal. : Pulichevidu;
Oriya : Paninoi;
P. : Bhanwar;
Tel. : Purititige.

Distribution : Throughout India, ascending to 1350m.

Description : Herbaceous twinner. Leaves cordate, hastate. Flowers in sessile clusters, pinkish purple.

Reproductive Cycle : August-October.

Habitat Ecology : Wasteplaces, agricultural fields, gardens grassy fields, sandy localities; common.

Material Examined : EBH-1121, 24-08-2002.

Plant/Parts Used : The Whole Plant. Leaves. Seeds.

Folk Uses : Fibres from plant used for making **ropes**. Leaves lopped as **fodder** for livestock. Seeds **edible**.

Uses in Literature : So far used earlier as fodder for milching cattles, edible in famine (leaves, stems, seeds); and for corns, epilepsy, headache, leprosy, rheumatism, ulcers and wounds (**Ambasta, 1986; Brahmam & Saxena, 1990; Jain, 1991; Rama Rao & Henry, 1996; Watt, 1972**).

Ipomoea muricata Jacq.+ (Pl. 37C).

Syn. : *Convolvulus muricatus* L., *Calonyction muricatum* G. Don, *C. bona-nox* var. *muricata* Chois.

Family : Convolvulaceae.

Vern. : Ghordan.

Regional Names

Beng.	:	Michai;
Hin.	:	Michai;
Bom.	:	Gariya;
Guj.	:	Garayo;
Mar.	:	Bhonvari;
Tam.	:	Kattutali.

Distribution : Kangra-Sikkim; very common.

Description : Twinner. Stem muricate. Leaves entire, cordate. Corolla tube linear. Capsule globose, apiculate.

Reproductive Cycle : June-August.

Habitat Ecology : Damp places.

Material Examined : EBH-578, 12-08-2001.

Parts Used : Leaves, Seeds.

Folk Uses : Considered **sacred** for offering during **worshipping**. Paste of seeds applied to check **falling of hairs**.

Active Constituents : Seeds yield behenic acid (3.78%).

Uses in Literature : Known to be used in India as cathartic, purgative; and for its edible pedicels (**Ambasta, 1986**).

Ipomoea quamoclit L.+ (Pl. 37D).

Syn : *Convolvulus pennatus* Desr., *C. quamoclit* Spreng., *Quamoclit vulgaris* Chois., *Q. pinnata* Boj.

Family : Convolvulaceae.

Vern. : Nagar Bel.

English, Sanskrit & Regional Names

Eng.	:	Cypress Vine, Indian Pink, Red Jasmine;
B.	:	Kamalata, Tarulata;
H.	:	Kamalata;
Kan.	:	Kempumallige, Kumalate;
Mal.	:	Suriyakanthi;
Mar.	:	Sitachekesa, Vishnukrantu;
Oriya	:	Kunjolota;
Tam.	:	Kembumalligai, Mayirmannikam;
Tel.	:	Kasiratnamu.

Distribution : Naturalised in the Himalaya upto 1200m.

Description : A slender hairless climber. Leaves pinnate with numerous narrow leaflets. Flowers long-stalked, crimson. Capsule ovoid.

Reproductive Cycle : August-October.

Habitat Ecology : Common in cultivated areas as a climber on shrubs and herbaceous plants.

Material Examined : EBH-1133, 03-10-2002.

Parts Used : Flowers. The Whole Plant.

Folk Uses : Paste of plant applied as an **antidote** to snake poison. Leaves cooked as **vegetable**. 2-5 ml decoction of flowers useful in **regulating menstrual disorders** (twice daily for 3-5 days).

Active Constituents : Root and aerial parts yield quamoclitic acid. Four new resin glycosides-quamoclins I, II, III and IV- isolated from jalapin fraction of seeds.

Uses in Literature : Recorded earlier as edible (leaves), refrigerant, sternutatory; and for breast pain, carbuncles, hemorrhoids, snake bites and ulcer (**Ambasta, 1986; Jain, 1991; Samwatsar & Diwanji, 1996b; Sinha, 1996; Watt, 1972**).

Jacaranda acutifolia Humb. & Bonpl.+ (Pl. 38A).

Syn : *J. mimosifolia* D. Don, *J. ovalifolia* R. Br.

Family : Bignoniaceae.
Vern. : Jackranda.

Distribution : Outer foothills of the W. Himalaya.

Description : Trees having pinnate leaves; blue flowers and nearly circular, flat, woody, dehiscent flowers.

Reproductive : April-June.

Habitat Ecology : Commonly planted by roadsides and railway-track sides.

Material Examined : EBH-452, 14-04-2000.

Parts Used : Bark. Leaves. Wood.

Folk Uses : Planted as an **avenue plant**. Wood used for **low-grade furniture, construction wood**, and **tool handles**. Finely powdered leaves dusted on wounds for quick healing. Decoction of bark and leaves (5 ml) given thrice a day till cure for **syphilis** and **gonorrhoea**.

Active Constituents : These are scutellarein, hydroxyquinone, jacaranose, jacaranone, verbascoside and phenylacetic-b-glucoside (leaves), lupenone, b-sitosterol and 7,2',3',4'-tetrahydroxyflavone-3-0-neohesperidioside (I) (bark).

Uses in Literature : Used earlier for blennorrhagia, pectoral, syphilis, tool handles, venereal diseases; and as vulnerary (**Ambasta, 1986; Jain & Sharma, 1996**).

Jasminum dispersum Wall.+ (Pl. 38B).

Family : Oleaceae.
Vern. : Chamaeli.

Distribution : Kashmir-S.W. China: 900-2500m.

Description : Climbing woody shrubs with imparipinnate leaves. Leaflets 3-5. Flowers white, numerous, fragrant. Fruit dark purple, elliptic.

Reproductive Cycle : April-November.

Habitat Ecology : Shrubberies; ravines.

Material Examined : EBH-221, 15-04-2000.

Parts Used : Leaves. Flowers.

Folk Uses : Flowers **offered** for religious worshipping. Green leaves chewed with 'gur' (crude sugar) to check **acidity** and **gastric problems**, and also its decoction (5-10 ml) given twice a day for **early healing of wounds**.

Biological Activity : EtOH (50%) extr. of aerial parts anticancerous and diuretic.

Uses in Literature : Recorded in India for its leaves (vegetable) and fruit (**Mao, 1993; Negi & Gaur, 1991**).

Major Ayurvedic Formulations : Jatyadi-taila.

Jasminum grandiflorum L.+ (Pl. 38C).

Syn: *J. aureum* Don., *J. officinale* L., *J. officinale* L. var. *grandiflorum* (L.) Stokes, *J. officinale* fa *grandiflorum* (L.) Kobuski

Family : Oleaceae.
Vern. : Malti.

English, Sanskrit & Regional Names

Eng. : *Common White Garden Jasmine, Spanish Jasmine, White Jasmine, Catalonian Jasmine;*

Sans. : Bahugandha, Gajavhaya, Punyagandha, Shankha-yuthika;

B. : Chameli, Jati;

Bo. : Chambeli;

G. : Chambeli,;

H. : Chaluba;

Kan. : Ajjige, Jaji, Jati, Mallige;

Kash. : Chirichog, Kiri;

Mar. : Pichakam, Pichakamulla;

P. : Bansu, Naugri, Somum;

Tam. : Manbadanam, Mullai, Pedarmalligai, Pichi;

Tel. : Jagi, Malati;

U.P. : Chameli.

Distribution : Afghanistan-S.W. China; S.E. Asia: 900-2500m. Cultivated.

Description : Climbing shrubs. Leaves imparipinnate, opposite. Leaflets 3-7, terminal longer than others. Flowers white, fragrant. Fruit with crimson juice.

Reproductive Cycle : April-September.

Habitat Ecology : Shrubberies; ravines.

Material Examined : EBH-482, 13-08-2000.

Parts Used : Leaves. Flowers. Roots.

Folk Uses : Flowers **offered** to deities, and its decoction (10 ml) good for **piles** and **skin eruptions** (twice daily for 5-8 days). Also, its extr. used as eye drops for curing **eye problems**. Besides, paste of fruits and roots considered good for treating **ringworm**. Leaves chewed for relief from **toothache**.

Active Constituents : Active properties of flowers are due to cis-3-hexenol, 2-vinylpyridine, indole, myrcene, linalool, geranyl linalool, a-terpineol, geraniol, linalyl acetate, nerolidol, phytol isophytol, farnesol, engenol, benzyl alcohol, p-cresol, methyl benzoate benzyl cyanide, benzyl acetate, methyl dihydrojasmonate methyl anthranitate, lis-jasmone, methyl N-methyl anthranitate, vanillin, cis-3-hexenyl benzoate, benzyl benzoate, methyl palmalte and methyl linolate.

Uses in Literature : Known to be used as an anthelmintic, aphrodisiac, audient, cooling, deobstrent, diuretic, psychoactive plant, social taboo (girls place flames in their hair when they become pregnant); and for apthae, attars, corns between the toes, dermatitis, emmenagogue, hair oils, headache, jasmine D1, otorrhoea, perfumed waters, religious offerings, sexual impotency, skin eruptions, toothache, treatment of ear discharge, ulceration of mouth (**Ambasta, 1986**; **Jain *et al.*, 1994**; **Kapur & Singh, 1996**; **Rana *et al.*, 1996**; **Shome *et al.*, 1996**; **Singh & Pandey, 1996**; **Srivastava *et al.*, 2003**; **Vihari, 1995**; **Watt, 1972**).

Major Ayurvedic Formulations : Jatyadimalhar, Jatyaditaila.

Jasminum humile L.+ (Pl. 38D).

Syn.: *J. chrysanthemum* Roxb., *J. revolutum* Sims., *J. bignoniaceum* Wall., *J. wallichainum* Lindl., *J. inodorum* Jacq., *J. pubigerum* D. Don var. *glabrum* DC

Family : Oleaceae.

Vern. : Chameli.

English, Sanskrit & Regional Names

Eng.	:	Yellow Jasmine, Italian Jasmine, Nepal Jasmine, Golden Jasmine;
Sans.	:	Svarnajuthica, Hemapushpika;
B.	:	Svarnajui;
Bo.	:	Hemapushpika;
H.	:	Peelajui, Peeli Chameli, Malto;
Kan.	:	Hasarumallige;
Mal.	:	Ponmallika;

Tam.	:	Semmalligai;
Tel.	:	Pachche-Adavimalle.

Distribution : Sub-Tropical Himalaya: 1500-3000m.

Description : Erect, stout shrubs. Leaves alternate. Leaflets 3-7, oblong or ovate. Flowers yellow. Fruit a juicy black berry.

Reproductive Cycle : May-September.

Habitat Ecology : Shrubberies, forests; common.

Material Examined : EBH-882, 11-08-2001.

Parts Used : Leaves. Flowers. Roots.

Folk Uses : Plant grown as an **ornamental.** Flowers considered **sacred** and **offered to deities**, and also in perfumery. Leaves and roots chewed for relief from **toothache**, and decoction of roots (5-10 ml) prescribed twice daily for 5-10 days to cure **skin ailments**.

Uses in Literature : So far used in India as antidote in poison cases, aromatic essential oil, astringent, tonic to heart; and for chronic sinuses, dyeing, earache, eye diseases, fistulas, headache, perfumery and cosmetics, ringworm affections, skin diseases, stomatitis, toothache, ulcerations, and wounds (**Ambasta, 1986; Chauhan, 1999; Chopra** ***et al.*****, 1965; Jain, 1991; Watt, 1972**).

Major Ayurvedic Formulations : Jatyadimalhar, Jatyaditaila.

Jasminum multiflorum (Burm. f.) Andrews+ (Pl. 39A).

Syn: *J. pubescens* (Retz.) Willd., *J. hirsutum* Willd., *J. multiflorum* Roth., *J. congestum* Wall., *J. bracteatum* Wight., *Mogorium pubescens* Lamk., *Nyctanthes multiflora* Burm. f., *N. pubescens* Retz.,

Family : Oleaceae.

Vern. : Chameli.

English, Sanskrit & Regional Names

Eng. : Common Jasmine, Downy Jasmine, Musk Jasmine;

Sans. : Atimukha, Attahasa, Attapuspaka, Bhringhabandhu, Bhringhsuhrita, Dalako-sha, Damana, Kunda, Madhya, Mahamoda, Makaranda, Manodna, Manohara, Manorama, Mughya, Muktapushpa, Tarapushpa, Vanahasa;

H. : Chameli, Kunda, Kundo, Kundphul;

Mal. : Gujari, Kattuchirakamulla, Kundam, Kurukuttimula, Kuruna;

Mar. : Mogra;

Oriya : Kontabelo;

Tam. : Magarandam, Malligai;

Tel. : Adavimalle, Gujari, Kundamu, Molla.

Distribution : Throughout India.

Description : Scandent, densely villous shrubs having calyx with yellow hairs and white flowers.

Reproductive Cycle : November-April.

Habitat Ecology : Shrubberies, forests; common.

Material Examined : EBH-216, 16-01-2000.

Parts Used : Leaves. Flowers.

Folk Uses : Flowers offered during **worshipping**, and in **perfumery**. Poultice of leaves prescribed for **suppuration of boils.**

Active Constituents : Leaves afford friedelin, lupeol, betulin, a-amyrin and betulinic, ursolic and oleanolic acids. Jasmolactones A, B, C and D, multifloroside, multiroside and 10-hydroxyoleoside-11-methyl ester, jasmultiside and multiflorin isolated from aerial parts.

Biological Activity : Jasmolactones B and C exhibit coronary vasodilating and cardiotropic activities.

Uses in Literature : Useful as antidote to snakebite; and for burns, chest pain, cough and cold, hounding dogs (vet), offerings and adornments, ulcers; and also in indolent ulcers, pinworms, stomachache, and urine blockage (**Ambasta, 1986; Dagar, 2003; Datt, 1996; Gogoi & Borthakur, 1991; Jain, 1991; Kapur & Nanda, 1996; Watt, 1972**).

Juglans regia L.+ (Pl. 39B).

Syn: *J. regia* L. var. *kumaonica* Cas. DC, *J. arguta* Wall.

Family : Juglandaceae.

Vern. : Akhrot.

English, Sanskrit & Regional Names

Eng. : Common Walnut Tree, European Walnut, Himalayan Walnut, Walnut Tree;

Sans. : Akhoda, Akschota;

Ass. : Kabsing;

B. : Akhrut;

G. : Akhrot;

H. : Akhor, Akhrot, Khor;

Kash. : Akhor, Krot;

Mar. : Akroda;

P. : Akhori, Charmagz, Kabotang;

Tam. : Akrottu;

Tel. : Akrotu.

Distribution : Kashmir-S. E. Tibet: 1500-3000m.

Description : Trees with imparipinnate, aromatic leaves and male and female flowers on the same tree. Drupe ovoid with green, thick, fleshy rind enclosing a woody, wrinkled two-valved nut.

Reproductive Cycle : April-October.

Habitat Ecology : Forests; widely cultivated in West Himalayas.

Material Examined : EBH-393, 11-08-2000.

Parts Used : Fruit. Rind. Bark. Roots.

Folk Uses : Wood used for **ornamental work**. Fruits **edible**, and its rind used as a **black dye**. 10-20g powdered kernels of fruits given with cow's milk before going to bed for checking **bed wetting**. Bark used for **scouring teeth** and **coloring the lips**, and also oil prepared by boiling its green bark with 'harad' (*Terminalia chebula*) and mustard oil, considered as an excellent **hair dye**. Paste of roots applied to check **skin-eruptions** and also its extr. good against **bleeding gums**.

Active Constituents : Active properties are due to 5-hydroxy-1,4-napthoquinone, berberine (0.08%) (stem-bark), cyclotris juglone (roots), 3,3'-bisjuglone isolated (root-bark).

Biological Activity : Juglone inhibits growth of liver ascites carcinoma and S-180 solid tumor in mice at 0.1 mg/ml *in vitro*.

Uses in Literature : Reported earlier to be used as an alexipharmic, alterative in scrofula, anthelmintic, aphrodisiac, astringent, cattle feed, cholagogue, detergent, edible (fruit), insect repellent for storage of grains, laxative, nerve tonic, psychoactive plant, tooth brush, tooth powder; and for bed wetting, calopacia, carving, chewing to reden the lips, confectionary (pastries), decolourising charcoal and activated carbon, diarrhoea, dispelling flies, drawing instruments, dyeing, eczema, foot and mouth disease (vet), frames, frost bite, furniture, gun-stocks, handlooms, herpes, hoof (vet), ice-creams, lacquer work, leucoderma, menorrhagia, musical instruments, oil for culinary purposes and illuminant, oil paints, ornamental works, paints, ploughs, printing inks, rheumatism, rickets, scrofula, shafts, sore throat, sores of toes, stupefying fish, syphilis, and tanning (**Ambasta, 1986; Aswal, 1996; Chauhan, 1999; Chopra *et al.*, 1956; Jain, 1991; Jain *et al.*, 1994; Kapur & Singh, 1996; Lalramnghinglova, 1996; Negi & Pant, 1990; Negi *et al.*, 1993; Rana *et al.*, 1996; Saini, 1996b; Singh 1991; Vishwanathan & Singh, 1996; Watt, 1972**).

Major Ayurvedic Formulations : Amrtipras-ghrit, Ashoka-taila, Dandasa (bark), Dantasodhak, Yavin.

Trade : 'Dandasa', Akahrot, Akhor, Akrut, Krot.

Kydia calycina Roxb.⁺ (Pl. 39C).

Syn: K. roxburghiana Wight., K. fraterna Roxb., K. pulverulenta Ham.

Family : Malvaceae.

Vern. : Pula.

Regional Names

Ass.	:	Bankopah, Pichhola;
B.	:	Bonkopas, Pola;
Bo.	:	Motipoturi, Varangata;
G.	:	Mhotihirwani, Nihotilirwani;
H.	:	Choupultea, Patha, Pathra, Potari, Pula;
Kan.	:	Bellaka,Nayibende;
Mal.	:	Vedunar, Velukku, Ventu;
Mar.	:	Bhendi, Iliya;
Oriya	:	Bankhopasia, Bharime;
P.	:	Pola, Pula;
Tam.	:	Vendai;
Tel.	:	Konda patri, Pachabotuka, Potri;
U.P.	:	Haldu.

Distribution : Tropical regions of India.

Description : Trees having palmately, 7-nerved, cordate, leaves and whitish-pink flowers..

Reproductive Cycle : July-September.

Habitat Ecology : Forests, shrubberies.

Material Examined : EBH-706, 03-08-2001.

Parts Used : Leaves. Wood.

Folk Uses : Leaves serve as **cattle feed**. Wood used for **house building** and **agricultural implements** and fibres from bark food for **coarse ropes**.

Active Constituents : Active properties of wood are due to 8-formyl-2,7-dihydroxy-5-isopropyl-1-methoxy-3-methylnaptha-lene (isohemigossypol 1-methyl ether), hibiscene C and hibiscoquinone.

Uses in Literature : Known to be used in various parts of India as detergent to wash hair, febrifuge, pulp for news print, veneers plywood's; and for agricultural implements, bobbins, body pain, boils, brush backs, carving, cheap grade pencils, coarse ropes, diabetes, fibre, fodder, house building, increasing saliva, inflammation, leucoderma, light packing cases, match boxes, oars, oedema, planks, ploughs, rheumatism, shoe-heels, shutter, skin diseases, splints, spoons, turnery, and ulcers (**Ambasta, 1986; Aminuddin & Girach, 1991; Rama Rao & Henry, 1996; Saini, 1996a; Satapathy & Brahmam, 1996; Watt, 1972**).

Trade : Pula (timber).

Lagerstroemia indica L.+ (Pl. 40A).

Syn: *L. elgans* Wall., *Velaga globosa* Gaertn.

Family : Lythraceae.

Vern. : Har-Shingar.

English, Sanskrit & Regional Names

Eng. : China Privet, Crape, Myrttle;
B. : Phurush, Telingachina;
Bo. : Chinai-Mendhi, Dhayti;
P. : Saoni, Dhaura;
Tam. : Pavalakurinji, Sinappu;
Tel. : Chinagoranta.

Distribution : Throughout India; usually cultivated in gardens.

Description : Trees having glabrous leaves and bright pink petals. Capsules nearly globose with calyx teeth erect on fruit.

Reproductive Cycle : June-October.

Habitat Ecology : Planted.

Material Examined : EBH-513, 28-09-2000.

Parts Used : Flowers.

Folk Uses : Grown as an **ornamental**. Decoction of its flowers (20-40 ml) given twice daily for 3-5 days to check **piles** and **sciatica**.

Active Constituents : Active properties of flowers are due to delphinidin-3-arabinoride, petnnidin-3-arabinoside, malridin-3-arasinoside, gallic acid, methyl gallate and ellagic acid.

Uses in Literature : So far reported as an astringent, febrifuge, hydragogue, narcotic, ornamental, purgative, stimulant; and for apathae of mouth (**Ambasta, 1986**; **Watt, 1972**).

Lantana camara L.⁺ (Pl. 40B).

Syn: *L. aculeata* L., *L. myxa* L., *L. camara* L. var. *aculeata* Moldenku

Family : Verbenaceae.
Vern. : Chudail buti, Phul lakdi.
English & Regional Names

Eng.	:	Lantana, Wild Sage;
G.	:	Ghanidalia;
H.	:	Desi-Lantana;
Kan.	:	Hesike, Kakke, Nata Hu Gida;
M.P.	:	Raimuniya;
Mal.	:	Arippu;
Mar.	:	Chadrung, Ghaneri;
Oriya	:	Naga-Airi;
P.	:	Desi Lantana;
Tam.	:	Arippu;
Tel.	:	Pulikampa.

Distribution : Native of Tropical America. Himalaya: ascending to 1500m.

Description : A shrub having numerous recurved prickles on branches. Leaves opposite, toothed. Flowers orange-yellow. Drupe black shining.

Reproductive Cycle : April-June.

Habitat Ecology : Semi-naturalised and common in cultivated areas.

Material Examined : EBH-913, 06-04-2001.

Parts Used : Stem. Leaves. Fruit.

Folk Uses : Grown as **hedges**. Ripe fruits **edible** and its stem used as **toothbrush**. Poultice of leaves considered good for **swellings** due to internal injuries, and also as an **insect repellent** for stored grains.

Active Constituents : These are lantadene A, lantadene B, lantalonic acid, lantic acid, ursonic acid, leucine, valine, tyrosine, glycine (leaves), geraniol, linalool, cedrene, caryophylene, farnerol, cis-nerolidol and trans-nerolidol (essential oil).

Biological Activity : Acid extr.of the shoot exhibits antibacterial activity. Antitumour activity of verbascoside *in vitro* might partly be due to inhibition of protein kinase.

Uses in Literature : Reported to be used as an antiseptic, antispasmodic, carminative, diaphoretic, edible (fruit), fish bait, for stored grains, mosquito repellent, poison (vet), stimulant, and toothbrush; and for anaemic patients, basketry, burns, chest diseases, chronic inflammation of skin, cold, cough, cuts, diabetes, diarrhoea, dysentery, dyspepsia (vet), fevers, fistulae, fits, green manure, infection of respiratory tract, itching, jaundice (vet), mouth ulcers, mumps, pulmonary problems, pustules, rabies, retaining humus in the soil, rheumatism, ringworm, sprains, swellings, tetanus, tumours, ulcers, vulnerary, whooping cough, and wounds (**Ambasta, 1986; Aminuddin & Girach, 1991; Bhattacharyya, 1996; Dwarakan & Ansari, 1996; Geetha *et al.*, 1996a; Girach & Aminuddin, 1995; Goud & Pullaiah, 1996a; Henry *et al.*, 1996; Hosagoudar & Henry, 1996b, c; Jain, 1991, 1996; Kurian, 1995; Lal *et al.*, 1996; Lalramn-ghinglova, 1996; Masih, 2003; Mishra *et al.*, 1996b; Saini 1996b; Watt, 1972**).

Leea aspera Wall. *ex* Roxb.+ (Pl. 40C).

Syn: *L. asiatica* (L.) Rid., *L..crispa* van Royan *ex* L., *L. edgeworthii* Sant., *L. herbacea* Wall., *L. pinnata* Andr.

Family : Leeaceae.

Vern. : Gan-Gola.

Regional Names

B.	:	Banchalita;
Mal.	:	Nellu, Nelugu;
Oriya	:	Alatikanopotro.

Distribution : Tropical Himalaya: ascending to 1650m.

Description : Robust, spreading shrubs with uppermost leaves usually simple pinnate and the lower ones 2-pinnate. Flowers yellow-green, in small, pubescent cymes. Fruit black, succulent.

Reproductive Cycle : July-August.

Habitat Ecology : Open slopes, screes.

Material Examined : EBH-1178, 19-07-2002.

Parts Used : Leaves. Fruit. Juvenile Stem.

Folk Uses : Fruits and juvenile stems cooked as **vegetable**. Leaves used as **fodder**, and its paste heals **cuts** and **wounds**.

Uses in Literature : Useful for cuts, rheumatism, snake bites, urinary complaints, wounds; and as an edible (**Jain, 1991**; **Sinha, 1996**; **Watt, 1972**).

Leptodermis lanceolata Wall.+ (Pl. 40D).

Syn. :*Hamiltonii suaveolens* Don.

Family : Rubiaceae.

Vern. : Padari.

Regional Names

- **Bo.** : Gidasawa, Gidesa;
- **M.P.** : Mahabel;
- **Mar.** : Janamirgiri, Pitondi;
- **P.** : Kanera, Kantalu, Muskei, Telunni;
- **Tel.** : Kondamurutidi.

Distribution : Temperate Himalaya; Kashmir-Bhutan: 1350-3300m.

Description : Shrubs. Leaves lanceolate, foetid when crushed. Flowers whitish-purple. Capsule cylindric.

Reproductive Cycle : June-October.

Habitat Ecology : Rocky slopes.

Material Examined : EBH-643, 13-09-2001.

Parts Used : Leaves. Seeds.

Folk Uses : 2-3g powdered seeds given twice daily with luke warm water for 20-25 days to cure **epilepsy**, a gargle of decoction of its leaves useful for **mouth sores** and **throat affections**.

Uses in Literature : Reported to be useful against boils, blisters in mouth and throat (**Jain, 1991**; **Maheswari *et al.*, 1986**).

Lettsomia thomsonii Clarke* (Pl. 41A).

Syn. : *Argyreia capitata* Brand., *A. nasirii* D. Austin, *A. thomsonii* (Clarke) Sensu Nair

Family : Convolvulaceae.

Vern. : Patdi.

Distribution : Throughout the Himalaya.

Description : Large climbers with hairy stem and leaves. Flowers pinkish white.

Reproductive Cycle : January-February.

Habitat Ecology : Forest edges, Forest clearing, Sal Forests.

Material Examined : EBH-680, 14-01-2000.

Parts Used : Leaves. Fruit.

Folk Uses : Fruits used as **pot herb**, and leaves as **cattle feed**.

Leucaena glauca Benth.+ (Pl. 41B).

Syn: *L. leucocephala* Lamk de Wit., *Mimosa lecocephala* Lamk., *Acacia frondosa* Willd., *A. caringa* Ham., *A. biceps* DC, *A. leucocephala* DC

Family : Fabaceae.

Vern. : Alseenia.

English & Regional Names

- **Eng.** : White Popinae, Lead Tree;
- **G.** : Lasobaval, Pardesi Baval, Vilaytibaval;
- **Mal.** : Takaranniram;
- **Oriya** : Rajokasundiri;
- **Tam.** : Tagarai-Virai.

Distribution : Throughout India.

Description : A small tree with 10-15 pairs of leaflets, whitish flowers and linear flat, shining pods.

Reproductive Cycle : September-November.

Habitat Ecology : Hedges, along boundaries of fields.

Material Examined : EBH-885, 17-10-2001.

Parts Used : Leaves.

Folk Uses : Considered an excellent **fodder** for milching cattle. Tree planted to check **soil erosion** and for **shade**.

Active Constituents : Leaves possess kaempferol-3-arabinoside, kaempferol-3-xyloside, quercetin-3-arabinoside quercetin-3-rhamnoside and quercetin-3-galactoside.

Uses in Literature : Used in India as source of protein, stimulant, tonic, vegetable; and for baskets, carotene, fancy bags, fibre, fodder, ornaments and poultry rations (**Ambasta, 1986; Gill & Nyawuame, 1994; Watt, 1972**).

Leycesetria formosa Wall.+ (Pl. 41C).

Family : Caprifoliaceae.
Vern. : Lahauli Kuja.

Distribution : Temperate Himalaya: 1750-3350m.

Description : Shrubs with hollow stems; ovate-lanceolate, entire leaves and white flowers with purple-tinged bracts. Berry glandular-hairy, dark purple.

Reproductive Cycle : May-November.

Habitat Ecology : Forests, shrubberies.

Material Examined : EBH-1235, 29-10-2002.

Parts Used : Buds. Twigs.

Folk Uses : Buds kept for **decorative** purposes, and its twigs considered an excellent **fodder** for **milching cattles**.

Uses in Literature : Used earlier for making flutes and whistles (**Ambasta, 1986**).

Litchi chinensis Sonner.+ (Pl. 41D).

Syn : *Nephelium litchi* Camb., *Dimocarpus* Lour.

Family : Sapindaceae.

Vern. : Litchi.

English & Regional Names

- **Eng.** : Litchi, Lychee;
- **Bo.** : Lichi;
- **H.** : Litchi;
- **Oriya** : Lishi;
- **Tam.** : Ilichi.

Distribution : Planted.

Description : A small tree with 4-12 leaflets and minute flowers, and red fruits.

Reproductive Cycle : April-June.

Habitat Ecology : Cultivated in gardens.

Material Examined : EBH-210, 20-05-2000.

Parts Used : Fruit.

Folk Uses : Fruit **edible** and its juice said to **relieve body heat.**

Active Constituents : Cyanidin-3-rutinoside, cyanidin-3-glucoside and malvidin-3-acetylglucoside isolated from fruits.

Uses in Literature : Described earlier in India as anodyne in neuralgic disorders and orchitis; and for edible (fruit), offerings (**Ambasta, 1986; Gogoi & Borthakur, 1991**).

Litsea polyantha Juss. (Pl. 42A).

Syn : *L. monopetala* (Roxb.) Pers., *Tetranthera monopetala* Roxb., *T.macrophylla* Wall., *T. alnoides* Miq., *T. fruticosa* Herb., *T. verticillata* Herb., *T. semecarpifolia* Wall.,*T. hexantha* Sieb., *Tomex pubescens* Herb.

Family : Lauraceae.

Vern. : Gwaoon, Paror.

English, Sanskrit and Regional Names

Eng.	:	Common Gray Mango Laurel;
Sans.	:	Gajapippali;
Ass.	:	Sualu, Muga;
B.	:	Barakurchita, Barakukar;
H.	:	Gwa, Kakuri, Kari, Karkawa, Katmarra, Katmedh, Katmonia, Kerauli, Marda, Meda, Papria, Patoia, Randkari, Sangran, Singraf, Singran;
Mar.	:	Ranamba;
Oriya	:	Kulya, Kurukochhodiya, Mahachhodiya;
P.	:	Gwa, Harien, Medalakri, Rian;
Tam.	:	Maida Lagadil, Muchaippeyetti, Pisinbattau;
Tel.	:	Meda, Nara, Naramamidi.

Distribution : Foot hills of Himalaya: ascending to 1000m.

Description : Trees with pubescent branches, alternate leaves having glabrous upper surface and tomentose beneath. Flowers white. Drupe ovoid, black when ripe.

Reproductive Cycle : March-July.

Habitat Ecology : Wastelands, roadsides, rocky wet slopes.

Material Examined : EBH-971, 19-07-2001.

Parts Used : Bark. Leaves. Wood.

Folk Uses : Decoction of bark (5-10 ml) useful against **diarrhoea** (twice daily for 3-5 days). Poultice of leaves good for **bone fractures**. Wood used for **agricultural implements, building works** and for **cheap furniture**.

Active Constituents : Polysaccharide isolated from leaves, contain xylose and arabinose in molar ratio of 1:4 and an arabinoxylan consisting of xylose and arabinose in a molar ratio of 1:2.

Uses in Literature : Known earlier as an astringent, feed for silkworms, vegetable; and for agricultural implements, boils, bone fracture (vet), candle manufacturing, chonic diarrhoea, chronic dysentery, cuts, furniture, giddiness, house building, plywood, relieving bodyache, rheumatism, sore (vet), ulcer (vet), and whitlow (**Ambasta, 1986; Chowdhury *et al.*, 1996; Jain, 1991; Jha & Varma, 1996; Rama Rao & Henry, 1996; Singh *et al.*, 1996b; Watt, 1972**).

Litsea sebifera Pers.[+] (Pl. 42B).

Syn.: *L. glutinosa* (Lour.) C. B. Robins, *L. chinensis* Lam., *L. tetranthera* Pers., *L. multiflora* Blume, *Neolitsea chinensis* (Lam.) Chun, *Tetranthera lauriflora* Blume, *T. mutiflora* Blume, *T. racemosa-umbellata* Blume, *T. roxburghii* Blume, *T. captitata* Herb., *T. apetala* Roxb., *Tomex tetranthera* Willd., *T. sebifera* Willd., *Sebifera glutinosa* Lour., *Laurus involucrata* Koenig.

Family : Lauraceae.
Vern. : Haryan.
English, Sanskrit & Regional Names

Eng. : Common Tallow Laurel;

Sans. : Adhavara, Bahurandhrika, Bhavya, Dhira, Jivani, Madhura, Manichhidra, Medarsa, Medodbhava, Purusadantika, Shalyada, Snehavati, Svalpaparni, Vibhavari;

Ass. : Bagnara, Heluka;

B. : Garur, Kukurchita, Maidalakri, Ratun;

Bo. : Chickana, Maidalakadi, Maidalakri;

H. : Garbijaur, Maidalakri, Medh, Menda, Singrauf;

M.P. : Menda;

Mar. : Maidalakadi, Miri;

Oriya	:	Jaisanda;
P.	:	Chamana, Gwa, Maidasak, Medachob, Medalakri, Medasak, Rian;
Tam.	:	Elumpurukki, Mushaippeyetti, Uralli;
Tel.	:	Naramamidi, Narra alagi.

Distribution : Throughout the warmer regions of India; lower Himalaya.

Description : Trees with subterminal, coriaceous leaves, which are pale when dry. Fruits pea-sized.

Reproductive Cycle : May-September.

Habitat Ecology : Near houses, fields.

Material Examined : EBH-1112, 14-08-2002.

Parts Used : Leaves. Fruit. Bark.

Folk Uses : Poultice of bark and that of *Prunus puddum* considered good for **rheumatic pains** and **bone fractures**. Decoction of its fruits prescribed as **refrigerant**, **aphrodisiac** and for **indigestion** (20-40 ml twice daily for 3-10 days). Paste of its leaves and stem said to heal **open wounds**.

Active Constituents : Chief ingredients are naringerin, naringen, kaempferol-3-and 7-glucosides, quercetin and its 3-thamnoside pelargonidin-3 and 5-glycosides, cystine, glycine, L-alanine, b-alanine, valine, tyrosine proline phenylalanine and leucine (leaves), two new alkaloids-sebiferine and litseferine (trunk-bark).

Uses in Literature : Known to be useful as anodyne, antidote to venomous animals, aphrodisiac, demulcent, edible, emollient, feebly balasamic, mild astringent, styptic; and for agricultural implements, boils, bone fracture (vet), bruishes, candles, chest pain, cold, cooling, cough, diarrhoea, dysentery, dyspepsia, fibres, floor boards, furniture, house building, incense industry, indigestion (vet), inflammation, leucorrhoea, oars, packing cases, pain in umbilicus region, rheumatic gouty joints, skin diseases, sprain, and wounds (**Ambasta, 1986; Chowdhury *et al.*, 1996; Jain, 1991; Rama Rao & Henry, 1996; Shah & Singh, 1990; Watt, 1972**).

Litsea umbrosa Nees+ (Pl. 42C).

Syn. : *L. umbrosa* Nees *var.* consimilis (Nees) Hook. f., *L. consimilis* Nees, *Neolitsea pallens* (D. Don) Momiyana & Hara, *Neolitsea umbrosa* (Nees) Gamble, *Tetradenia pallens* Don, *T. umbrosa* Nees, *T.pallens* Don

Family : Lauraceae.

Vern. : Pangra.

Regional Names

Kash. : Chirindi;

P. : Chindi, Chirudi.

Distribution : Temperate & Sub-Tropical Himalaya: 1500-2700m.

Description : Glabrous trees having lower surface of leaves covered with white bloom. Flowers pale-yellow.

Reproductive Cycle : March-May.

Habitat Ecology : Forms understory of forests.

Material Examined : EBH-1213, 17-05-2002.

Parts Used : Leaves. Stem Wood.

Folk Uses : Leaves kept **as cushion** to keep animals dry during rainy season. Wood employed for making **agricultural implements.**

Active Constituents : Bark yields b-amyrin and b-sitosterol.

Uses in Literature : Reported to be used as an illuminant, edible, oil for burning; and for itching, skin affections (**Ambasta, 1986; Jain, 1991; Kapur, 1996; Watt, 1972**).

Lonicera japonica Wall.⁺ (42 D).

Syn. :*L. macrantha* DC, *Caprifloium macranthum* Don, *C. japonicum* Don.

Family : Caprifoliaceae.

Vern. : Lushai.

English Name

Eng. : Japanese Honey Suckle.

Distribution : C. Nepal-Bhutan. China. S. E. Asia. 1500-2400m.

Brief Description : Shrubs with cordate-oblong leaves; white corolla and black berry.

Reproductive Cycle : May-June.

Habitat Ecology : Shrubberies, planted.

Material Examined : EBH-929, 11-05-2001.

Plant/Parts Used : The Whole Plant.

Folk Uses : Commonly grown as an **ornamental** shrub.

Active Constituents : Plant contains lonicerin (leaves), (–) cis-2,6,6-trimethyl-2-vixyl-5-hydroxytetrahydropyran, lina-lool, pinene, hexane-1, cis-3-hexa-nol-1, cis and trans-2-methyl-2-vinyl-5-(a-hydroxyisopropyl) tet-rahydrofuran, geraniol, a-terpineol, benzyl alcohol, b-phenyl ethyl alcohol carvacrol, eugenol (flower essential oil), loniceroside A, B, epivogeloside, secologanin dimethyl acetal, coganin and vogeloside (aerial parts).

Biological Activity : Aq. extr. of the plant material inhibits *in vitro* culture of Jinke 68-1 virus, enterovirus ECHO and Herpes virus (Luc *et al.*, 1960).

Uses in Literature : Known to be used in Indian literature as antipyretic, diuretic, stomachic; and for cordage, dysentery (**Ambasta, 1986**; **Jain, 1991**; **Lalramnghinglova, 1996**; **Yang & Blunden, 1993**).

Lonicera quinquelocularis Hardw.+ (Pl. 43A).

Syn.: *L. diversifolia* Wall., *L. royleana* Wall.

Family : Caprifoliaceae.
Vern. : Chaunsh.

English and Regional Names

Eng.	:	The Himalayan Honey Suckle;
Kash.	:	Pakhur, Tata bateri;
P.	:	Bakru, Phut;
U.P.	:	Bhati, Taknoi.

Distribution : Himalaya: 1800-3200m.

Description : A shrub with ovate-lanceolate leaves, 2-lipped, yellowish-white corolla and ovoid, white berry.

Reproductive Cycle : April-September.

Habitat Ecology : Forests, cultivated areas; common.

Material Examined : EBH-495, 28-08-2000.

Parts Used : Leaves. Branches. Wood.

Folk Uses : Leaves serve as an excellent **fodder**. Wood used as **fuel** and branches for making **walking sticks**.

Active Constituents : Aerial parts yield sitosterol, hexacosanol, octacosanol, n-triacontanol, nonacosane and L-inositol.

Uses in Literature : So far used as laxative; and for carvings, fibres, fodder, food utensils and turnery (**Ambasta, 1986**; **Kapur, 1996**; **Kapur & Singh, 1996**; **Watt, 1972**).

Loranthus ligustrinus Wall.* (Pl. 43B).

Syn: *Helix anthera* Ligustrina (Wall.) Danser

Family : Loranthaceae.

Vern. : Reenain, Reeni.

Distribution : Outer Himalayan hill ranges between 600-1400m.

Description : A parasitic shrub having pink corolla and ovoid, truncate fruits.

Reproductive Cycle : April-September.

Habitat Ecology : Common parasite on trees.

Material Examined : EBH-

Parts Used : Leaves. Bark. Aerial Plant Parts.

Folk Uses : Leaves used as **fodder**. Massage of oil, obtained by boiling it with powdered bark, gives relief from **rheumatism** and **arthritis**. Tablets (1g) prepared by grinding aerial plant parts with '**desi ghee**' useful in **fractures of the bones** (3 tablets thrice a day for 15 days).

Luffa cylindrica (L.) Roem. (Pl. 43C).

Syn: *L. aegyptica* Mill., *L. pentandra* Roxb., *L. racemosa* Roxb., *L. clavata* Roxb., *L. acutangula* W. & A.., *L. petola* Seringe, *L. cattu-picinna* Seringe, *l. parvala* Wall., *L. gosa* Wall., *L. hederacea* Wall., *L. satpatia* Wall., *Bryonia cheirophylla* Wall., *Momordica cylindrica* L., *M. luffa* L.

Family : Cucurbitaceae.

Vern. : Ghiya.

English, Sanskrit & Regional Names

Eng. : Gourd Towel, Spinge Gourd, Vegetable Sponge, Washrag Sponge;

Sans. : Aibhi, Brihatkoshtaki, Dhamargava, Dhirghapatolika, Ghoshaka, Hastiparna, Hastighosha, Hartikoshataki, Mahakoshataki, Mahaphala, Rajakoshataki, Sapitaka;

Ass.	:	Bhatkarel, Bhol;
B.	:	Dhindul, Hastighasha;
Bo.	:	Ghosali, Gonsali, Parula, Turi;
G.	:	Galaka, Martigonsali, Turia;
H.	:	Ghiaturai, Nenua, Purula;
Kan.	:	Tuppahirekai;
Mal.	:	Kattupechal;
Mar.	:	Ghadaghosali, Ghosali, Parosi;
P.	:	Ghiatori, Ghiaturia, Ghigandoli;
Tam.	:	Pichukku, Pikku;
Tel.	:	Guttibira, Netibira, Nunebira.

Distribution : Throughout India.

Description : A climber having 5-lobed, dotted leaves; yellow flowers and cylindrical green fruits.

Reproductive Cycle : May-September.

Habitat Ecology : Cultivated.

Material Examined : EBH-1187, 23-07-2002.

Parts Used : Fruits.

Folk Uses : Fruits cooked as **vegetable**, and used in culinary preparations, '**pakoras**', '**kofta**', '**badies**', etc.

Active Constituents : Fruits mainly contain major saponins–lucyosides A, E, F, I, J, K, L and M.

Uses in Literature : Known to be used as an anthelmintic (vet), carminative, cathartic, diuretic, edible, emetic, galactagogue, purgative, vegetable; and for abnormal stoppage of the menses, anaemia, cold (vet), cough (vet), diarrhoea, dyspepsia (vet), expulsion of worms, fibre, indigestion (vet), iron deficiency, killing white ants, liver diseases, piles, prickly heat, and scabies (**Ambasta, 1986**; **Jain,**

1991; Jha *et al.*, 1996; Kulkarni & Kumbhojkar, 1993; Kurian, 1995; Satapathy & Brahmam, 1996; Singh & Maheswari, 1992; Watt, 1972).

Lyonia ovalifolia (Wall.) Drude+ (PL. 43D).

Syn : *Pieris ovalifolia* (Wall.) D. Don

Family : Ericaceae.
Vern. : Bhelar.
Regional Names

P.	:	Ailan, Angyar;
U.P.	:	Ailan, Angyar, Ayar.

Distribution : S. E. Asia. 1500-3000m.

Description : A small tree with long-petioled leaves; white flowers and globose capsule.

Reproductive Cycle : April-May.

Habitat Ecology : Common in oak forests, shrubberies; and usually grow in association with *Rhododendron arboreum.*

Material Examined : EBH-1228, 30-04-2002.

Parts Used : Leaves.

Folk Uses : Paste of leaves heals **skin eruptions** and relieves **itching.**

Active Constituents : Putatine, b-sitosterol-b-d-glucoside, ursolic acid, lyofolic acid, hexacasane, hexacosanol, hyperoside, b-sitosterol, taxaxerol, and quercetin isolated from leaves.

Uses in Literature : Known to be useful as edible (flowers and fruits eaten), insecticidal, poison (vet); and for boils, cutaneous troubles, pimples, scabies (vet), skin eruptions, worms, wounds (**Ambasta, 1986; Ansari, 1991; Jain, 1991; Kapur & Nanda, 1996; Lalramnghinglova, 1996; Manandhar, 1990; Mao, 1993**).

Machilus odoratissima Nees+ (Pl. 44A).

Syn. : *Persea odoratissima* (Nees) Kosterm

Family : Lauraceae.
Vern. : Bhadral.
Regional Names

H.	:	Kawala;
P.	:	Badroor, Dalchini, Kalban, Kaula, Mithpatta, Taura.

Distribution	:	Himalaya: 1500-2100m.
Description	:	An evergreen tree with long-pointed, elliptic-lanceolate leaves; solitary pale-yellow flowers; long, ellipsoid, dark-brown fruits.
Reproductive Cycle	:	March-April.
Habitat Ecology	:	Forests.
Material Examined	:	EBH-1103, 03-04-2002.
Parts Used	:	Wood.
Folk Uses	:	Wood used for manufacturing cheap **furniture**, and in **house building**.
Uses in Literature	:	Reported earlier to be good for house building, rearing silk worms, and tea-boxes (**Ambasta, 1986**; **Watt, 1972**).
Trade	:	Machilus.

Mallotus phillipinensis (Lam.) Muell.-Arg. (Pl. 44B).

Syn. : *Rottlera tintorii* Roxb., *R. aurantiaca* Hook & Arn., *R. affinis* Hassk., *R. montana* Wall., *R. mollis* Wall., *Croton philippensis* Lamk., *C. punctatus* Retz., *C. coccineus* Vahl., *C. montanus* Willd., *C. distans* Wall., *C. cascarilloides* Rauesch ex Steud.

Family : Euphorbiaceae.
Vern. : Kaambal.

English, Sanskrit & Regional Names

Eng.	:	Monkey Face Tree, Kamala Tree;
Sans.	:	Bahupushpa, Pikaksha, Raktachurnaka;
Ass.	:	Gangai, Jagguru;
B.	:	Kamlagundi, Tung;
Bo.	:	Kamala, Ruhin;
G.	:	Kapilo;
H.	:	Kambhal, Rohni, Rula;
Kan.	:	Kunkumadara;
Kash.	:	Kambil;
Mal.	:	Chenkolli, Piponnakam, Poonagam;
Mar.	:	Shendri;
Oriya	:	Bosontogundi, Punango;
P.	:	Kamal, Rulya;
Tam.	:	Kabilappodi, Kurangumanjanatti;
Tel.	:	Adavigubbtuda, Nagarakta.

Distribution : India, China, S. E. Asia, Australia; upto 1800m.

Description : Trees. Leaves ovate-lanceolate and their lower surface densely dotted with minute red glands. Male flowers-yellow. Capsule covered with crimson powder.

Reproductive Cycle : April-October.

Habitat Ecology : Often gregarious; moist forests.

Material Examined : EBH-334, 25-07-2000.

Parts Used : Leaves. Fruits. Seeds. Stem Wood.

Folk Uses : Paste of young leaves useful against **itching, ringworm** and other **skin ailments**. 1-3g powdered fruits given with luke warm water on empty stomach for **expelling worms** from the body, and to cure **jaundice**. To lower the body heat, hands and feet are washed in water containing decoction of its leaves (one glass) and a pinch of rock salt. **'Kungu'** (a dye) extracted from its seeds is used in religious rites.

Active Constituents : Chief active principles are rottlerin, isorottlerin, corotoxigenin, coroglaucigen (seeds), betulin-3-acetate, lupeol, lupeol acetate, sitosterol, bergenin (heartwood), rottlerin, isoallorottlerin and red compound (leaves).

Biological Activity : Aq. extr. of red resinous powder (0.03g) exhibits lithotropic effect in adult albino rats.

Uses in Literature : Known to be used as an alternative to sendur, anthelmintic, anthelmintic (vet), antioxidant for ghee, aphrodisiac, blood purifier, cathartic, contraceptive, dye, fodder, fuel, manure, purgative, styptic, tonic for pregnant women, red dye from it adorned by women on forehead as a sign of being married, vermifuge; and for anorexia, boils, controlling allergic infections, downer cow syndrome (vet), dysentery, fever, fits, giddiness, hemiplegia, hepatic, herpes, house decoration, hypocaemia (vet), jaundice, leprous eruptions, match boxes, parasitic skin diseases, rapid drying of varnish (seeds used), rheumatism, snake bites, sores, stomachache, tannins, thoracic pains, ulcers, vermilion, and wound (vet) (**Ambasta, 1986**; **Banerjee & Pal, 1996**; **Chhetri, 1994**; **Hosagoudar & Henry, 1996c**; **Jain, 1968, 1991**; **Kapur & Singh, 1996**; **Mukherjee & Namhata, 1990**; **Rama Rao & Henry, 1996**; **Rana** ***et al.*****, 1996**; **Shome** ***et al.*****, 1996**; **Singh & Prakash, 1994**; **Singh, 1996a**; **Tirkey** ***et al.*****, 2003b**; **Tiwari & Tiwari, 1996**; **Vihari, 1995**; **Watt 1972**).

Major Ayurvedic Formulations : Dhanvantara-ghrita, Herbenol cream, Kampillaka churna, Kampillaka-taila, Kash-isadighrit, Krimighatini-vati, Krimighatini-vati, Krimikuthar-ras, Misraka-sneha, Tinctura Kamala.

Trade : Kamala powder, Kanikaa, Kainda.

Malvaviscus conzatii Greenm.+ (Pl. 44C).

Family : Malvaceae.

Vern. : Gudhal.

Description : A shrub with serrated leaves; red orange flowers and staminal column.

Reproductive Cycle : Major part of the year.

Habitat Ecology : Hedges; near houses.

Material Examined : EBH-37, 13-01-1999.

Parts Used : Flowers.

Folk Uses : Flowers considered **sacred** and offered to various deities. Also, 5-10 ml of its juice prescribed in **impotency**, once daily for few months.

Active Constituents : Betaine, β-sitosterol and its glucoside, allantoin, kaempferol and their 3β-glucoside and 3β-sophoroside isolated from flowers.

Biological Activity : Betaine exhibits sperm immobilising action *in vitro*.

Mangifera indica L.+ (Pl. 44D).

Syn. : *M. domestica* Gaertn.

Family : Anacardiaceae.

Vern. : Aam, Amb.

English, Sanskrit & Regional Names

Eng. : Cuckoo's Joy, Cupid's Favourite, Mango Tree, Spring Tree;
Sans. : Adisairrabha, Manmathalaya, Parapushtamahotsava;
Ass. : Ghariam;
B. : Ambra;
Bo. : Am, Amba;
G. : Ambo, Amri;
H. : Am, Amb;
Kan. : Mavu;
Mal. : Amram, Gomanne, Nattumavu;
Mar. : Amba;

Oriya	:	Ambo, Boula, Chuto;
P.	:	Am, Amb, Mawashi;
Tam.	:	Adishelarayam, Kilimukkuma;
Tel.	:	Gujjumamidi, Makandamu.

Distribution : Commonly planted.

Description : Trees with shining leaves; and yellow, odorous flowers and compressed, yellowish drupe.

Reproductive Cycle : March-July.

Habitat Ecology : Planted along roadsides, field borders, local pathways.

Material Examined : EBH-1007, 09-7-2002.

Parts Used : Leaves. Fruit. Seeds. Wood.

Folk Uses : Tree considered **sacred**, and worshipped in various religious ceremonies. Leaves good for **scouring** teeth. Wood used for **house construction**. Ripe fruits **edible**, and its pulp used for making **'Aam papad'** (Traditional Indian candies) and **restorative tonic**; and unripe ones **pickled** and their decoction useful in **sunstroke**. Powdered seeds (1 teaspoonful) taken with cow's milk on empty stomach for expelling **intestinal worms**, and as snuff for checking **nose bleeding**. Also, powdered mixture of its seeds and those of **'jamun'** (*Eugenia jambolana*) in divided dose, given twice a day for 2-3 months to cure **diabetes, blood dysesntery** and **piles**.

Active Constituents : Various constituents isolated are homomangiferin and characterised as 2-b-D-glucopyranosyl-3-methoxy-1, 6, 7-trihydroxyxanthone, leucine, tyrosine and valine (leaves), protocatechuic acid, catechin and mangiferin (bark), alanine, glycine, g-aminobutyric acid, kinic and shikimic acids (leaves and bark), b-sitosterol, (+) catechin, (±)catechin, (-) epicatechin and leucocyanidin (seed coat), stearic acid isolated (seeds), exudate methyl, ethyl, propyl, butyl, amyl and iso-butryl alcohols (leaf oil), isomyl alcohol, a-pinene, b-pinene, myrcene, limonene and fenchone (pericarp oil), and methyl propyl, isopropyl, butyl and isomyl alcohols (seed oil).

Biological Activity : Gallic acid (1 mg/ml) and quercetin (4 mg/ml) showed marked antiinfluenza virus activity.

Uses in Literature : Used earlier as an abortifacient, anthelmintic, antiscorbutic, antisyphilitic, appetiser, astringent, cremation wood, diuretic, edible, febrifuge, laxative, magico-therapeutic to vanish sty, porridge, restorative tonic, sacred (Lord Buddha reposed under this tree), stomachic, symbolic (abode of god), vegetable; and for abdominal pain, agricultural implements, anasarca, anthrax (vet), antifertility, arthritis, asthma, blood cancer, cholera, colic (vet), constipation (vet), cough, cuts, diabetes, diarrhoea, diarrhoea (vet), dropsy, drums and other musical instruments, dyeing, dysentery containing slight mucous, expelling placenta, extraneous affections, eye complaint, gingivitis, gum, healthy and long hair, house construction, liver disorder, malaria, menorrhagia, piles, plague, pneumonia (vet), pregnant woman to beget fair complexioned babies, preparing lakes, rabies (vet), rheumatism, scabies, skin diseases, style, sunstroke, swelling of throat (vet), tanning, throat diseases, toothache, ulcers, ulcerated tongue, uterine haemorrhage, wounds, writing inks, and also in marriages and festivals for decoration (**Ambasta, 1986; Aminuddin & Girach, 1991; Borthakur & Gogoi, 1994; Chowdhury *et al.*, 1996; Ghosh *et al.*, 1996; Gill *et al.*, 1993, 1997; Gogoi *et al.*, 2003; Goud & Pullaiah, 1996a; Henry *et al.*, 1996; Hosagoudar & Henry, 1996a, c; Jain, 1991; Jain *et al.*, 1997; Jha *et al.*, 1996; Kapur *et al.*, 1996; Lalramnghinglova, 1996; Manandhar, 1996; Masih, 2003; Mishra *et al.*, 1996b; Mohanty & Padhy, 1996; Pandey & Varma, 2003; Sahu, 1996; Saini, 1996a; Shah & Singh, 1990; Sikarwar, 1996; Singh & Prakash, 1994; Tirkey *et al.*, 2003b; Watt, 1972**).

Trade : Mango.

Melia azedarach L.+ (Pl. 45A).

Syn : *M. sempervirens* Sw., *M. bukayan* Royle

Family : Meliaceae.

Vern. : Drek.

English, Sanskrit & Regional Names

Eng.	:	Barbados Lilac, Bead Tree, Indian Lilac, Persian Lilac;
Sans.	:	Brihan Nimba, Mahadroksha, Vishamushtika;
Ass.	:	Thamage;
B.	:	Ghornim, Mahanim;
Bo.	:	Bakayan, Vilayatinim;
G.	:	Bakan limbodo;
H.	:	Bakayan, Deikna, Mahanimb;
Kan.	:	Arevebu, Hutchu Bevu;
Mal.	:	Malaveppu;
Mar.	:	Bakananimb, Limbara;
P.	:	Bakain, Chein, Kacheu;
Tam.	:	Malaivembu, Sigarinimbam;
Tel.	:	Turukavepa, Vettivapa.

Distribution : Throughout India: 600-2700m.

Description : Trees having bi-or tripinnate leaves; lilac flowers, honey scented; and yellow, wrinkled drupe.

Reproductive Cycle : May-October.

Habitat Ecology : Planted in villages, along roadsides in Himalayas.

Material Examined : EBH-363, 14-10-2000.

Parts Used : Leaves. Bark. Seeds. Wood.

Folk Uses : Decoction of aerial plant parts given for **purifying blood**. Poultice of bark beneficial in **joint pains**. Juice of leaves prescribed for expelling **intestinal worms**, and applied locally for killing **lice**. Also, 25ml of its decoction taken on empty stomach for 10 days for the cure of

skin ailments. Powdered seeds and jaggery (1:1) in divided dose, taken one hour before meals for 40 days, for curing **piles**. Wood good for **agricultural implements** and **building works**.

Active Constituents : These are kaemferol-3-L-rhamno-D-glucoside, rutin (leaves), kuline, tetracyclic triterpenoids-kulactone, kulalactone, and kulinone (tree-bark).

Biological Activity : Azedarachin A and 12-O-acetylazedarachins A and B exhibit antifeedant activity against larvae of *Spodoptera exigua*.

Uses in Literature : Known earlier to be used as an alexipharmic, anthelmintic (vet), antilithic, antiseptic, astringent, blood purifier, brain tonic, cathartic, deobstruent, diuretic, diuretic (vet), edible, emmenagogue, febrifuge, fodder, fuel wood, laxative, psychoactive plant, resolvent, stomachic, symbolic abode of god, tonic, vermifuge, vermifuge (vet); and for bile influences in cattle, cold, curry, dilatation of the pupil, dyeing, enlarged lymph nodes and glands, fever (vet), furniture, gout, hair fall, headache, hypertension, hysteria, leprosy, making idols, malarial fever, nausea, necklaces, nervous headache, paintings (Mithila), preserving dried fruit and clothing from insects, promoting conception, rheumatism, rosaries, scrofula, skin diseases, small pox (vet), spasm, spleen enlargement, sprain (vet), swelling (vet), tuberculosis, tumours, ulcer (vet), and wounds **(Ambasta, 1986; Borthakur & Sarma, 1996; Chhetri, 1994; Goud & Pullaiah, 1996b; Jain, 1991; Jain *et al.*, 1994; Jha & Basak, 1994; Jha *et al.*, 1991; Kapur & Singh, 1996; Kurian, 1995; Lalramnghinglova, 1996; Rana *et al.*, 1996; Sharma, 1996; Siwakoti & Varma, 1996; Srivastava *et al.*, 2003; Watt, 1972)**.

Trade : Persian lilac.

Melothria heterophylla (Lour.) Cogn.+ (Pl. 45B).

Syn. : *Zehneria umbellata* Thw., *Solena amplexicaulis* (Lamk,) Gandhi., *S. heterophylla* (Lour.) Cogn.

Family : Cucurbitaceae.

Vern. : Gal kakdu, Jangli kheera.

Regional Names

B.	:	Kudari;
Bo.	:	Gomtta, Gometi;
H.	:	Tarali;
Mal.	:	Njerinjanpuli;
Oriya	:	Karakai, Makirla, Matka;
P.	:	Bankakra;
Tam.	:	Pulivanji;
Tel.	:	Thiyyadonda.

Distribution : Throughout India.

Description : Perennial scandent climber with yellow flowers; striped fruits and white seeds.

Reproductive Cycle : May-October.

Habitat Ecology : Forests, shrubberies.

Material Examined : EBH-556, 02-07-2001.

Parts Used : Leaves. Aerial Plant Parts. Root. Fruit.

Folk Uses : Decoction of leaves helpful in curing **malaria** (20 ml taken twice daily for 7 days). Poultice of roots good for **skin eruptions**. Fruits **eaten** raw. Also, used in the preparation of local wine '**sur**'.

Active Constituents : Main constituents of the plant are lignoceric, tricosanoic and behenic acids.

Uses in Literature : Reported to be used as coolant, invigorating purgative, stimulant, tonic, vegetable; and for antifertility, cuts, diabetes, dysuria, earache, fever, inflammation, jaundice, labour, paralysis, snake bite, sore eyes, sores of small-pox, spermatorrhoea, stomachache, syphilis, and unconsciousness (**Ambasta, 1986; Jain, 1991; Kapur & Srivastava, 1996; Negi & Gaur, 1991; Pandey *et al.*, 1996; Sinha, 1996b; Tiwari & Tiwari, 1996**).

Michelia champaca L.+ (Pl. 45C).

Syn : *M. rufinervis* DC., *M. doldsopa* Ham., *M. aurantiaca* Wall., *M. rheedi* Wight

Family : Magnoliaceae.

Vern. : Danti.

English, Sanskrit & Regional Names

Eng. : Champak, Golden Champa, Yellow Champa;

Sans. : Anjana, Atigandhka, Bhramavatithi, Bhringmohi, Chambunala, Champeya, Deepapushpa, Gandhaphali, Hemapushpa, Hemavhaya, Kamabana, Katu, Kumara, Kusuma, Kusumadhivata, Nagapushpa, Pati-Champaka, Ugragandha, Vanadapika, Vavlabdha;

Ass. : Phulchopa, Phulsopa, Titasoppa;

B. : Champa, Champaka;

Bo. : Champa, Chapha;

G. : Champo, Pitochampo, Raechampo, Sonchampo;

H. : Champ, Champa, Champac, Champaca;

Kan. : Sampige, Kola-Sampige;

Mal. : Champakam;

Mar. : Kudchampa, Pivalachampa, Sonachampa;

Oriya : Chompa, Chompoko;

P. : Chamba, Chamoti, Chamuti;

Tam. : Amariyam, Sambagam, Shampangi, Vandumarmalar;

Tel. : Champakmu, Champeyamu, Gangaravi, Heman-gamu, Kanjanamu, Sampamgi.

Distribution : Planted in outer Himalayas upto 2000 m.

Description : An evergreen tree. Leaves, tapering at ends. Flowers yellow, scented. Carpels ovoid.

Reproductive Cycle : April-October.

Habitat Ecology : Planted around houses in villages and temples.

Material Examined : EBH-152, 20-05-2000.

Parts Used : Fruits.

Folk Uses : Fruits prescribed in case of **respiratory ailments**, and also as an **appetiser.**

Biological Activity : Magnosprengerine and salicifoline possess muscle relaxant and hypotensive activity.

Uses in Literature : Known in India as an abortifacient, antiperiodic, antispasmodic, astringent, carminative, diuretic, emmenagogue, expectorant, febrifuge, edible (fruit), purgative, stimulant, stomachic, symbolic (sacred in offering), tonic, vermifuge; and for asthma, boils, bronchitis, canoes, carbuncle, carriage work, cephalalgia, cholera, cold, colic, colouring garments, cough, delirium, door panels, dysentery, dyspepsia, furniture, gonorrhoea, gout, headache, house building, inflammation, infusion, mania, maniacal excitement, menorrhagia, mucus in urine, native drums, nausea, oil seeds, opthalmia, perfumery and attar preparations, planking, pthisis, sculptures, sores, spermatorrhoea, timber, ulcer (**Ambasta, 1986; Banerjee, 1996; Jain, 1991; Khanna *et al.*, 1996b; Lalramnghinglova, 1996; Rao, 1996; Saini, 1996a; Vihari, 1995; Watt, 1972**).

Trade : Champak.

Milletia extensa Benth. *ex* Baker+ (Pl. 46A).

Syn. : *M. auriculata* Baker *ex* Brandis, *Robinia macrophylla* Roxb., *Pongamia macrophylla* Grah., *Otostema macrophylla* Benth., *Amerimnum pellidum* Hamilt.

Family : Fabaceae.

Vern. : Gauj.

Regional Names

H. : Ganj, Gauj;
Mal. : Vallimurutali;
Oriya : Arkawla, Kissi;
Tel. : Kondatangedutige.

Distribution : Himalaya.

Description : A woody climber. Leaflets glabrescent above and pale beneath. Corolla dull-white. Pod straight, very hard.

Reproductive Cycle : July-September.

Habitat Ecology : Open slopes, ravines.

Material Examined : EBH-675, 19-07-2001.

Plant/Parts Used : Leaves. Whole Plant.

Folk Uses : Leaves used as **fodder**, and dried plant parts kept in storage containers for **repelling insects**.

Active Constituents : These are auriculatin, aurmillone, millettin (seeds), aurmillone, isoaurmillene (pods).

Uses in Literature : Reported to be used as a fermenting agent, fibre yielding, fodder (vet), insect repellent, piscicide, symbolic (magic cure for fever); and for cattle's sores, collage (vet), fish poison, killing bed bugs (vet), spleen diseases, toothache, and washing wounds (**Ambasta, 1986; Aminuddin & Girach, 1991; Jain, 1991; Rao, 1996; Sikarwar, 1996; Siwakoti & Varma 1996**).

Momordica charantia L. (Pl. 46B).

Syn : *M. humilis* Wall., *M. muricata* DC., *M. senegalensis* Lamk., *Cucumis africanus* Bot.

Family : Cucurbitaceae.

Vern. : Karela.

English, Sanskrit & Regional Names

Eng.	:	Bitter Gourd;
Sans.	:	Brihadavalli, Karavellaka;
Ass.	:	Kakiral;
B.	:	Baramasiya, Potikakar;
Bo.	:	Karla;
G.	:	Karela, Karola;
H.	:	Karela, Karola;
Kan.	:	Hagal;
Mal.	:	Kaipavalli, Pavakacheti;
Mar.	:	Karale;
Oriya	:	Karena;
P.	:	Karela, Karila;
Tam.	:	Pavakkachedi;
Tel.	:	Mattakakata, Tellakakara.

Distribution : Throughout India.

Description : Climbers. Leaves orbicular, glabrous. Petals yellow. Fruits rostrate.

Reproductive Cycle : July-October.

Habitat Ecology : Cultivated.

Material Examined : EBH-555, 09-07-2000.

Parts Used : Fruit.

Folk Uses : Unripe fruits consumed as **vegetable** or **pickled**; 20 ml of its juice taken on empty stomach in morning for reducing excessive **blood sugars**, and curing **skin diseases**; and as an **anthelmintic**, **blood purifier**, and also efficaceous against **lice** when applied externally.

Active Constituents : Active properties of fruits are due to momoridocosides G, F1, F2, I, K and L.

Biological Activity : Positive hypoglycaemic activity of fruits are due to charautin-b-sitosterol-glucoside and stigmast-5,25-diene-3b-0-glucoside.

Uses in Literature : So far recorded as an abortifacient, anthelmintic, astringent, blood purifier, carminative, emetic, febrifuge, laxative, stomachic, tonic, vegetable; and for bilious affection, blood cancer, burning of the soles of the feet, colic, cooling, cutaneous diseases, diabetes, diseases of spleen and liver, dissipating melancholy and gross humours, dysentery, eczema, gout, headache, haemorrhoids, hypertension, jaundice, leprosy, malaria, night-blindness, oedema, peptic ulcer (vet), poisonous bite (vet), poultry diseases, psoriasis, rheumatic pains, scabies, snake bite, sores, water worms, wound maggots (**Ambasta, 1986; Bajpayee & Dixit, 1996; Ghosh *et al.*, 1996; Gill *et al.*, 1993; Jain, 1991; Jha *et al.*, 1996; Kothari & Moorthy, 1996; Kulkari & Kumbhojkar, 1993, 1996; Lalraminghinglova, 1996; Mahato & Mahato, 1996; Negi & Gaur, 1991; Rama Rao & Henry, 1996; Shome *et al.*, 1996; Singh & Pandey, 1996; Siwakoti & Varma, 1996; Watt, 1972**).

Major Ayurvedic Formulations : Mahavisagarbha Taila.

Trade : Karela.

Moringa pterygosperma Gaertn. (Pl. 46C).

Syn. : *M. oleifera* Lamk., *M. zeylanica* Pers., *M. polygana* DC, *Hyperanthera moringa* Vahl., *H. decandra* Willd., *Guilandina moringa* L.

Family : Moringaceae.
Vern. : Sunaney.
English ,Sanskrit & Regional Names

Eng. : Drum Stick Tree, Indian Horse Raddish;
Sans. : Damasamalu, Tikshnandhaka;
Ass. : Saijna, Sohjna;
B. : Sajina;
Bo. : Mangai, Sanga;

G.	:	Midhosaragaro, Seyla;
H.	:	Mungna, Saonjna;
Kan.	:	Nugge;
Mal.	:	Muringa, Muringa Elai, Murinna, Tishnagandha;
Mar.	:	Murinna, Tishagandha;
Oriya	:	Achajhada, Managachajhada;
P.	:	Saonjna;
Tam.	:	Karunjanam, Tavuselam;
Tel.	:	Mochakamu, Sajana.

Distribution : Widely cultivated throughout foothills of W. Himalaya:1100 m.

Description : Trees. Leaves usually 3-pinnate. Leaflets opposite, pale beneath. Flowers pale-white, honey scented. Pod pedunculous, seeds winged at the angles.

Reproductive Cycle : February–April.

Habitat Ecology : Hilly slopes, forests; not common.

Material Examined : EBH-1170, 03-02-2001.

Parts Used : Leaves. Flowers. Pods. Inflorescence. Seed Oil. Fruits. Gum.

Folk Uses : Young leaves and fruits consumed as **vegetable**. 2-3g powdered **inflorescence** (1/2 kg) in combination with 25g black pepper and 5g asofoetida, given once a day for 15 days to check **gastric ailments**. Seeds prescribed for preventing **conception** and a massage of its oil is good in **arthritis**. One teaspoonful of powdered gum along with dried coconut, almonds, pistachio, given to women for overcoming **post labour stress**. Sweet gum also given to children for checking **dysentery**. And also prescribed as an **aphrodisiac** in combination with **'shilajeet'**, **'musli'** and gum **'katira'**.

Active Constituents : Active properties are due to aldotriuronic acid, arabinose, galactose, glumronic acid, rhamnose, mannose, xylose (gum), aspartic acid, glutamic acid, glycine, threonine, alanine, valine, leucine, isoleucine, histidine, lysine, arginine, pheayl alanine, tryptophan, cysteine, methimine (leaves), alanine, arginine, glycine, serine, threonine, valine, glutamic and aspartic acids (flowers and fruits), 4-hydroxymellein, vanillin, octacosanoic acid, b-sitosterol and b-sitosterone (stem), benzyl isothiocyanate (seeds), and four auxin-like factors isolated (fast growing fruits).

Biological Activity : 4(a-L-rhamnosyloxy) benzyl isothiocyanate exhibits anti-bacterial activity against *Mycobacterium phlei* (40 mM/1) and *Bacillus subtilis* (56 mM/I).

Uses in Literature : Known earlier as an abortifacient, anthelmintic, antilithic, antipyretic, antirheumatic, antiscorbutic, aphrodisiac, appetiser, cardiac stimulant, carminative, cholagogue, cooking oil, diuretic, edible (fruit and pods), emetic, flatulent colic, illuminant, lubricants, psychoactive plant, rubefacient, stimulant, stomachic, symbolic (in folk proverbs), tonic, vesicant; and for acute traumic pain, anthrax (vet), articular pains, asthma, baldness, blood pressure, boils (vet), calculus affections, calico printing, catarrhal affections, cholera, circulating stimulants, cold, cosmetics, cough, cuts, debility of nerves, delivery, dental caries, dermatitis, dispersing swellings, dog bite, dropsy, dyeing, dysentery, dyspepsia (vet), ear complaints, eczema, enlarged spleen or liver, epilepsy, expelling placenta, eye (vet), eye aliments, fits (vet), foot sores, glandular swellings, gout, gum boils, headache, hoarseness, hydrocele, hysteria, increasing flow of bile, injuries, intermittent fevers, internal and deep seated inflammation, leprosy, madness, mats, menstrual complaints, otalgia, paper or cordage, paralysis, patches, perfumeries, pneumonia, postnatal, pustules, reducing labour pains in pregnant women, rheumatism (vet), scabies, scrofula, scurvy, small pox, soreness of mouth and throat, swelling (vet), syphilis, tetanus, toothache, typhoid, ulcer (vet), urinary complaint, venomous bites, and wounds (**Ambasta, 1986; Banerjee, 1996; Chauhan, 1999; Geetha *et al.*, 1996b; Hembrom, 1996; Henry *et al.*, 1996; Hosagoudar & Henry, 1996b; Jain, 1991; Jain *et al.*, 1994; Jha *et al.*, 1996; Kulkarni & Kumbhojkar, 1993; Molla & Roy, 1996; Mukherjee & Namhata, 1990; Negi & Gaur, 1991; Prasad *et al.*, 1996; Rama Rao & Henry, 1996; Saini, 1996a; Shome *et al.*, 1996; Singh, 1996b; Siwakoti & Varma, 1996; Srivastava *et al.*, 2003; Vishwanathan, 1997; Vishwanathan & Singh, 1996; Watt, 1972**).

Major Ayurvedic Formulations : Ekangavira-rasa, Ratnagiri-rasa, Sobhanjanadi-lepa, Syamadi-churna, Visatundika-taila.

Morus alba L.+ (Pl. 46D).

Syn. : *M. serrata* Wall., *M. tatarica* L.

Family : Moraceae.

Vern. : Toot.

English, Sanskrit & Regional Names

Eng.	:	Mulberry or White Mulberry;
Sans.	:	Tula;
B.	:	Tut;
Bo.	:	Chinni;
G.	:	Shetur;
H.	:	Chun;
Kan.	:	Hipnerle;
Kash.	:	Tut;
Mar.	:	Ambat, Tut;
Oriya	:	Tuticoli;
P.	:	Tut, Tutri;
Tam.	:	Pattuppuchi;
Tel.	:	Pippalipand Chettu, Reshmechattu;
U.P.	:	Shahtoot.

Distribution : Cultivated extensively in drier areas of W. Himalaya upto 3300m.

Description : Trees with lanceolate, toothed leaves; fruits black when ripe.

Reproductive Cycle : March-May.

Habitat Ecology : Common along agricultural fields.

Material Examined : EBH-507, 11-05-2000.

Parts Used : Roots. Fruit. Leaves.

Folk Uses : Fruits **edible** and considered **refrigerant**. Raw leaves chewed to overcome **spitting** habit. 1-2g mixture of powdered leaves and black pepper considered beneficial against **piles** (once a daily for 15 days), and also an excellent **fodder** for rearing silk worms. Powdered roots (½ kg), mixed with 25g **'Ashwagandha'** and 25g black pepper for **flatulence** and **rheumatic ailments** (one teaspoonful, twice a day for 15 days).

Active Constituents : Main ingredients are mulberrin, mulbernochromene, cyclomulberrin, cyclomulberrochromene (root), heptanoic (0.047), caprylic (0.21), pelargonic (0.047), capric (0.07), myristic (0.105), palmitic (11.85), palmitoleic (0.188), methyl heptadecanoic (0.14), stearic (4,46), oleic (12.67), linoleic (68.3),nonodecanoic (0.641), linilenic (0.98%), acids (fruit), two phytoalexins-moracin C and moracin D (shoot), and astragalin (leaves).

Biological Activity : Mulberrifuran A exhibits antibacterial activity against *Staphylococcus aureus* and *Fusarium roseum* whereas Moran A exhibits hypoglycaemic effect in normal and alloxan induced hyperglycaemic mice.

Uses in Literature : Reported earlier as an anthelmintic, diaphoretic, edible, fodder, purgative, refrigerant, vermifuge; and for agricultural implements, boats, buildings, dyspepsia, feeding silk worms, fever, fibre (bark), furniture, melancholia, paper, silk worm, and sore throat (**Ambasta, 1986; Jain, 1991; Negi & Gaur, 1991; Saini, 1996a; Watt, 1972**).

Trade : Mulberry.

Mucuna pruriens (L.) DC+ (Pl. 47A).

Syn. : *M. prurita* Hook., *M. utilis* Wall., *Caryopogon pruriens* Roxb., *Dolichos pruriens* L.

Family : Fabaceae.

Vern. : Draagal, Gazal bel.

English, Sanskrit & Regional Names

Eng.	:	Cowhage, Cowitch;
Sans.	:	Durabhigraha, Galrabhange;
B.	:	Akolshi, Bichchoti;
Bo.	:	Kuhili;
G.	:	Kanchan, Kivanch;
H.	:	Goncha, Kivachh, Kaunch;
Kan.	:	Nasukunni, Hasagunigida;
Mal.	:	Naykkurma;
Mar.	:	Kuhili;
Oriya	:	Alokushi;
P.	:	Gunchagaji, Kanaucha;
Tam.	:	Attumabuttar, Punaiyavarai;
Tel.	:	Pravrishayani, Telladuradagondi.

Distribution : Throughout India.

Description : Climber with membranous leaflets, purplish corolla; and turgid, pale brown pods.

Reproductive Cycle : July-September.

Habitat Ecology : On trees; ravines, slopes.

Material Examined : EBH-1193, 23-07-2003.

Parts Used : Seeds.

Folk Uses : Roasted seeds considered to be useful in **joint pains**. 2-3g of powdered seeds having seed coat removed after boiling in water, taken with cow's milk for 3 weeks to overcome **impotency** and as **nervine tonic**.

Active Constituents : Seeds contain mucadine, mucunadinine, mucuadininene, prurienidine mucunin, mucunadine, nicotine, L-dopa, glutarthiane lecithin, gallic acid, a glucoside and aminoacids.

Biological Activity : Plant extr. shows marked hypotensive and hypocholestrolaemic effects in normal rats.

Uses in Literature : Recorded to be useful as an anthelmintic, aphrodisiac, appetizer, astringent, diuretic (vet), edible (pods), emmenagogue, lactagogue (vet), local stimulant, mild vesicant, nervine tonic, psychoactive plant, purgative, stimulant, vegetable, vermifuge (vet) ; and for anasarca, blisters (vet), boils (vet), bone fracture, cancer, cough, digestion, dropsy, dysentery, elephantiasis, eye diseases, facial paralysis, fertility, fever, general debility, hemiplegia or paralysis on one side of the body and hallucinations accompanied with high fever, impotency, intestinal (vet), leucorrhoea, madness, nervous disorders (vet), night dreams, regulating menstruation, renal affections, ringworm, scabies (vet), scorpion and honey bee bites, seminal weakness, snake bite, sores, spermatorrhoea, syphilis, ulcer (vet), urinary complaints, worm (vet), and wounds (vet) (**Ambasta, 1986; Aminuddin & Girach, 1991; Borthakur & Sarma, 1996; Gill & Nyawuame, 1994; Girach *et al.*, 1997; Jain, 1991; Jha & Varma, 1996; Jha *et al.*, 1991; Kurian, 1995; Radhakrishnan *et al.*, 1996; Rama Rao & Henry, 1996; Rana *et al.*, 1996; Sahu, 1996; Satapathy & Brahmam, 1996; Singh, 1991; Singh & Pandey, 1996; Singh & Zaheer, 1996; Sinha, 1996; Siwakoti & Varma, 1996; Srivastava *et al.*, 2003; Watt, 1972**).

Major Ayurvedic Formulations : Agastya-haritaki, Ashwagandha-churn, Badam-pak, Brimhana-gutika, Dhatupaustic-churn, Kamdev-ghrit, Kawanch-churna, Kawanch-pak, L-dopa, Masbaladi-pachan, Muslipak, Sperman-tablets, Strenax-tablets, Tentax-forte, Vaji-kasana-ghrita, Vanari-gutika.

Trade : Seeds sold as 'kaunch beej'.

Murraya koenigii Spreng.+ (Pl. 47B).

Syn. : *Bergera koenigii* L.

Family : Rutaceae.

Vern. : Dandhelu, Ghandla, Kadi patta.

English, Sanskrit & Regional Names

Eng.	:	Curry Leaf Tree;
Sans.	:	Alakavhaya, Krishnanimba, Surabhinimba;
Ass.	:	Bishahari, Narasingha;
B.	:	Barsunga, Kariaphyli;
Bo.	:	Goranimb, Karrinim;
G.	:	Kadhilimbdo;
H.	:	Barsunga, Gandhela, Katnim;
Kan.	:	Rribevu;
Mal.	:	Kariveppilei
Mar.	:	Jhirang, Karepacha, Kudianim;
Oriya	:	Basango, Sanomorosinghi;
P.	:	Bowala, Gandanim;
Tam.	:	Karuvembu, Karuvepallai;
Tel.	:	Karepeka, Karivepaku.

Distribution : Throughout India, ascending to 1650m.

Description : Shrubs with gland dotted leaves which are crowded towards the end of branches. Flowers white. Berry wrinkled, black.

Reproductive Cycle : March-June.

Habitat Ecology : Often planted.

Material Examined : EBH-1109, 16-05-2003.

Parts Used : Leaves. Seeds. Fruit.

Folk Uses : Plant grown as an **ornamental** around temple courtyards. Leaves used as **condiment** for flavouring curries, and its poultice considered useful against

hernia; and also a dose of 2-3g powdered leaves prescribed with cow's milk (twice a day for 3 weeks) to cure **piles** and **diabetes**. 25-30 ml decoction of leaves and fruits (once daily) promotes **flow of urine**. Seeds eaten raw for **purifying blood**.

Active Constituents : These are mahanimbine, koenimbine, koenigicine (fruits and leaves), girinimbine, muconicine, cyclo mahanimbine and mahanimibidine (leaves).

Biological Activity : Koenoline exhibits cytotoxic activity against ICB cell culture. Also roots are antiamoebic, and hypotensive; and aerial parts spasmolytic.

Uses in Literature : Known to be used as an anthelmintic, carminative, edible (vegetable), febrifuge, laxative, stimulant for eruptions, stomachic, tonic, vegetable; and for acute abdominal pain, agricultural implements, bites of poisonous animals, bleeding, bone fracture, burning pains, chronic liver disorders associated with alcoholism and dyspepsia, culinary purposes, diarrhoea, dysentery, fertility (vet), flavouring curries (leaves), grooms as ornamental, herpes, joint pains, malarial fever, postpartum, purifying blood, renal pains, rinder pest (vet), scurvy, skin eruptions, snake bite, swellings in various tract of the body, toning stomach, tuberculosis, vomiting, and worms (**Aminuddin & Girach, 1996; Borthakur, 1996a; Chauhan, 1999; Gill *et al.*, 1997; Gogoi & Borthakur, 1991; Gogoi *et al.*, 2003; Jain, 1991; Kurian, 1995; Lal *et al.*, 1996; Mohanty & Padhy, 1996; Parmar & Kaushal, 1982; Rama Rao & Henry, 1996; Singh, 1996b; Siwakoti & Varma, 1996; Watt, 1972**).

Myrica esculenta Buch.-Ham. *ex* D. Don⁺ (Pl. 47C).

Syn. : *M. nagi* auct. non Thunb., *Myricaria sapida* Wall. Tent., *M. integrifolia* Roxb., *M. missions* Wall. Cat., *M. farquhariana* Wall., *M. rubra* Sieb. & Zucc., *Nugeia japonica* Gaertn.

Family : Myricaceae.

Vern. : Kaphal.

English, Sanskrit & Regional Names

Eng. : Box Myrtle;
Sans. : Aranya, Bhadra, Kaitarya, Mahakumbhi, Purusha;
Ass. : Nagatenga;
B. : Katphal;
Bo. : Katphal;
H. : Kaiphal;
Mal. : Marulu;
Mar. : Kayaphala;
P. : Kahela, Kahi, Kaphal;
Tam. : Marulam;
Tel. : Kaidaryama.

Distribution : Kashmir-Bhutan. Burma. China. S.E. Asia 1000-2300m.

Description : Trees. Leaves oblong lanceolate, leathery. Male flowers spikes reddish. Fruit red.

Reproductive Cycle : November-April.

Habitat Ecology : Cultivated areas.

Material Examined : EBH-213, 17-01-2000.

Parts Used : Fruits. Leaves. Bark.

Folk Uses : Fruits **consumed** as rich source of vitamin C and considered **carminative**. Leaves chewed as an effective remedy for **mouth ulcers**. A pinch of powdered bark taken as snuff for checking **headache** and **sinusitis**; and also added in the preparation of local wine, **'sur'**. Also gargles of the mixture of its bark and that of *Cassia fistula* (1:1) recommended as a remedy for **tonsilitis**. For curing **hydrocoel**, 1-2g mixture of its bark roasted in 'Alsi oil' (*Linum usitatissimum* L.), 10g 'Ajwain' (*Trachyspermum amoni* L.), 10g ginger (*Zingiber officinale*) given twice a day for 15-20 days. Wood used as **fuel**.

Active Constituents : Stem-bark yields b-sitosterol, taraxerol, myricadiol, myricanol, myricanone, myricanol glucoside, myricanol and myricanone.

Uses in Literature : Reported to be used earlier as an antiseptic, astringent, carminative, edible (fruit), fish-poison, refreshing drink, resolvent, sedative, stimulant, stomachic, tooth brush, tonic; and for affections of the chest and throat, anaemia, asthma, catarrh, cholera, chronic cough, chronic dysentery, cold, dental problems, diarrhoea, diuresis, dyeing, earache, eye diseases, fever, gonorrhoea, gum troubles, headache, heating, inflammation, lung affections, malaria, menorrhagia, nasal congestion, paralysis, pectoral, phlegm, piles, poisons suffering from overdoses of mercury, putrid sores, rheumatism, tanning, tumours, ulcers, and urinary discharges (**Ambasta, 1986; Ansari, 1991; Borthakur, 1996b; Chauhan, 1999; Chopra *et al.*, 1956; Dey, 1980; Dobremez, 1976; Jain, 1991; Lal *et al.*, 1996; Lalramnghinglova, 1996; Negi & Gaur, 1991; Negi *et al.*, 1993; Parmar & Kaushal, 1982; Sinha, 1996; Watt, 1972**).

Major Ayurvedic Formulations : Brihat-phal, Devdar-vadi-kwath, Irimedadi-tel, Kathphakdi-churna, Kathphaladi-kwath, Katphala-kvatha, Katphala-lepa, Katphala-taila, Khadiradi-gutika, Pushganug-churn, Shiroghna-churna.

Trade : Fresh fruits sold in the local markets. However, traded under the name of 'kaiphal' or 'katphala chhal'.

Myrsine africana L.[+] (Pl. 47D).

Syn. : *M. glabra* Gaertn., *M. scabra* Gaertn., *M. retusa* Ait., *M. rotundifolia* Lamk., *M. bifaria* Wall., *M. potama* Don

Family : Myrsinaceae.

Vern. : Chhota Mehndu.

Regional Names

H. : Chapra;

Kash. : Gugil;

P. : Bandaru, Khukan, Shamsad.

Distribution	:	Temperate Himalaya: 900-1800m.
Description	:	Shrubs. Leaves sessile, sharpely toothed. Flowers white, clustered. Mature berries globular becoming dark purple or black.
Reproductive Cycle	:	April-October.
Habitat Ecology	:	Forests, shrubberies, open slopes; often gregarious.
Material Examined	:	EBH-884, 28-09-2001.
Parts Used	:	Leaves. Fruits. Gum. Wood.
Folk Uses	:	A pinch of gum prescribed with luke warm water for overcoming **menstrual pain**. Fruits **anthelmintic** and **refreshing**. Leaves used as **fodder** and its decoction beneficial in **purifying blood**. Wood considered as excellent **fuel**.
Active Constituents	:	Fruit contains embelfin (to the extent of 3%) and quercitol.
Biological Activity	:	Alc. extr. of twigs and leaves exhibits inhibition of walker carcinosarcoma 256 in rats.
Uses in Literature	:	Used earlier as an alterative, anodyne, anthelmintic, astringent, bitter tonic, blood purifier, diaphoretic, hedges, laxative in dropsy, powerful cathartic, stomachic, vermifuge; and for biliousness, body strength, colic pain, cough, diarrhoea, dysmenorrhoea, flatulence, ringworm, and skin affections (**Ambasta, 1986; Chauhan, 1999, Jain, 1991; Watt, 1972**).
Major Ayurvedic Formulations	:	Agnitundivati, Amritarishta, Bidangarishta, Chyawanprash, Dasmoola-kwath, Dasamoolarista, Mahanarayan-taila, Pushanug-churna, Sanjivani-vati, Sudarshan-churna, Vidangadi-churna.
Trade	:	'Kakarsinghi' (roots).

Nerium indicum Mill. (Pl. 48A).

Syn. : *N. odorum* Solander, *N. odoratum* Lamk., *N. latifolium* Mill.

Family : Apocyanaceae.

Vern. : Kaner.

English, Sanskrit & Regional Names

Eng.	:	Indian Oleander, Sweet Scented Oleander;
Sans.	:	Asvamaraka, Pratihasa, Vishavrykshanka;
Ass.	:	Karavi, Karbira, Karvir;
B.	:	Karabi;
Bo.	:	Ganira, Kanhera;
G.	:	Kanera;
H.	:	Kaner, Karber, Kuruvira;
Kan.	:	Kanagalu;
Kash.	:	Gandeela, Gandula;
M.P.	:	Arali, Kanave Eram, Kattalari;
Mal.	:	Chuvannarali, Satraprasam;
Mar.	:	Kaneri;
Oriya	:	Konera, Koniyoro, Krobyo;
P.	:	Ganhira, Kanhira;
Tam.	:	Asuvabari, Irattaichegappayalari, Vellalari;
Tel.	:	Ettaganneru, Kasturipatha.

Distribution : N. W. Himalaya: 600-1500m.

Description : Shrubs with long, pointed, leathery leaves. Flowers pink, fragrant. Fruit paired, cylindric.

Reproductive Cycle : April-November.

Habitat Ecology : Foothills of W. Himalaya.

Material Examined : EBH-991, 17-05-2002.

Parts Used : Leaves. Stem-bark. Root-bark.

Folk Uses : A pinch of dried leaves taken as snuff for relief from headache. Powdered root-bark (2g) given twice daily with honey for 30 days in **paralysis**. Crushed stem-bark in combination with fennel (*Foeniculum vulgare*) and sugar (1:1) prescribed as **blood purifier**.

Active Constituents : These are quercitin, isoquercetin, neriumol, nerifol, oleandrin, neriodin, ursolic acid, adynerin, ursolic acid, adynerin, a compound similar to rutin; b-D-diginoside, b-D-digitaloside, digitoxigenin (leaves), neriodorin, nerioderin, karabin (root, bark, seeds), scopoletin, and scopoline (bark).

Biological Activity : Alc. extr. of root, bark, leaf and flower exhibits antibacterial activity. Also, oleandrin and oleandrigenin-b-D-diginopyranoside showed cardiopkinetic and diuretic activities.

Uses in Literature : Reported to be used in India as an abortifacient, anthelmintic, attenuant, cardiac, carminative, diaphoretic, febrifuge, fodder, gunpowder, insect repellent, insecticide, powerful heart poison, rat poison, resolvent, symbolic (sacred), vermic (vet); and for asthma, blisters, boils, bronchitis, copious lachrymation, dysentery, epilepsy, eye diseases, gum troubles, hooka tubes, leprosy, menorrhagia, opthalmia, ornamental garlands, pthisis, ringworm, skin eruptions like herpes, snake bite, swellings, ulcers, worms, and wounds (**Ambasta, 1986; Chauhan, 1999; Dastur, 1970; Girach *et al.*, 1996; Jain, 1991; Kapur & Singh, 1996; Khanna *et al.*, 1996a; Kurian, 1995; Saini, 1996a; Watt, 1972**).

Major Ayurvedic Formulations : Karaviradhya-taila, Kasi-sadi-taila, Laghu-visgarbha-taila.

Opuntia dillenii Haw.+ (Pl. 48C).

Syn. : *O. stricta* (Haw.) Haw. var. *dillenii*, *Cactus indicus* Roxb.

Family : Cactaceae.

Vern. : Chu-Chuar, Thor.

English, Sanskrit & Regional Names

Eng. : Prickly Pear, Slipper Thorn;

Sans. : Bahudagdhika, Kandashaka, Nagaphana;

B. : Nagphana, Phenimana;
G. : Chorhathola;
H. : Haththathoira, Nagphani;
Kan. : Papaskalli;
Mal. : Nagmullu, Palakkalli;
Mar. : Chapal, Nagphana, Samar;
Oriya : Nagophenia, Poturiyabiju;
P. : Chhittarthohar;
Tam. : Manjarnagadali, Pattanadugalli;
Tel. : Nagajemudu, Nagamullu.

Distribution : Planted and naturalised.

Description : Shrubs. Branches thick, flat and dotted with hairy tufts of short spinescent bristles. Flowers orange-red, large.

Reproductive Cycle : April-June.

Habitat Ecology : Edges of fields.

Material Examined : EBH-1323, 23-05-2003.

Plant/Parts Used : Whole Plant. Stem. Flowers. Latex.

Folk Uses : Plant grown as hedge. Flower juice beneficial for the cure of **asthma**. Latex applied externally for arresting **baldness**. Ash of burnt plant massaged on affected parts to cure **pain** due to **injury**. Plant consumed as **vegetable** (after removing thorns and boiling in water) for checking **flatulence** and **high blood pressure**. Soot formed by burning a piece of new silk cloth dipped in its latex and mustard oil against a bronze plate, recommended as an eye-liner for curing **eye ailments**.

Active Constituents : Flowers yield isoquercitin, b-sitosterol, opuntiol, opuntiol acetate and a glucoside narcissin.

Biological Activity : Ether extr. of the stem exhibits some antibiotic properties.

Uses in Literature : Used earlier as an antipyretic, carminative, demulcent, edible (fruits), expectorant, hedge, laxative, purgative, refrigerant, stomachic, symbolic in religious ceremonies of 'Santhals' (Spines); and for allaying heat, asthma, biliousness, boils, bronchitis, chest pains, controlling spasmodic cough, dyeing, enlargement of spleen, expelling worms, fibre as paper material, food, gonorrhoea, hastening formation of pains in the body, inflammation, leucoderma, loss of consciousness, opthalmia, pains, piles, pimples, promoting flow of urine, scorbutic ulcers, skin diseases, sores, swellings, syphilis, tumours, ulcers, urinary complaints, and whooping cough, (**Ambasta, 1986**; **Jain, 1991**; **Kurian, 1995**; **Parmar & Kaushal, 1982**; **Singh, 1996b**; **Watt, 1972**).

Oroxylum indicum Vent.⁺ (Pl. 48D).

Syn. : *Spathodea indica* Pers., *Calosanthes indica* Blume, *Bignonia petandra* Lour., *B. indica* L.

Family : Bignoniaceae.

Vern. : Arlu, Tataein.

English, Sanskrit & Regional Names

Eng.	:	Indian Trumpet Flower;
Sans.	:	Adhvantashatrava, Svarnavalkana;
Ass.	:	Bhatgila, Dingari, Kering, Toguna,;
B.	:	Nasona;
Bo.	:	Phalphura, Tetu;
G.	:	Podval;
H.	:	Assarsauna, Sauna;
Kan.	:	Bunepale, Sonepatta, Tigdu;
Mal.	:	Palakapayani;
Mar.	:	Taitu;
Oriya	:	Phompania;
P.	:	Miringa, Tatmorang;
Tam.	:	Paluiyudaichi, Vangam;
Tel.	:	Dundilamu, Mundakaparnamu.

Distribution : Throughout India, ascending to 1350m.

Description : Trees. Flowers dark red. Capsule thick, margins incurved. Seeds winged all round.

Reproductive Cycle : May-August.

Habitat Ecology : Forests, shrubberies.

Material Examined : EBH-708, 03-07-2001.

Parts Used : Pods. Seeds. Stem.

Folk Uses : Seeds check **constipation** and also ingredient of '**chywanprash**'. For **bone fracture** in children, a single dose of tablet formed by grinding a piece of 4 cm stem, 3 barley grains and one black pepper, has been recommended with luke warm water on empty stomach. In adults, this dose is formed by increasing the number of barley grains and black pepper to 5 and 2, respectively. Powdered pods (2-4g) given to check **dysentery, stomachache** and **fever** (twice a day till cure). In **veterinary**, pods are fed orally along with 'rai' (*Brassica juncea* Hook) and 'ajwain' (*Trachyspermum amoni* L.) for **discomfort** or **pain in stomach** and **spleen affections**.

Active Constituents : Active principles are baicelin its 6-glycuronide and 7-glucoronide, scutellarein, its 7-glucuronide, oroxylin A, chrysin and scutellarein-7-rutinoside (leaves and stem-bark), prunetin and b-sitosterol (heartwood), glucuronide-oroxidin, synthesis of biacatein-6-glucuronide and 7-glucuronide (seeds), fixed oil (25.0) and crude protein (7.9%).

Uses in Literature : Known to be used earlier as anodyne, antidote to snake bite, antiseptic, astringent, black dye, carminative, diaphoretic, ingredient of 'chywanprash', psychoactive plant, purgative (seeds), stomachic, symbolic (garlands of idols), tonic, vermic, vegetable (corolla, young shoots, unripe fruits); and for affections of the brain, blood dysentery, body pain, bone fracture (vet), cholera, chronic dysentery, colic, crack in nipples, diarrhoea, dropsy, dysentery (vet), dyspepsia, dysuria, enlarged spleen, epilepsy, headache, inducing conception, inflammatory diseases of chest, leprosy patients, liver ailments, madness, miscarriage, oedema, otorrhoea, pains, piles, pimples, polyuria, post-delivery pain, remittent and purpueral fevers, rheumatism, rutting (vet), scabies, scorpion sting, small pox, snake bite, sore backs of herds and bullocks, sores of nails and soles (vet), tanning, tuberculosis, ulcers, urticaria, vitality, and wounds (vet) (**Ambasta, 1986; Borthakur, 1996b; Borthakur & Sarma, 1996; Brahmam & Saxena, 1990; Brahmam *et al.*, 1996; Chopra *et al.*, 1956; Chowdhery, 1996; Girach, 1992; Girach & Aminuddin, 1995; Jain, 1991; Jain *et al.*, 1994; Kothari & Moorthy, 1996; Mehrotra *et al.*, 1995; Mohanty & Padhy, 1996; Molla & Roy, 1996; Mukherjee & Namhata, 1990; Pandey *et al.*, 1996; Rama Rao & Henry 1996; Reddy *et al.*, 1996; Satapathy & Brahmam, 1996; Siwakoti & Varma, 1996; Srivastava *et al.*, 2003; Upadhye *et al.*, 1994; Vishwanathan & Singh, 1996; Watt, 1972**).

Major Ayurvedic Formulations : Amritarishta, Brhatpanca-mulayadi-kvatha, Chywanprasha, Dashmool-ghrita, Dashmula-rishta, Dashmul-taila, Dasmool-kwath, Mahanarayan-taila, Pancha-mooladi-kwatha, Pushayanugchurn, Shyonaaka-putpale, Shyonaaka-siddha-ghrta.

Osbeckia stellata Wall. (Pl. 49A).

Syn. : *O. crinita* Nand., *Melastoma crinata* Roxb.

Family : Melastomataceae.
Vern. : Bandar Ki Tiur, Kumbhakavach.

Distribution : E. & N. Himalaya: 1200-2400m.

Description : Shrubs having 4-angled stems, opposite leaves with bristles. Flowers pink purple. Capsule oblong-ovoid, opening at the top by pores.

Reproductive Cycle : July-October.

Habitat Ecology : Shrubberies, under wood thickets, cultivated areas.

Material Examined : EBH-748, 24-09-2001.

Parts Used : Leaves. Stem.

Folk Uses : Tender stem edible. Paste of crushed leaves used as a substitute to Henna for **colouring palm, sole** and **nails of bride** ceremoniuosly before marriage.

Biological Activity : Antiviral activity confirmed (**Jain *et al.*,1994**).

Uses in Literature : Used earlier as an antidote to snake bite, poison (vet); and for chyluria, cough, digestion, dysentery, nose bleeding, stomachache, toothache, and wounds (**Jain, 1991; Lalramnhinglova, 1996; Manandhar, 1990; Rao, 1981a**).

Osyris arborea Wall.+ (Pl. 49B).

Syn. : *O. wightiana* Wall., *O. quadripartita* Salz. *ex* Decne.

Family : Santalaceae.
Vern.: Banchai.
Regional Names

Kan. : Baingani, Kuriganda,;
Mar. : Lotal, Popoli.

Distribution : Sub-Tropical Himalaya: 350-2000m.

Description : A shrub. Leaves nearly sessile, oblong-ovate. Flowers yellowish-green. Drupe globose, red when ripe.

Reproductive Cycle : August-April.

Habitat Ecology : Screes, open slopes, rare in cervices.

Material Examined : EBH-835, 23-07-2001.

Parts Used : Stem-bark. Leaves.

Folk Uses : Decoction of leaves drunk as a **beverage**, and its paste applied on cuts and wounds for quick healing. Powdered bark used for preparing **'sund'**, given to expecting women for overcoming **weakness**, and for restoring **post-delivery** health.

Uses in Literature : Recorded earlier as edible (fruit), powerful emetic, substitute for tea (leaves); and for dislocated bone, mother after child birth, and for preparing hot beverage like tea (**Ambasta, 1986**; **Manandhar, 1980b**; **Mao, 1993**; **Negi & Gaur, 1991**; **Watt, 1972**).

Pavetta indica L.$^{+}$ (Pl. 49C).

Syn. : *Ixora pavetta* Roxb.

Family : Rubiaceae.
Vern. : Papadi.
English, Sanskrit & Regional Names

Eng.	:	Indian Pellet Shrub, White Pavetta;
Sans.	:	Kakachedi, Papata, Tiriakphala;
Ass.	:	Gobor-Hitha, Samsuka;
B.	:	Jui, Kukrachura;
Bo.	:	Papat;
G.	:	Papat;
H.	:	Kankra, Karnikara, Kathachampa, Papari;
Kan.	:	Pappadi, Pavati;
Mal.	:	Mallikamuti, Pavatta;
Mar.	:	Papadi;
Oriya	:	Boniki, Katophingi, Paniphingi, Sua;
Tam.	:	Araniya, Karanai, Kattukkaranai, Pavattai;
Tel.	:	Duyipapata, Kondapapata, Lakkapapidi, Manjipapata, Namapapata, Nunepapata, Tapra;

Distribution : Throughout India upto 1500m.

Description : A shrub with leathery elliptical leaves. Flowers white, fragrant. Fruit black, globular.

Reproductive Cycle : July-October.

Habitat Ecology : Forests, shady ravines.

Material Examined : EBH-1081, 16-10-2002.

Parts Used : Flowers. Leaves.

Folk Uses : Plant grown as an **ornamental** shrub. Its toasted leaves applied externally to relieve **haemorrhoidal pain**, and 8-10 flowers pounded with milk recommended as an **aphrodisiac** tonic.

Biological Activity : Diuretic activity confirmed (**Jain *et al.*, 1991**).

Uses in Literature : Known to be used earlier as diuretic, edible (flowers and fruits), laxative, tonic in jaundice; and for blisters, dropsical affections, dysentery, epilepsy, haemorrhoidal pain, headache, intestinal worms especially roundworms in children, rheumatism, ulcerated nose, ulcers, urinary diseases, visceral obstructions, and on boils to remove pus (**Ambasta, 1986; Chowdhery, 1996; Girach *et al.*, 1997; Henry *et al.*, 1996; Jain *et al.*, 1991; Kharkongor & Joseph, 1981; Rama Rao & Henry, 1996; Sinha, 1987, 1996; Siwakoti & Varma, 1996; Watt, 1972**).

Phoenix sylvestris Roxb.+ (Pl. 49D).

Syn.: *Elate sylvestris* L., *Katou indel* Ham.

Family : Arecaceae.

Vern. : Khajrey.

English, Sanskrit & Regional Names

Eng. : Date Sugar Palm, Indian Wine Palm, Wild DatePalm;

Sans. : Bhumikharjurika, Kakakarkati, Skandhappala;

B.	:	Kajar, Kejur;
Bo.	:	Khajuri, Khjura
G.	:	Kajuri, Kharak;
H.	:	Khaji, Salma, Thakil;
Kan.	:	Ichalu;
Mal.	:	Intappana, Kattinta;
Mar.	:	Boichand, Sindikajuri;
Oriya	:	Khorjuri, Khorjuro;
P.	:	Khaji, Khajur;
Tam.	:	Ichamanai, Periyayinju;
Tel.	:	Peddayita.

Distribution : H.P.-W. Nepal. India. Burma: 150-1500m.

Description : Trees. Leaves pinnate with spiny stalk. Male flowers in white clusters. Fruits yellowish-reddish brown.

Reproductive Cycle : May-June.

Habitat Ecology : Open hills, rocky slopes.

Material Examined : EBH-119, 07-05-1999.

Parts Used : Fruits. Leaves. Roots.

Folk Uses : Tender buds and fruit **edible** as a **cooling tonic**. Leaves used for making **mats**, **brooms**, and for **thatching**. A dose of 2-3g mixture of (1/2 kg) powdered root in combination with 25g 'jayphal' (*Myristica fragrans* Houtt.) 25g 'jwati', and 25g cloves given twice a day with butter for 10-15 days to cure **piles**.

Uses in Literature : Useful in India as cardiotonic, constipative, cooling beverage, edible (buds and fruits), restorative, nervine tonic, psychoactive plant; and for abdominal complaints, ague, bakery, beverage, brooms, diarrhoea, dysentery, fattening, fevers, gleet, gonorrhoea, gum trouble, heart complaints, intoxicating alcoholic drink, loss of consciousness, basketry, mats and bags, menstrual complaint, nervous debility, paper making, pickling, relief in pain during pregnancy, ropes, thatching, toddy tapped from apical region of the trunk, toothache, venereal disease, and vomiting (**Ambasta, 1986; Goud & Pullaiah, 1996a; Jain, 1991; Jain *et al.*, 1994; Ragupathy & Mahadevan, 1991; Rama Rao & Henry, 1996; Ravishankar & Henry, 1992; Sahu, 1996; Singh, 1996b; Srivastava, 2003; Vishwanathan & Singh, 1996; Watt, 1972**).

Phyllanthus parvifolius Buch.-Ham. *ex* D. Don* (Pl. 50A).

Syn.: *P. juniperinus* Wall., *P. praetervisus* Muell.

Family : Euphorbiaceae.
Vern. : Bhui-Amla, Bhumi-Akhra, Dharti Kaphal.

Distribution : Temperate Himalaya. 1600-2200m.

Description : Shrubs. Leaves oblong-ovate. Flowers brownish-purple, minute. Fruits purplish-black.

Reproductive Cycle : June-October.

Habitat Ecology : Fairly common in forests, also sometimes on open rocky slopes.

Material Examined : EBH-799, 28-06-2002.

Plant/Parts Used : Whole Plant. Leaves. Stem. Root. Fruits.

Folk Uses : Raw leaves and fruit chewed with black pepper (*Piper nigrum* L.) for curing **ulcers** of the mouth; and leaves added to water for **purifying** it. Also, 10 ml decoction of root, stem and leaves given twice daily for 10-15 days to cure **typhoid**.

Picea smithiana (Wall.) Boiss.⁺ (Pl. 50B).

Syn. : *P. morinda* Link., *Abies smithiana* Wall. *ex* Aitch., *A. spinulosa* Griff., *A. kutrow* Royle, *Pinus smithiana* Lamb.

Family : Pinaceae.

Vern.: Rai.

English & Regional Names

Eng.	:	West Himalayan Spruce;
H.	:	Rai, Rau, Re, Riar, Kachal, Kachhlu, Sall, Tosh.

Distribution : N. W. Himalaya: 2100-3600m.

Description : Conical trees. Leaves spirally arranged, needle-like. Cone pendulous, shining dark brown.

Reproductive Cycle : April-May.

Habitat Ecology : Associated with *Cedrus deodara, Pinus wallichiana, Abies pindrow* and *Quercus spp.*

Material Examined : EBH-447, 14-04-2000.

Parts Used : Wood. Cones. Resin. Bark.

Folk Uses : Cones used for making **decorative art pieces**. Oil extracted from distillation of its resin is externally used for healing **wounds** and **fresh cuts**. 2-3g mixture of ½ kg powdered bark in combination with 25g black pepper (*Piper nigrum* L.) and 30g fennel (*Foeniculum vulgare*) given with cow's milk on empty stomach for 5-8 days to **purify blood**.

Active Constituents : These are amentoflavone, cupressuflavone and agathisflavone (leaves).

Uses in Literature: Reported earlier as an activator for saw-dust boards, bath salts, deodrant, rubefacient, stimulant; and for room sprays (**Ambasta, 1986; Kapur, 1996**).

Trade : Spruce.

Pinus roxburghii Sarg.[+] (Pl. 50C).

Syn.: *P. longifolia* Roxb. *ex* Lambert

Family : Pinaceae.

Vern. : Chir, Cheel.

English, Sanskrit & Regional Names

Eng.	:	Himalyan Long Leaved Pine, Three Leaved Pine;
Sans.	:	Bhadradaru, Putikashtha;
Ass.	:	Telia;
B.	:	Saralakashtha;
G.	:	Deodara;
H.	:	Chir, Dhupasarala;
Kash.	:	Salla;
Mal.	:	Charalam, Putikashtam;
Mar.	:	Saraladeodara;
Oriya	:	Sorolokatha;
P.	:	Anandher, Chir, Drabchir;
Tam.	:	Suruldevadari;
Tel.	:	Devadaru, Saralu.

Distribution : Afghanistan-Bhutan: usually 1000-2000m.

Description : Tree; leaves acicular, in cluster of three, 22-30 cm long. Cones ovoid-conical, solitary.

Reproductive Cycle : March-May.

Habitat Ecology : Forming extensively forests; gregarious.

Material Examined : EBH-106, 11-05-2000.

Parts Used : Leaves. Bark. Seeds. Resin. Wood.

Folk Uses : Pine needles used for making **brooms, corrugated packaging boxes**, and its juice prescribed to cure patients suffering from **tuberculosis**. Also, this on decomposition after mixing with cow dung, form an excellent **manure**. Seeds **edible**. Wood used as **torch**, and for **furniture** and **house building**. Paste of bark effectively used for treating **crab-bites**. Cones employed for making decorative **art pieces.** Turpentine oil good as an **anthelmintic**, and added to sacred fire for performing magico-religious rites.

Active Constituents : Principle constituents are friedelin, ceryl alcohol and b-sitosterol (bark), b-pinene and car-3-ene (turpentine oil), pine oil (needles), a-pinene, b-pinare, car-3-ene, abietic acid and isopimaric acid (turpentine from xylem resin).

Biological Activity : The oil from 'chir' pine needles is reported to have antibacterial activity against a number of organisms including *Staphylococcus aureus.*

Uses in Literature : So far reported as an antiseptic, carminative in flatulent colic, colouring matter, diaphoretic, diuretic, edible (seeds), expectorant, fuel wood, insecticide, psychoactive plant, rubefacient in various thematic affections such as lumbago, arthritis and neuralgia, stimulant; and for abscess, bangles, baskets, boils, bone fracture, buboes, burning of body, charcoal, chronic bronchitis, constipation, cough, cracks in sole of feet, denaturants, fevers, furniture, gangrene of lungs, general care of hovers, gonorrhoea, house building, inflammation of abdomen, itching, leprosy, mange (vet), match industry, minor haemorrhages on tooth-sockets, musical instruments, nose, painting, perfumery, phthisis, pine oil in various pharmaceuticals, preparation of plasters-ointments and fumigants, rafters, resin, rheumatism, seatworm infestations, shingles, skin diseases, snake bite, sport-goods, sprain, sprain (vet), swelling, synthetic pine oils, tanning, tea chests, torches, turpenting, tympanitis, typhoid, and ulcers (**Ambasta, 1986; Chauhan, 1999; Jain, 1968, 1991; Jain *et al.,* 1994; Sensarma, 1991; Srivastava *et al.,* 2003; Watt, 1972**).

Major Ayurvedic Formulations : Chandanaadi-vati.

Trade : Birozaa, Chil, Chir.

Pinus wallichiana A. B. Jackson+ (Pl. 50D).

Syn. : *P. excelsa* Wall. *ex* D. Don, *P. griffithii* Mc Clell, *P. chylla* Lodd.

Family : Pinaceae.

Vern. : Kail.

English & Regional Names

Eng.	:	Bhutan Pine, Himalayan BluePine, Five Leaved Pine,
H.	:	Kail;
Kash.	:	Beyar, Biar, Kaiar, Tser, Yara, Yiro, Yur;
P.	:	Andal, Biar, Chiti, Darchil, Kachir, Kail, Keiiri, Lhim, Lhimster, Lim, Palsam, Partal, Shim, Somshing, Yari, Yero;

Distribution : Himalaya: 1800-4300m.

Description : A tree. Leaves needle-like, blue green, borne on short shoots in clusters of 5.

Reproductive Cycle : April-June.

Habitat Ecology : Forming forests in drier areas over a wide range of altitudes, also found in wet areas in secondary forests.

Material Examined : EBH-1220, 29-05-2003.

Parts Used : Wood. Leaves.

Folk Uses : Wood extensively used for **furniture** and **building works**.

Active Constituents : Turpentine from xylem resin contains a-pinene (90.0), isomers of undecane, dodecane, tridecane abietic, isopimaric and lambertianic acids.

Biological Activity : Needle oil shows positive antibacterial activity.

Uses in Literature: Used earlier for abscess, bone fracture (vet), dislocation of joints, plastering walls, ulcer, unconsciousness, and wounds (**Chauhan, 1984**; **Jain, 1991**).

Trade : Blue Pine, Kail.

Pistacia integrimma J. L. Stewart *ex* Brand.+ (Pl. 51A).

Syn.: *P. kinjuk* Stocks, *P. chinensis* Bunge,
P. chinensis Bunge. var. *integrimma* (Stewart) Zohary,
Rhus integrimma Wall., *R. kakra-singee* Royle

Family : Pistaceaceae.

Vern. : Kakad Singhi.

English, Sanskrit & Regional Names

Eng.	:	Crab's Claw;
Sans.	:	Chandraspada, Kasavinashini;
Ass.	:	Kakaisrnghi;
B.	:	Kakra;
G.	:	Kakra;
H.	:	Kakra;
Kan.	:	Kakadsasingi, Karkatasrini;
Kash.	:	Gurgu, Kakkar;
Mal.	:	Karkatasringhi;
Mar.	:	Kakra;
Oriya	:	Kakadashringi, Kakadashrungi;
P.	:	Kakkar, Kakkeram.
Tam.	:	Karkata Singhi;
Tel.	:	Kakarsingi, Karkatashrungi.

Distribution : Afghanistan - W.Nepal: 600-2400m.

Description : Trees. Leaves odd or even pinnate. Flowers red, without petals. Drupe glabrous, wrinkled, grey when ripe.

Reproductive Cycle : March-May.

Habitat Ecology : Riversides, cultivated areas.

Material Examined : EBH-755, 07-04-2001.

Parts Used : Pods.

Folk Uses : 2-3g powdered fruits alone or in combination with roasted rind of *Punica granatum*, given either with honey or with cow's milk twice daily for 3-5 days to cure **cough**, **asthma**, and as an **expectorant**. And its paste in mustard oil relieves **pain** in muscles. One teaspoonful of the mixture prepared by adding 10g of its pod, 5g 'magan' (*Piper longum*), 5g black pepper (*Piper nigrum*), 5g 'kadu' (*Picorhhiza kurrao*), 5g 'patish' (*Aconitum heterophyllum*) 5g 'geru' and 100g honey, prescribed thrice daily for 15 days for curing **convulsive coughs**. Also, mixture prepared by adding equal quantity of fruits, coriander (*Coriandrum sativum* L.) and fennel (*Foeniculum vulgare*) taken for 8-10 days with lukewarm water has been found to be good for relief from gripping **ovarian pain** and as an **anthelmintic**. For performing '**havan**' during religious ceremonies, its pods added to sacred fire along with 'ghee' and 'jaggery'.

Active Constituents : Active principles isolated from fruits are pistacin and psitacinin.

Biological Activity : Gall oil antibacterial, antiprotozoal, antispasmodic and CNS depressant.

Uses in Literature : Known to be used as an antidote to scorpion and snake bites, carminative, expectorant, fodder, ingredient of 'chywanprash', tonic; and for asthma, carvings, chronic pulmonary affections, construction and beautification purposes, cough, diarrhoea, diseases of respiratory tract, dyeing, dysentery, dyspeptic vomiting, fever, furniture, inlay work, irritability of the stomach, panels, phthisis, picture frames, ploughs, psoriasis, respiratory troubles, roping, spinning wheels, tanning, turnery, and want of appetite (**Ambasta, 1986; Chauhan, 1999; Jain, 1991; Kapur & Nanda, 1996; Mehrotra *et al.*, 1995; Watt, 1972**).

Major Ayurvedic Formulations : Balachaturbhadra-churna, Brihat-taalishaadi-chu-rna, Balacatur-bhadrika-churna, Chayavanprasha, Dasti-moolarishta, Kantakatari-avaleha, Karkadi-churna, Kasni, Shringyadi-churna.

Trade : Kaakarsingi.

Plumbago zeylanica L.⁺ (Pl. 51B).

Syn. : *P. auriculata* Blume., *Thela alba* Lour.

Family : Plumbaginaceae.
Vern. : Chitrak.
English, Sanskrit & Regional Names

Eng.	:	Ceylon Lead Wort, White Flowered Leadwort;
Sans.	:	Analanania, Chitraka, Jataveda, Vaishavan-Ara;
B.	:	Chita, Chitruk;
Bo.	:	Chitrack;
G.	:	Chitaro, Gholochitro;
H.	:	Chitarak, Chitra, Chiti;
Kan.	:	Chitramula, Vahni;
Mal.	:	umpokotuveli;
Mar.	:	Chitraka, Chitramula;
Oriya	:	Chitramulo, Krishanu;
P.	:	Chitrak;
Tam.	:	Adigarradi, Vellaikkodiveli;
Tel.	:	Chitramulamu, Tellachitramulamu.

Distribution : Throughout India, ascending to 1650m.

Description : A rambling undershrub with glandular leaves and white corolla. Capsule enclosed in the persistent calyx, opening transversely near the base.

Reproductive Cycle : April-September.

Habitat Ecology : Moist slopes, edges of forests.

Material Examined : EBH-807, 22-07-2001.

Parts Used : Juvenile Stem. Roots.

Folk Uses : Young stems are edible as **potherb**. Roasted bark (1/2 teaspoonful) taken with cow's milk on empty stomach for the cure of **leucorrhoea**. Paste of roasted roots, a pinch of alum and turmeric in coconut milk applied externally to check **boils, ringworms** and **itching**. For performing **abortion**, a small piece of root is kept in uterus, and its amulet tied around waist for relief from pain in **sciatica**.

Active Constituents : Chief constituents are plumbagin, 3-chlorplumbagin and a new substance-3, 3'-biplumbagin, a new binaphthaquinone-chitranone, zeylinone, isozeylinone, elleptinone, droserine, zeylanone, isozeylanone, plumbagic acid (root) and isoschinandone, sitosterol, plumbazeylone, a new dihydronaptho-quinone-dihydroserone-tyrosine (4.5), threonine (4.1), alanine (3.5), histidine (3.4), glycine (1.7), methionine (0.5) and hydroxyproline (0.2%) (aerial parts).

Biological Activity : Plumbagin (0.75 mg/kg) active against P388 lymphocytic leukaemia and at 4 mg/kg exhibits antibacterial and antifungal activity against a wide variety of bacteria and fungi. Besides aq. and alc. ext. has been observed to be a powerful aphicide.

Uses in Literature : Known to be useful as an abortifacient, acrid poison, antiperiodic, aphicide, appetiser, diuretic, expectorant, expellent of phlegmatic tumors, psychoactive plant, stimulant, sudorific, symbolic (worn on arm in body pain), uterine tonic (vet); and for abdominal pain, abscess, anasarca, arthritis, black-water fever, boils, bone fracture (vet), chest pain, contraction of uterus after child birth, diarrhoea, diarrhoea (vet), dropsy, dysentery, dyspepsia, ecbolic action in delayed labour, eczema, elephantiasis, enlargement of spleen, hydrocele, infection in horses, influenza, inhibiting/preventing conception, leprosy, malaria, muscular pain, night blindness, pains after birth, paralysis, piles, rheumatic pains, severe headache, skin diseases (vet), skin disorder, snake bite, spleen complaint, sprains, stomach disorders, stomatitis (vet), vesicant properties, and wounds **(Ambasta, 1986; Balasubramanian & Prasad, 1996; Bhandary *et al.*, 1996; Borthakur, 1993; Brahmam & Saxena, 1990; Brahmam *et al.*, 1996; Chauhan, 1999; Gill *et al.*, 1993; Girach *et al.*, 1996; Goud & Pullaiah,**

1996b; Hembrom, 1996; Jain, 1991; Jain & Sharma, 1996; Jain *et al.*, 1994; Jha & Varma, 1996; Kothari & Moorthy, 1996; Manandhar, 1987b; Rama Rao & Henry, 1996; Ranjan, 1996; Reddy & Raju, 1996; Sensarma, 1991; Sharma & Borthakur, 2003; Sikarwar, 1996; Singh, 1991; Singh *et al.*, 1996b; Siwakoti & Varma, 1996; Swami & Gupta, 1996; Srivastava *et al.*, 2003; Varma, 1997; Watt, 1972).

Major Ayurvedic Formulations : Agni-tundi-vati, Amlayadi-churna, Arogyavardhini-vati, Chaudraprabha-vati, Chitraka, Chitrakadi-churna, Chitrakadi-guttika, Citrakadi-vati, Chitrak-ghrita, Chitrak-haritaki, Devaradi-kwath, Navayas-lauha, Panakolasava, Panchkola-chorna, Citrakadi vati, Pippalyasava, Punarnavamandur, Sarvajwar-lauha, Sudarshan-churna, Trimalla-churna, Yog-raj-guggulu.

Trade : 'Chitrak' (roots).

Plumeria rubra L. (Pl. 51C).

Syn. : *P. acuminata* Roxb., *P. acutifolia* Poir.

Family : Apocynaceae.
Vern. : Champa.
English, Sanskrit & Regional Names

Eng.	:	Frangipani, Jasmine Tree, Pagoda Tree;
Sans.	:	Devaganagalu, Gosampige;
Ass.	:	Goalanchi, Gulanchi, Gunanch;
B.	:	Dalanaphula, Goruchampa;
Bo.	:	Champa, Chameli, Dolochapa, Gutachin, Khadchampo, Khairchampa, Sonchampa;
G.	:	Rhadochampo;
H.	:	Champa, Chameli, Goburchamp, Golainchi, Gulachin;
Kan.	:	Devaganigile, Kadu-Sampige;
Mal.	:	Arali, Vellachampakam, Velattalari;
Mar.	:	Khairchampa, Ruruchahpa, Sonchampa;
Oriya	:	Golochi, Kutokompa;
Tam.	:	Ilattalari, Kallimandarai, Kuppiyalari, Navillavalari, Perungalli;
Tel.	:	Arhataganneru, Nuruvarahalu, Vadaganneru, Veyyiverahalu.

Distribution : Many parts of India.

Description : A small tree with leaves prominently nerved beneath. Flowers cream-coloured, usually with a yellow centre.

Reproductive Cycle : June-September.

Habitat Ecology : Cultivated.

Material Examined : EBH-1250, 16-07-2003.

Plant/Parts Used : The Whole Plant.

Folk Uses : Grown as an **ornamental** plant.

Active Constituents : These are plumerinine (stem), fulvoplumierin fulvoplumierin, allamcin, allamandin, 2,5-dimethoxy-p-benzoquinone, plumericin, and lirio-dendrin (bark).

Biological Activity : Fulvoplumierin, allamcin, allamandin, 2,5-dimethoxy-p-benzoquinone, plumericin and liriodendrin exhibit cytotoxic activity against a panel of cell lines including murine lymphocytic leukaemia (P-388) and a human cancer cell type, viz. Colar, breast, fibrosarcoma, lung, melanoma.

Uses in Literature : So far reported as an abortifacient, ague, antihepatic, cathartic, emmenagogue, febrifuge, purgative, rubefacient, vegetable (flower); and for allaying thirst, asthma, cough, decoration, diarrhoea, dropsical and venereal affections, gonorrhoea, gum troubles, intermittent fevers, itching, leprosy, musical instruments, rheumatism, scabies, sores, sprain (vet), swelling, syphilis,and wounds (vet) (**Ambasta, 1986; Gogoi & Borthakur, 1991; Hosagoudar & Henry, 1996b, c; Jain, 1991; Jha *et al.*, 1996; Rama Rao & Henry, 1996; Siwakoti & Varma, 1996; Watt, 1972**).

Pogostemon benghalensis **Burm. f.⁺ (Pl. 52A).**

Syn. : *P. plectranthoides* Desf., *Origanum benghalensis* Burm. f., *O. indicum* Roth., *Mentha secunda* Roxb., *M. fruticulosa* Roxb.

Family : Lamiaceae.

Vern. : Kali Basunti.

Regional Names

B.	:	Bakoha, Jin, Jui-Lata;
Bo.	:	Pangla;
Oriya	:	Dumobadottoka;
Tel.	:	Kasurijang.

Distribution : H. P.-E. Nepal, upto 1300 m.

Description : Strongly-scented, pubescent shrubs. Leaves opposite. Flowers white-tinged.

Reproductive Cycle : December-April.

Habitat Ecology : Common on wastelands, roadsides.

Material Examined : EBH-180, 29-01-1999.

Parts Used : Roots.

Folk Uses : Ash of roots useful in **whooping cough**, and its poultice effective against **joint pains**.

Uses in Literature : Reported in Indian literature as antihaemorrhagic, drink, edible (eaten in times of scarcity), febrifuge, stimulant, stomachic, styptic; and for bleeding, body pain, boils, cuts, detergent, diarrhoea, diarrhoea (vet), green manure, and wounds (**Ambasta, 1986; Brahmam & Saxena, 1990; Datt, 1996; Jain, 1991; Kothari & Moorthy, 1996; Rama Rao & Henry, 1996; Siwakoti & Varma, 1996**).

Polygonum chinense L.+ (Pl. 52B).

Syn. : *P. auriculatum* Meissn., *P. polycephalum* Wall., *P. brachiatum* Poir, *P. pratens* Don, *P. corymbosum* Willd., *P. ovatum* Heyne, *P. adenotrichum* Wall., *Ampelygonum chinense* Lindl., *P. panduraeforme* Ham., *Coccoloba crispata* Ham.

Family : Polygonaceae.

Vern. : Tarod.

Regional Names

Ass.	:	Kalnap, Kuki;
Kan.	:	Bilchini-Ganigalu;
Mar.	:	Paral.

Distribution : Hilly districts throughout India, upto 2650m.

Description : Nearly erect shrubby climbers. Leaves rough with minute prickles, oblong lanceolate. Flowers white. Nut 3-angled, dull black.

Reproductive Cycle : September-October.

Habitat Ecology : Hilly and mountain areas.

Material Examined : EBH-1222, 23-09-2003.

Parts Used : Leaves. Fruits.

Folk Uses : Leaves and ripe fruits **edible**.

Active Constituents : Leaves possess kaempferol, quercetin, batafolin, kaempferol-7-0-glucoside, kaempferol-3-0-glucoronide, b-sitosterol and ellagic, gallic, 3-0-methylellagic acids.

Uses in Literature : So far used in India as antiscorbutic, edible (leaf shoots, fruit), fodder, good nectar and pollen source for bees, tonic, vulnerary; and for dysentery, fever, giddiness, stomachache, (**Ambasta, 1986; Borthakur, 1996; Hosagoudar & Henry, 1996b; Jain, 1991; Jamir & Rao,**

1990; Kapur & Nanda, 1996; Lalramnghinglova, 1996; Radhakrishanan *et al.*, 1996; Singh *et al.*, 1996b; Sinha, 1996)

Populus ciliata Wall. (Pl. 52C).

Syn.: *P.pyriformis* Royle, *P. rotudifolia* Griff.

Family : Salicaceae.

Vern. : Phalsa.

English & Regional Names

Eng.	:	Himalayan Poplar;
H.	:	Pahari-Peepal, Tilaunja;
P.	:	Dudfrass, Falish, Kramali.

Distribution : Pakistan-S.W. China: 2100-3600m.

Description : Trees. Leaves ovate, teeth ciliate. Flowers in drooping racemes. Capsule ovoid.

Reproductive Cycle : April-July.

Habitat Ecology : Stream sides; drier areas in coniferous forests.

Material Examined : EBH-776, 17-06-2001.

Parts Used : Leaves. Bark. Wood.

Folk Uses : Decoction of bark along with black pepper useful as **blood purifier**. Leaves used as **fodder** and wood for cheap **furniture** and **building constructions.**

Uses in Literature : Reported earlier to be used as blood purifier, fodder, fuel, stimulant, tonic; and for bone fracture, constructions, cricket bats, fruit crater, furniture, limbs, match boxes, packing, splints, turnery, veeners, and water troughs (**Ambasta, 1986; Jain, 1991; Kapoor & Singh, 1996; Kapur & Nanda, 1996; Singh *et al.*, 1996a; Watt, 1972**).

Populus nigra L.+ (Pl. 52D).

Syn. : *P. nigra* var *italica* (Munchn) Koechne.

Family : Salicaceae.

Vern. : Poplar, Popular.

English & Regional Names

Eng. : Black Poplar, Lombardy Poplar;

Kash. : Frasht;

P. : Buins, Do, Farsh, Frast, Kramadi, Makkal, Pakshubut;

Distribution : N. W. Himalaya.

Description : A large-sized tree, forming a narrow cylindric crown.

Reproductive Cycle : April-July.

Habitat Ecology : Planted along roadsides.

Material Examined : EBH-474, 14-04-2000.

Parts Used Examined : Wood. Gum.

Folk Uses : Tender shoots given as **fodder**. Wood used in match-stick and **plywood** industries, **building works** and as **fuel**. 4-5g mixture prepared by pounding 250g of gum in combination with 10g '**shilajeet**', given on empty stomach with cow's milk for 15 days as an **aphrodisiac** and to overcome **sexual dysfunctions**.

Active Constituents : Various constituents are salicin, salicortin d, new phenol, glycoside-nigracin (bark), 2-methylpropyl, 2-methyl-2-propenyl-2-methylbutyl-, 3-methylbutyl-, 2-methyl-2-butenyl-, 4-methylpeatyl-, 5-methylhexyl and 3,7-dimethyl-2,6-octadienyl caffeates (bud exudates).

Uses in Literature : Reported earlier as coacher, depurative, fodder; and for bone fracture, carving, colds, cross-beams, furniture, haemorrhoids, hair growth, matches, poles, round boxes for grapes, small boxes, timber, truck, and barrio trays (**Ambasta, 1986; Jain, 1991; Singh** ***et al.*****, 1996a; Watt, 1972**).

Premna latifolia Roxb.[+] (Pl. 53A).

Syn. : *P. mucronata Roxb.*

Family : Verbenaceae.

Vern. : Vikar.

Regional Names

B.	:	Goharo;
H.	:	Agnium, Bakarcha, Jhatela;
Mal.	:	Knappa;
Oriya	:	Gondhono;
P.	:	Bankar, Ganhila;
Tam.	:	Erumaimunai, Munnai;
Tel.	:	Peddanella kure.

Distribution : Temperate Himalaya: upto 1400m.

Description : A small tree with subcoraceous leaves and broad-ovate, greenish-white flowers. Fruit globose, dark purple.

Reproductive Cycle : July-September.

Habitat Ecology : Common on roadsides.

Material Examined : EBH-624, 29-07-2001.

Parts Used : Leaves. Stem-bark.

Folk Uses : 20 ml decoction of leaves, 'mulathi' (*Glycyrrhiza glabra* L.) and 'dhansoa' (*Anethum sowa Roxb. ex* Flem.) in equal quantity given to cure **diabetes** (twice daily for 15 days). Bath of medicated water, obtained by boiling it with stem-bark, helps in curing **urticaria** and **skin affections**. Juice of leaves rubbed as an **emollient** on affected parts.

Active Constituents : Principle constituents are a new furanoid- 2,4-di (3′,4′-methylenedioxyphenyloxy) furan (premna-latin), b-sitosterol and its glucoside, 5-hydroxy-4′-methoxyflavone-7-O-arabinopyra-nosyl-rhamnoside and 5-hydroxy-4′-methoxyflavone-7-O-trioside, two iridoids-7-deoxyloganic acid (=bisdesoxydi-hydromonotropein) and geniposidic acid (leaves) and pimarene-sandarcopimar-15-en-1a, 8b, 11a-triol (VIII) (bark).

Uses in Literature : Reported earlier in India as diuretic, firewood, fodder, vegetable (leaves and tender shoots); and for blisters, boils, charcoal, colic, colic (vet), dropsy, flavouring, gonorrhoea, indigestion, indigestion (vet), ringworm, syphilis, and ulcers (**Ambasta, 1986; Watt, 1972**).

Prinsepia utilis Royle[+] (Pl. 53B).

Family : Rosaceae.

Vern. : Bhekal.

Regional Names

H. : Bekhra, Cherara, Dhatila, Jhatela;

P. : Bekhul, Gurinda, Kharnglara, Kharnguru, Phulwara, Tatua.

Distribution : Temperate Himalaya: 1200-2700m.

Description : Thorny shrubs. Leaves leathery. Flowers white. Drupe purple.

Reproductive Cycle : June-October.

Habitat Ecology : Forests. Shrubberies. Hedges.

Material Examined : EBH-263, 04-06-2000.

Plant/Parts Used : Roots. Stem. Branches. Seeds. Fruits. Whole Plant.

Folk Uses : Commonly grown as **hedges**. Fruits **edible**. Stem excellent for making **walking sticks**. To ward off **evil**

eyes and **spirits**, a branch smeared with cow's dung is hung on main door of the house. A blue **dye** extracted from its roots. Seed oil **edible** and used as an **illuminant** and **rubefacient** for checking **headache**. Also, massage of oil boiled with pounded plant parts useful in **joint pains**.

Active Constituents : These are new lignan-prinsepiol-along with (-) epicatelhin and b-sitosterol-b-glucoside (stems), 1.8% myristic, 15.2% palmitic, 4.5% stearic, 0.9% lignoceric, 32.6% oleic and 45.6% linoleic acids (seed oil).

Uses in Literature : Known to be used as an edible, illuminant, insect repellent, fuelwood, rubefacient, tonic; and for bodyache, burns, cooking oil, cuts, hydrogenation, pain due to fatigue, rheumatism, soap making, walking sticks, and wounds (**Ambasta, 1986; Chauhan, 1999; Jain, 1991; Kapur & Srivastava, 1996; Panda, 1996; Rana** ***et al.*****, 1996; Watt, 1972**).

Major Ayurvedic Formulations : Bhekla oil (its properties resemble closely to that of olive oil).

Prunus amygdalus Baill.⁺ (Pl. 53C).

Syn. : *P. communis* Fritsch, *Amygdalus communis* L., *P. dulcis* (Mill.) D. A. Webb.

Family : Rosaceae.

Vern. : Badam.

English, Sanskrit & Regional Names

Eng.	:	Almond Tree;
Sans.	:	Badama, Netroamaphala;
B.	:	Bilati-Badam;
G.	:	Badam;
H.	:	Badam;
Kan.	:	Badami;
Mal.	:	Vatam-Kotla;
Mar.	:	Badam;
Oriya	:	Badamo;
P.	:	Badam;

Tam.	:	Parsivadumai, Vadamkottai;
Tel.	:	Badamamu, Parsibadamu.

Distribution : Cultivated in cooler parts of India.

Description : Trees. Leaves conduplicate in bud. Flowers pink, appearing before the leaves. Drupe velvety.

Reproductive Cycle : May-September.

Habitat Ecology : Planted.

Material Examined : EBH-386, 07-08-2000.

Parts Used : Seeds. Fruits.

Folk Uses : Seeds **edible**. Its decoction prepared in milk along with grates said to be good **aphrodisiac** preparation; and paste used as a lotion to cure **burns**. Seed oil given to check **constipation**, and massaged in case of **joint pains** and **skin ailments**. Also, 5-10 seeds (soaked in water overnight) prescribed with 'misri' (larger sized pellets of sugar) and butter, in the morning on empty stomach as **brain tonic** and for giving **body tone** and **vitality**.

Active Constituents : Prunasin, daucosterin, sitosterol (bitter seeds), daucosterin, sitosterol (sweet almonds) a new biflavone-4'''-methoxy-2'',3''-dehydroflavenol-4',4''-oxymethyl-5,7-dimethoxy flavone, coumaric acid, 2-methylnonacosan-3-one, n-octacosanol, n-triacentane, b-sitosterol, procyamidin dimer (II), (+) catechin and (-) epicatechin (seed coat).

Uses in Literature : Reported earlier as an alternative, astringent, attenuant, demulcent, detergent, discutient, diuretic, edible (kernels), laxative, lithnotropic, nervine tonic, poultice in irritable sores, stimulant, symbolic-offering, tooth powder; and for amenorrhoea, confectionary, cosmetic preparations, cough, diabetes, gums, intestinal pain, marquetry, obstructions of liver and spleen, painful menstruation, peptic ulcers, perfumery, pharmaceutical, poultice in irritable sores, skin eruptions, sores, and turnery (**Ambasta, 1986**; **Jain, 1991**; **Shome *et al.*, 1996**; **Watt, 1972**).

Prunus armeniaca L.[+] (Pl. 53D).

Syn. : *P. americana* Marsh.

Family : Rosaceae.

Vern. : Khumani.

English & Regional Names

Eng. : Apricot, Common Apricot, Mishmus, Moon of the Faithful;
H. : Chilu, Khubani, Zardalu;
Kash. : Cherkisk;
P. : Allukashmiri, Jardalu, Mandata.

Distribution : N.W. India.

Description : Trees. Leaves convulate in bud. Petals white. Drupe large, downy.

Reproductive Cycle : May-August.

Habitat Ecology : Cultivated.

Material Examined : EBH-390, 28-06-2000.

Parts Used : Fruit. Seeds. Leaves. Flowers.

Folk Uses : 20 ml decoction of leaves recommneded for 7 days on empty stomach for expelling **intestinal worms**. Ripe fruits **edible** and its paste applied to heal wounds; and also employed to **clean brass utensils**. Seed oil used as **laxative, tonic** and for **cooking**; also massaged over affected parts to cure **joint pains**.

Active Constituents : Chief constituents isolated are quercitin-3-rhamnoglucoside (leaves), linallol, g-daclactone, hexanal, (E)2-hexenal, (E,E)2,4- decadienal, (E)2-nonenal, g-dodecalactone and b-ionone (dried fruits), lilac alcohol, lilac aldehyde, lilac acetate, 6-hydroxy-2,6-dimethyl-2,7-octadienal and 2 (Z) 6-dimethyl-2,7-octadien-1,6-diol (flowers), 24-dehydrox-cholesterol, cholesterol, estrone, a-estradiol (estradiol-b-17-oL) (seeds), myrsitic palmitic stearic, oleic and linoleic acids (Seed kennels).

Uses in Literature : Used earlier as an anthelmintic, antidiarrhoeal, antipyretic, cattle feed, demulcent, emetic, edible (fruit, oil), firewood, fodder cakes, illuminant, laxative, manure, nervine tonic, psychoactive plant, refrigerant, stimulant; and for allaying thirst, blood clotting, burning, earache, gum of commercial importance, liver troubles, jam, oil for massage, opthalmia, pharmaceutical and cosmetic industry, piles, and wounds (**Ambasta, 1986; Jain, 1991; Jain *et al.*, 1994; Kapur & Singh, 1996; Parmar & Kaushal, 1982; Rajendran *et al.*, 1996; Rana *et al.*, 1996**).

Prunus cerasoides D. Don+ (Pl. 54A).

Syn. : *P. puddum* (Wall.) Roxb. *ex* Brandis,
Cerasus puddum Seringe.

Family : Rosaceae.

Vern. : Paaja, Pajja.

English, Sanskrit & Regional Names

Eng. : Himalayan Cherry;
Sans. : Padmagandhi, Padmavriksha;
Ass. : Dieng-Soh-Iong-Krem;
B. : Padmak;
Bo. : Padma-Kasta;
G. : Padmak, Padmakathi;
H. : Padam, Phaja;
Mar. : Padmakastha;
P. : Amalguchh, Pajja.

Distribution	:	Himachal Pradesh to S.W. China:1200-2500m.
Description	:	Trees. Leaves glossy, long-pointed, toothed. Flowers long-stalked, pink. Fruits ovoid, long, yellow or red.
Reproductive Cycle	:	November-May.
Habitat Ecology	:	Forests; often planted by wayside resting places.
Material Examined	:	EBH-99, 08-01-1999.
Parts Used	:	Leaves. Flowers. Stem. Bark. Branches. Fruit.
Folk Uses	:	Plant considered **sacred** and it is essential for both bride and groom to take bath under the plant's branch at their respective places. Leaves serve as good **fodder** for livestock. Ripe fruits **edible** and considered as **blood purifier**. Poultice of pieces of inner bark boiled along with dried 'ginger' (*Zingiber officinale*), 'jayphal' (*Myristica fragrans* Houtt.), and 'cloves' applied to heal **fractured** and/or **dislocated bones.**
Active Constituents	:	Main ingredients are chyrsophanol, emodin and their 8β-D-glucosides, orientalone, physcion, puddumin A, padmakastein and its derivatives, β-sitosterol behenate, tectochrysin, genistein, leucocyanidin and 4'-glucoside of genkwanin (stem bark), naringenin, apigenin, β-sitosterol, sakuranetin, prunetin and genkawanin (stems).
Biological Activity	:	Spasmolytic activity found positive (**Jain** ***et al.*****, 1991**).
Uses in Literature	:	Reported to be used as an analgesic, carminative, conceptive, edible (fruit), expectorant, febrifuge, psychoactive plant, tonic; and for bone dislocation and fractures, building, burning sensation of the body, cold, cough, cuts, diarrhoea, furniture, hair oil, hemicrania, hoof (vet), injury (vet), internal injuries, marriages, pipe

stalks, pregnancy, religious ceremonies, rheumatic pains, rosaries, seminal weakness, sprains, stomach problem, tanning, venereal disease, walking sticks, and wounds (**Ambasta, 1986**; **Ansari, 1991**; **Chauhan, 1999**; **Chopra** ***et al.*****, 1956**; **Dey, 1980**; **Jain, 1991**; **Jain** ***et al.*****, 1991, 1994**; **Lal** ***et al.*** **1996**; **Lalramnhinglova, 1996**; **Sinha, 1996b**; **Uniyal, 1968**; **Watt, 1972**).

Prunus domestica L.[+] (Pl. 54B, C).

Syn. : *P. insititia* L., *P. communis* Huds. var. *insititia* (L.) Hook. f.

Family : Rosaceae.

Vern. : Aloo Bukhara, Plum.

English & Regional Names

Eng. : Plum Tree, Prum Tree, Common Plum;

B. : Alubukhara;

G. : Alubukhara;

H. : Alubukhara.

Distribution : W. Himalaya.

Description : A cultivated branched tree without spines. Rececptacle hollow. Flowers stalked, pink. Fruits ovoid.

Reproductive Cycle : April-August.

Habitat Ecology : Planted.

Material Examined : EBH-299, 11-08-2000.

Parts Used : Fruits. Wood.

Folk Uses : Ripe fruits **edible**. 3-4g mixture of powdered pulp (1/2 kg), fennel (25g) and black pepper (25g), given once a day for curing **piles** and **dysentery**. Dried branches serve as good **fuelwood**.

Active Constituents : Chief constituents are prudomestin, leucoantho-cyanidin, kaempferol, kaempferide, aronadendrin, dihydrokaempferide, pinocembrin, isosakuranetin, chrysin, naringenin, quercitin (heartwood), phloracetophenone, 4-0-methyl-phloracetophenon and its 6-0-glucoside (I) and frakinol (bark) galacturmic acid, arabinose, galactose, xylose and rhamnose (fruits).

Biological Activity : Naringenin possesses significant antiulcer activity in experimentally induced ulcers in pylorusligated rats (132 mg/kg) rats.

Uses in Literature : Useful as an aperient, astringent, demulcent, digestive, edible (fruit), laxative, nutritious, refreshing, refrigerant; and for bilious states of the system, cabinet work, cathartic decoctions, cold, confectionary, desserts, fever, illuminating purposes, inlay and turnery, jams, skeleton of paper match boxes (**Ambasta, 1986**; **Watt, 1972**).

Prunus persica Benth. & Hook. f.+ (Pl. 54D).

Syn. : *Amygdalus persica* L., *A. colinus* Wall., *Persica vulgaris* Miller, *P. saligna* Royle

Family : Rosaceae.
Vern. : Adu, Aru.

English & Regional Names

Eng.	:	Peach Tree, Nectarine;
H.	:	Aru;
Oriya	:	Pishu;
P.	:	Aru, Bembi, Chirmaru, Katherti.

Distribution : N.W. Himalaya, cultivated.

Description : Trees. Leaves conduplicate in bud. Petals pink. Drupe downy with irregulary furrowed stones.

Reproductive Cycle : February-May.

Habitat Ecology : Commonly planted near habitations.

Material Examined : EBH-110, 14-02-1999.

Parts Used : Leaves. Fruit. Gum. Seed-Oil.

Folk Uses : Ash of burnt leaves heals **burn injuries**. Gum used as an **adhesive** for pasting papers. Seed oil used for **cooking** and **body massage**. Fruits **edible**, **pickled.**

Active Constituents : Kaempferol (leaves), 6-0-(b-glucoronopyranosyl)-D-galactose, 6-0-(4-0-methyl-b-D-glucoronopyran-osyl)-D-galactose and 2-0-(b-D-gluconopyransyl)-D-mannose, rhamnose, xylose, arabinose and galactose (gum), proteins (PR-A and PR-B) containing 18 amino acids residues (seeds), kaempferol, dihydrokaempferol, hentriacontave, hentriacoritanol, naringenin, quercetin, b-sitosterol and its glucoside (heart wood).

Biological Activity : Alc.extr. of embryo and immature seeds exhibits analgesic and anti-inflammatory activities in rats.

Uses in Literature : Reported earlier as an abortifacient, anthelmintic, antiemetic, antiscorbutic, cattle feed, demulcent, dessert, diuretic, dye, eczema, edible, fertilizer, insecticidal, illuminant (oil), psychoactive plant, purgative, table fruit, tonic, vegetable; and for beverages, biliousness, blood diseases, building purposes, canning, carbonizing, charcoal, cough, deftness, earache, eczema, foot and mouth disease, headache, hoof rot of cattle, itching and scabies, lessoning thirst, leucoderma, peach brandy, pickling, piles, removing bad smell of mouth and sputum, rheumatism, scabies, stomach troubles, and whooping cough (**Ambasta, 1986; Jain, 1991; Jain *et al.*, 1994; Kapur & Singh, 1996; Lal *et al.*, 1996; Parmar & Kaushal, 1982; Sharma, 1996; Watt, 1972**).

Psidium guajava L.[+] (Pl. 55A).

Family : Myrtaceae.

Vern. : Amrood.

English, Sanskrit & Regional Names

Eng. : Guava tree;

Sans. : Aprithaktvacha, Ovidhavija;

Ass.	:	Madhuriam;
B.	:	Goaachhi, Peyara;
Bo.	:	Jam, Perala;
G.	:	Jamrud, Peru;
H.	:	Amrud, Safed safari;
Kan.	:	Pera, Koyya;
Mal.	:	Koyya, Malakkapera;
Mar.	:	Jamba, Tupkel;
Oriya	:	Bodojambo, Julabojamo;
P.	:	Anjirzarad, Amrud;
Tam.	:	Segappugnoyya, Vellaikoyya;
Tel.	:	Ettajama, Tellajama.

Distribution : Widely cultivated throughout India.

Description : Small trees. Leaves opposite, ovate-oblong. Petals white. Fruit globose or pear-shaped. Pulp yellow with white or deep pink.

Reproductive Cycle : July-September.

Habitat Ecology : Cultivated.

Material Examined : EBH-1240, 22-07-2003.

Parts Used : Fruit. Leaves.

Folk Uses : Fruits **edible** and considered good for **diabetic** and **anaemic** patients. Roasted fruits taken on empty stomach for **colic** and **piles**. Leaves good for **scouring teeth** and its paste cures **bleeding gums** and **pyrrohoea**; and also its gargles gives relief from **aching tooth**.

Active Constituents : Chief constituents are sesquiguavaene, ellagic acid, ellagitannin-guavin b and guavin A, C and D, galacturonic acid, galactose, arabinose with traces of xylose, rhamnose (fruits), and quercitin-3-a-arabinopyranoside (leaves).

Biological Activity : Spasmolytic activity of leaf is mainly due to prescence of quercetin. (+) gallocatechin exhibits antimutagenic activity against UV-induced mutation in *Escherichia coli.*

Uses in Literature : Known to be used in India as an antibiotic, antiemetic, antinarcotic to overcome the influence of liquor, aperient, astringent, colouration material in 'Mithila' paintings, edible, astringent, laxative, tongue cleaner, tonic; and for carbuncle, cholera, chronic fever, chronic infantile diarrhoea, chutneys, colic, cooling, cough and cold, deep cuts, dyeing, dyspepsia, engraving, gonorrhoea, headache, inflammation, intestinal haemorrhage, jam, jaundice, ketchup, lac-turnery, malaria, marmalades, menstrual disorders, mouth wash for swollen gums, piles, pulmonary disorders, pustules in mouth, scurvy, sores, spear handles and instruments of various kinds, sprains, stomachache, stones, tanning, ulcer, and wound (vet) (**Ambasta, 1986; Bhandary *et al.*, 1996; Borthakur, 1993; Chowdhury *et al.*, 1996; Gill *et al.*, 1993; Girach & Aminuddin, 1995; Hosagoudar & Henry 1996a, b; Jain, 1991; Jain *et al.*, 1997; Jha & Basak; 1994; Lalramnghinglova, 1996; Panda, 1996; Shome *et al.*, 1996; Singh & Maheswari, 1992; Siwakoti & Varma, 1996; Tirkey *et al.*, 2003b; Verma *et al.*, 1995; Vishwanathan & Singh, 1996; Watt, 1972**).

Punica granatum L.+ (Pl. 55B).

Syn. : *P. nana* L.

Family : Punicaceae.

Vern. : Daadu.

English, Sanskrit & Regional Names

Eng. : Pomengrate Tree;

Sans. : Dantabijaka, Pindapushpa, Vrittaphala;

Ass. : Dalim;

B. : Dalimgachh, Dalimphal;

Bo. : Anara, Dalimba;

G. : Dadam, Daramphal;

H. : Anar, Dharimb;

Kan.	:	Daalimbe;
Mal.	:	Pumatalam, Uruyampalam;
Mar.	:	Dalimba;
Oriya	:	Dalimbo;
P.	:	Anar, Danu, Daruna;
Tam.	:	Kalumalmandalai, Tusagam;
Tel.	:	Dadimanu, Dalimma, Puvvudanimma.
U.P.	:	Anar, Rumman.

Distribution : W. Himalaya: 700-2700m.

Description : Small trees, armed with axillary thorns. Petals red. Fruits woody, globular, brownish-orange.

Reproductive Cycle : May-September.

Habitat Ecology : Gregarious on dry lime soils.

Material Examined : EBH-710, 11-07-2001.

Parts Used : Fruit. Seeds. Leaves.

Folk Uses : Fruits **edible** and as a rich source of vitamin C. 1-2g powdered roasted rind and seeds of *Achyranthes aspera* (1:1) taken with honey for **cough**. Pounded leaves with a pinch of rock salt useful for **tonsilitis**. 2-3g mixture of powdered seeds (20g) in combination with 5g 'soa' (*Anethum sowa* Roxb.) 5g 'kadu' (*Picorrhiza kurroa*), 5g 'patish' (*Aconitum heterophyllum*) 5g 'magan' (*Piper longum*) taken with honey, twice a day till cure for **cough, stomachache** and **giddiness**. For curing **liver ailments**, 2-3g mixture of its powdered seeds (100g) and those of 10g 'soa' (*Anethum sowa* Roxb.), 10g 'black pepper' (*Piper nigrum*) and 10g rock salt, prescribed thrice a day for 15-20 days. For **persistent cough**, chutney of its seeds grinded along with coriander, 'soa' and leaves of 'pudina' *Mentha spp* is prescribed.

Active Constituents : Constituents isolated are cyanidin-3-gluoside delphinidin-3-glucoside and 3,5-diglucoside (seed coat), punicalin, punicalagin, punicafolin, granatin A, granatin B, conlagin, strictinin, 1,2,4,6-tetra-0-galloyl, b-D-glucose, a new hydrolysable tannin 2-0-galloyl-4,6(S,S)-galloyl-D-glucose (I) (leaves), 28% gallotannic acid together with a yellow colouring matter (rind), glucose, fructose, maltose, starch, carotene, thiamine, riboflavin, nicotinic acid, vitamin, pectin and amino acid (edible portion of the fruit)(Anon., 1990).

Biological Activity : Aerial parts CNS depressant and diuretic. Flower buds antifungal and antibacterial. Chloroform extr. of its stem and root anthelmintic. Fruit extr. exhibits high activity against *Micrococcus pyogenes* var. *aureus, Escherichia coli, Salmonelia typhosa, S. monterideo, S. mschottmulleri* and *Shigella paradyesenteriae*. Aq. extr. of roots completely inhibits activity of *Mycobacterium tuberculosis*-607.

Uses in Literature : Reported earlier as an anthelmintic, anti-vomiting agent, astringent, cardiac, colouration material in 'Mithila' paintings, condiment, edible, expectorant, good source of vitamin C, laxative, souring agent, stomachic, styptic and cicatrizing, symbolic (offering), tonic; and for bleeding after child birth and miscarriage, brain affections, bronchitis, chest troubles, cholera, congestion of the liver, conjunctivitis, controlling cholesterol level, cough, diarrhoea and dysentery, dyeing, dyspepsia, enlarged spleen, epistaxis, eye complaint, fever, gastric, general debility, gums, tone to the stomach, horses in dysentery, increasing sexual vigour, inflammation of the urinary tract and haemorrhage, iron, jaundice, kidney stones, leucorrhoea, malaria, menstrual disorders, pimples, refreshing drink, sore throat, tannin, uterine diseases, and wine, (**Ambasta, 1986; Anonymous, 1990; Chopra *et al.*, 1956; Dagar & Dagar, 1996; Dey, 1980; Ghosh *et al.*, 1996; Hosagoudar & Henry, 1996a, c; Jain, 1991; Jain *et al.*, 1997; Jha & Basak, 1994; Kapur & Singh, 1996; Khanna *et al.*, 1996b; Kulkarni & Kumbhojkar, 1993; Kurian, 1995; Lalramnghinglova, 1996; Mahato & Mahato, 1996; Mohanty & Padhy,**

1996; Parmar & Kaushal, 1982; Rana *et al.*, 1996; Sharma & Borthakur, 2003; Shome *et al.*, 1996; Singh, 1996b; Singh & Pandey, 1996; Tirkey *et al.*, 2003b; Watt, 1972).

Major Ayurvedic Formulations : Bhaskara-lavana, Corticis, Dadhika-ghrta, Dadimashtakadi-churna, Dadimaadi-churna, Dadima-ghrita, Dadimastaka, Dadima-ghrta, Dadimaya-taila, Decoctum-granati, Eladichurn, Fructus, Lavanbhaskar-churna, Sukra-matrka-vati.

Trade : Daruni, Pomegranate Rind.

Pyrus communis L.+ (Pl. 55C).

Syn. : *P. pyrifolia* Nakai var. *culta* Nakai, *P. sinensis* Hort. (non L.)

Family : Rosaceae.

Vern. : Nashpati.

English & Regional Names

Eng.	:	Pear, Wild Pear;
H.	:	Nashpati;
Kash.	:	Batang, Kistabahira;
P.	:	Batank, Nashpati, Tangi;
Tam.	:	Perrikkay;
Tel.	:	Berikaya;
U.P.	:	Bagugosha.

Distribution : Cultivated in N. W. Himalaya.

Description : Trees. Leaves upto 6.5 cm long.Flowers white. Fruit pyriform, yellow.

Reproductive Cycle : June-August.

Habitat Ecology : Cultivated.

Material Examined : EBH-195, 21-07-1999.

Parts Used : Fruit.

Folk Uses : Fruits **relished** and **pickled**, and its juice considered **refreshing** and **cooling.** For promoting **appetite**, fruits cooked with a pinch of rock salt, 'safed jira' and 'black pepper' is said to be helpful.

Biological Activity : Lupeol, b-sitosterol, betulin, betulinic acid, hydroguinone and arbutin possessed bacteriostatic activity against *Escherichia coli, Salmonella typhi, Shigella flexneri* and *Staphylococcus aureus.*

Uses in Literature : Known to be used in India as an edible, psychoactive plant; and for acute fever, beveıages, carving, engraving purposes, malaria, mathematical and drawing instruments, turning, and wines (**Ambasta, 1986; Gill** *et al.*, **1993; Jain** *et al.*, **1994; Watt, 1972**).

Pyrus malus L.+ (Pl. 56A).

Syn.: *Malus communis* Desf., *M. sylvestris* Hort. non Mill., *M. domestica* Borkh., *M. pumila* Mill.

Family : Rosaceae.

Vern. : Seb, Seon.

English, Sanskrit & Regional Names

Eng. : Apple Tree, Crab Apple;
Sans. : Seba;
Bo. : Seb;
H. : Seb, Seo, Sev;
Kan. : Sebu, Sevu;
P. : Cho, Chung, Kashu, Seo, Tsunt;

Distribution : Cutivated in the cooler parts of India.

Description : Tree. Leaves simple, entire, ovate. Flowers pink. Fruit large-sized, shortly peduncled.

Reproductive Cycle : May-October.

Habitat Ecology : Planted.

Material Examined : EBH-431, 23-09-2001.

Parts Used : Fruit.

Folk Uses : Fruits **edible, pickled** and even cooked as **vegetable**, and also considered beneficial for people **addicted to alcohol**, if eaten regularly for 30 days.

Active Constituents : Fruit yields vomifoliol-4-O-b-D-xylopyranoside-6-O-b-D-glucoside, quercetin-3-0-a-D-galactoside, 3-0-xyloside, 3-0-a-L-arabinofuranoside, 3-0-rhamnoside and 3-0-b-D-glucoside, rectin, and phloretamide.

Uses in Literature : So far known as an anthelmintic, edible (fruit), heart tonic, hypnotic, refrigerant; and for apple sauce, cider, cog-wheels, jam, jelly, knobs, mallet heads, marmalade, tool handles, and turnery (**Ambasta, 1986**; **Watt, 1972**).

Pyrus pashia Ham.+ (Pl. 56B).

Syn. : *P. variolosa* Wall., *P. verruculosa* Bertol., *P. nepalensis* Herb. Ham. *ex* Hort.

Family : Rosaceae.

Vern. : Kainth, Segal.

Regional Names

Ass. : Chalthei, Soh-Jhur, Soh-Shur;

H. : Mehal, Mol;

P. : Batangi, Kaenth, Shehgal, Tang.

Distribution : Afghanistan-S. W. China, Burma: 750-2700m.

Description : Trees. Leaves crenate, long-pointed. Flowers white, tinged with pink. Fruit yellow, globular.

Reproductive Cycle : May-August.

Habitat Ecology : Shrubberies, cultivated areas.

Material Examined : EBH-879, 29-05-2002.

Parts Used : Leaves. Twigs. Fruits. Wood.

Folk Uses : Poultice of leaves good for **boils**. Young twigs chewed as a remedy for **mouth ulcers**. Ripe fruits **edible** and considered **analgesic**, and a dose of 5-10g mixture comprising of its fruits (250g-without seeds), black pepper (25g) and 'soa' (25g) (*Anethum sowa* Roxb.) prescribed twice daily for 10-15 days as **blood purifier**. Wood used for **agricultural implements**.

Active Constituents : Main ingredients are friedelin, b-sitosterol (bark), n-triacontane, myricyl alcohol as chloride and sitosterol (leaves).

Biological Activity : Aerial parts anticancerous and CNS depressant.

Uses in Literature : Known to be used earlier for bobbins, catarrh (vet), combs, edible purposes, eye complaint, injury (vet), mouth infection, textile mills, tobacco pipes, walking sticks; and as fodder and fuel (**Ambasta, 1986; Jain, 1991; Lal** ***et al.*****, 1996; Rana** ***et al.*****, 1996; Singh, 1996b; Watt, 1972**).

Quercus glauca Thunb.+ (Pl. 56C).

Syn. : *Q. annulata* Sm., *Q. phullata* Ham., *Q. dentosa* Lindl., *Q. laxiflora* Lindl.

Family : Fagaceae.

Vern. : Banni.

English & Regional Names

Eng. : Blue Japanese Oak, The Green Oak;

P. : Banku, Banni, Bran, Imbri.

Distribution : S. E. Asia: 800-3000m.

Description : Trees. Leaves stalked, oblong-lanceolate, long-pointed, saw toothed. Nut ovoid, pointed, brown.

Reproductive Cycle : March-July; fruiting the following year.

Habitat Ecology : Moist forests, ravines not only gragarious; often associated with *Rhododendron arboreum* and *Lyonia ovalifolia.*

Material Examined : EBH-62, 14-04-1999.

Parts Used : Leaves. Fruit. Bark. Wood.

Folk Uses : Wood good for **agricultural implements, walking sticks** and as **fuel**. Leaves used as **fodder** and also kept in woollens as **repellent**. Fruits and bark considered useful in **asthma**. Also, 2-3ml of fruit juice taken with milk in the morning as a remedial measure for **piles**.

Active Constituents : Active properties are due to querglanin, isoquerglanin, cyclobalanone, 24-methylenecycloartanone (leaves), sitosterol, stigmast, stigmast-4-en-3-one, friedelin, epifried elinol, glutinol, b-amyrin acetate and lupeol (trunk wood).

Uses in Literature : Known in India as edible, fodder; and for bridges, door parts, rafters, sledge runners, tannin, and window frames (**Ambasta, 1986**; **Jain, 1991**; **Watt, 1972**).

Quercus leucotrichophora A. Camus[+] (Pl. 56D).

Syn. : *Q. incana* Roxb., *Q. dealbata* Wall., *Q. lanata* Smith., *Q. lanata* Don var. *incana, Q. oblongata* Don.

Family : Fagaceae.

Vern. : Baan.

English & Regional Names

Eng. : Grey Oak, Himalayan White Oak, Kumaon Oak, Ban Oak;

H. : Ban, Banji;

Kash. : Iri, Shiddar, Silaspuri;

P. : Ban, Daghunban, Kharpatarserai.

Distribution : Pakistan-C. Nepal.Sri Lanka. Burma: 1200-2400m.

Description : Trees. Leaves stalked, ovate-lanceolate, spinous towards the tip. Nut ovoid.

Reproductive Cycle : April-October.

Habitat Ecology : Gregarious and common in forests; often associated with *Rhododendron arboreum* and *Lyonia ovalifolia.*

Material Examined : EBH-277, 19-05-2000.

Parts Used : Root. Leaves. Bark. Wood.

Folk Uses : Wood considered good for **house building, agricultural implements** and **charcoal**. Bark constituent of local wine, **'sur'**, and for preparing **black ink**, and also its poultice prepared in 'til oil' (*Sesamum indicum* L.) useful against **joint pains.** Ash of its leaves mixed with fodder, helpful in curing **gastro-intestinal** disorders in animals. Acorns **edible** and said to cure **gonorrhoea.**

Active Constituents : Leaves afford quercetin and its 3-0-dissacharide.

Biological Activity : Hypotensive activity and effect of CNS found positive (**Jain *et al.*, 1991**).

Uses in Literature : Recorded in India as an astringent, diuretic in gonorrhoea, edible (acorns), fodder; and for agricultural implements such as ploughs, handles and beams, asthma, building, charcoal, confectionary, diarrhoea, hardboards, indigestion, scabies, skin diseases, tanning, urinary complaints (**Ambasta, 1986; Jain, 1991; Jain *et al.*, 1991; Kapur & Nanda, 1996; Negi & Gaur, 1991; Uniyal & Chauhan, 1973; Watt, 1972**).

Quercus semicarpifolia Sm. (Pl. 57A).

Syn. : *Q. obtusifolia* Don, *Q. cassura* Don

Family : Fagaceae.

Vern. : Kharsu.

English & Regional Name

Eng. : Brown Oak of Himalaya., Kharsu Oak;

P. : Banchar, Kharen, Klarshu;

Distribution	:	Afghanistan-S. W. China: 2100-3800m.
Description	:	Tree with entire, rusty, ellepticto-oblong leaves. Nut solitary, black when ripe.
Reproductive Cycle	:	May-August.
Habitat Ecology	:	Forming forests, gregarious and common.
Material Examined	:	EBH-203, 29-06-2000.
Plant/Parts Used	:	Plant. Wood.
Folk Uses	:	Wood used for **agricultural implements, house building, furniture** and as **firewood**. Plant considered excellent for rearing **silk worms**.
Active Constituents	:	Bark possesses friedelin, b-sitosterol, quercetin-3-0-L-rhamnoside, pelargonidin-3-D-glucoside and taraxerol.
Uses in Literature	:	Used in India for bed steads, building purposes, carrying poles and ploughs, charcoal, door-frames, feeding silkworms, furniture, hardboards, ploughs, tannin; and as firewood, and fodder (**Ambasta, 1986; Watt, 1972**).

Randia dumetorum (Retz.) Lam.$^{+}$ (Pl. 57B).

Syn.: *R. nutans* DC & W. & A., *R. brandisii* Gamble, *R. tomentosa* W. & A. non Blume, *R. longispina* DC & W. & A., *R. floribunda* DC & W. & A., *R. rotteri* W. & A., *R. stipulosa* Miq., *R. spinosa* Blume, *R. malabarica* Wall., *Gardenia nutans* Roxb., *G. spinosa* L., *G. longispina* Roxb., *G. floribunda* Roxb., *G. dumetorum* Retz., *G. stipularis* Rottl. & Willd., *G. glabra* Br., *G. propinqua* Br., *G. latifolia* Schelecht., *Posoqueria dumetorum* Roxb., *P. nutans* Roxb., *P. longispina* Roxb., *P. floribunda* Roxb., *Ceriscus malabaricus* Gaertn., *Catuneregum nutans* (DC.) Tiruv., *Xeromphis spinosa* (Thunb.) Keay

Family : Rubiaceae.

Vern. : Rada.

English, Sanskrit & Regional Names

Eng.	:	Common Emetic Nut;
Sans.	:	Bastishodhana, Ramachhardanaka;
Ass.	:	Bahmona, Gural, Mon;
B.	:	Madan;
Bo.	:	Gehala, Gelaphal;
G.	:	Medhola;
H.	:	Arar, Madanphal;
Kan.	:	Banegara, Kara, Mangri;
Kash.	:	Kirkla, Kokoa;
Mal.	:	Karalikkaya, Kattunaranna;
Mar.	:	Gelaphela, Monigeli;
Oriya	:	Potova;
P.	:	Mandkolla, Mindhal;
Tam.	:	Murkkarai;
Tel.	:	Madanamu, Mangara.

Distribution : Sub-Tropical Himalaya upto 1200m.

Description : Spiny trees. Leaves ovate. Flowers white, fragrant. Fruit globose, greenish-purple.

Reproductive Cycle : April-September.

Habitat Ecology : Common in Sal forests, shrubberries.

Material Examined : EBH-725, 20-07-2001.

Parts Used : Leaves. Seeds. Fruit. Thorn. Wood.

Folk Uses : Fruits **eaten** after roasting and used to wash **woollen clothes**, and also 1-2g of its powder, prescribed once daily for 7-14 days to cure **sexual debility**. Powdered seeds (1/2 teaspoonful) mixed with honey given twice daily on 4th day of menstrual cycle to promote **conception** in women. A chutney prepared from its leaves along with those of *Premna latifolia*, and seeds of *Punica granatum* and a pinch of salt prescribed for **regulating menstruation**. Wood considered suitable for making **walking sticks**.

Active Constituents : These are dumetoronins A,B,C,D,E & F, b-D-galactopyra-nosyl (1→3)-deanolic acid, lupeol, two new triterpenoid glycosides (fruit), a-amyrin, b-sitosterol, oleanolic acid, ursolic acid, mannitol (stem heartwood), and randialic acid-A and randialic acid-B (bark).

Uses in Literature : Reported earlier as an abortifacient, anodyne, anthelmintic, antispasmodic (for teething ailments of children), astringent, calmative, colour intensifier, diaphoretic, emetic, expectorant, fish-poison, fodder, insect repellent, nauseant, nervine, edible (ripe fruit), sacred tree, sedative, synergesics in insecticide preparation, wood for walking sticks; and for acne, bobbins, bodyache during fever, boils, cotton-reels, cramps, diarrhoea, dyeing, dysentery, fever, fruits (eaten after roasting or cooking), jaundice, lactation, malaria, mathematical instruments, pimples, ploughs, rheumatism, sores, umbrella-handles, and in soap industry (**Ambasta, 1986; Chauhan, 1999; Chopra *et al.*, 1956; Jain, 1991; Lalramnghinglova, 1996; Pandey *et al.*, 1996; Singh & Prakash, 1996; Watt, 1972**).

Major Ayurvedic Formulations : Madanphala-churna, Pippalayadi-churna.

Trade : Madanphal.

Reinwardtia indica Dumort.+ (Pl. 57C).

Syn : *R. trigyna* (Roxb.) Planch., *R. tetragyna* Planch., *R. repens* Planch., *Linum trigynum* Roxb., *L. repens* Don., *Macrolinum trigynum* Reichb., *Kittelocharis trigyna* Alef.

Family : Linaceae.

Vern. : Piyan.

Regional Names

Bo. : Abai;

H. : Basanti;

P. : Balbasant, Gudbatal, Karkum;

U.P. : Basant.

Distribution : Pakistan-S.W. China. India. Burma. S. E. Asia; upto 1800mt

Description : Shrubs. Leaves entire, ovate-lanceolate. Flowers bright yellow. Capsule globose.

Reproductive Cycle : November-May.

Habitat Ecology : Edges of terraced fields, grazing grounds; common.

Material Examined : EBH-756, 14-02-1999.

Parts Used : Leaves. Flowers.

Folk Uses : Flowers **offered** for **worshipping**, and its paste along with leaves applied for healing **cuts** and **injuries**.

Uses in Literature : Known to be useful earlier as an edible (flowers), poison; and for maggot infested wounds, 'founder' disease in cattle, whitlow, and also in paralysis, typhoid, wound (vet) (**Ambasta, 1986; Jain, 1991; Lal *et al.*, 1996; Manandhar, 1990; Watt, 1972**).

Rhamnus persicus Laws.+ (Pl. 57D).

Syn. : *R. pentapomica* Parker

Family : Rhamnaceae.

Vern. : Nikki-Kander.

Regional Names

H. : Chirla;

P. : Kander, Kukai, Warak.

Distribution : Temperate Himalaya: 650-1700m.

Description : A small tree. Leaves acute at the apex and minutely tomentose beneath. Flowers few. Fruit obovoid-globose.

Reproductive Cycle : April-October.

Habitat Ecology : Fairly common in shady oak forests.

Material Examined : EBH-1202, 14-04-2001.

Parts Used : Leaves. Wood.

Folk Uses : Wood used for **agricultural implements**, and leaves as **fodder**.

Uses in Literature : So far reported in India for edible purposes,dyeing, turnery, and as fodder (**Ambasta, 1986**).

Rhamnus triquetra Wall.+ (Pl. 58A).

Syn. : Ceanothus triquetra Wall.

Family : Rhamnaceae.

Vern. : Katheru.

Regional Name

P. : Gardhan, Gudlei, Phagora, Phulla, Rangrek.

Distribution : N.W. Himalaya: 1000-1700m.

Description : Shrub with tomentose shoots. Flowers greenish- white. Fruits obovoid, 3-lobed.

Reproductive Cycle : May-August.

Habitat Ecology : Common; forest areas.

Material Examined : EBH-1203, 12-08-2001.

Parts Used : Fruit. Bark. Wood.

Folk Uses : Fruits **edible** and wood used as fuel. Decoction of its bark along with a pinch of 'black pepper' taken as a **blood purifier** and for **body tone** (once daily for 7-10 days).

Active Constituents : These are physcion, emodin, chrysophanol, kaempferol, â-sitosterol and its glucoside (whole plant), emodin and frangulin (heartwood).

Biological Activity : CNS depressant activity (emodin) and anti-inflammatory (kaempferol 7-methyl ether) activities found positive.

Uses in Literature : Known to be used as an astringent, blood purifier, deobstruent, tonic (vet); and for boils and skin diseases (**Jain, 1991**).

Rhododendron arboreum Sm.[+] (Pl. 58B).

Syn. : *R. puniceum* Roxb., *R. cinnamomeum* Wall., *R. album* Sweet, *R. campbelliae* Hook. f., *R. windsorii* Nutt., *R. nilagiricum* Zenk.

Family : Ericaceae.

Vern. : Braah, Brass.

English & Regional Names

- **Eng.** : Tree Rhododendron, Rose Tree;
- **B.** : Baras;
- **H.** : Burans;
- **Mal.** : Kattupuvarasu;
- **P.** : Ardawal, Bras, Mandal;
- **Tam.** : Alingi, Billi.

Distribution : Temperate Himalaya: 1500-3600m.

Description : Trees. Leaves oblong-lanceolate. Flowers blood red. Capsule scaly, curved.

Reproductive Cycle : March-May.

Habitat Ecology : Forests, shrubberies; common.

Material Examined : EBH-285, 11-04-2001.

Parts Used : Flowers. Wood.

Folk Uses : Wood used for making **agricultural implements, tool handles, plates** ('Kunal'), and also serves as excellent **gun powder**. Flowers raw **eaten** or made into culinary preparations like, **'pakoras'** for serving with tea and also for making **chutney** and **squash** which are used against **nose bleeding**, passing of **blood in urine** and as **cardio-tonic** and **refrigerant**. Besides, flowers form an important constituent of local wine, '**sur**'.

Active Constituents : These are taraxerol, ursolic acid, betulinic acid (bark) and ursolic acid, quercetin, resin (flowers).

Biological Activity : Aq. ethenolic and methanolic extrs. of flowers exhibit significant anti-inflammatory activity against carrageeuin, PG(E2), histamine, and 5HT induced rats hind paw oedema in decreasing order of activity.

Uses in Literature : Reported earlier as refreshing drink, sacred, vegetable; and for amoebic dysentery, boxes, chutneys, diarrhoea, fever, gum-stocks, headache, jam, kukri, nose bleeding, park-saddles, plywood, posts, rheumatism, smoking pipe, tool handles, wounds, decorations of temples, clearing fish bone stuck in throat, poisoning rats, and wood for building and plates (**Ambasta, 1986; Chauhan, 1984; Jain, 1991; Lal *et al.*, 1996; Manadhar, 1990; Panda, 1996; Singh, 1996b; Sinha, 1996; Watt, 1972**).

Major Ayurvedic Formulations : Rhododendron juice.

Rhus cotinus L.+ (Pl. 58C).

Syn. : *R. vellutina* Wall., *R. laevis* Wall., *Cotinus coggyria* Scop.

Family : Anacardiaceae.

Vern. : Tung.

English & Regional Names

Eng. : Indian-Sumach, Smoke Tree, Venetian-Sumach, Wig Tree;

H. : Batra;

Kash. : Chariant, Darengri;

P. : Manu, Paan, Tung, Tunga.

Distribution : W. Sub-Tropical Himalaya: 1000-1700m.

Description : Small trees. Leaves simple. Flowers in drooping hairy panicles. Drupe hairy, unequally lobed.

Reproductive Cycle : March-August.

Habitat Ecology : Forests, shrubberies.

Material Examined : EBH-349, 22-03-2002.

Parts Used : Branches. Wood.

Folk Uses : Branches used for **scouring teeth** and wood for **carvings.**

Active Constituents : The plant contains (–) fustin.

Uses in Literature : Described earlier for basket making, cabinet work, in laying, picture frames, tanning, carvings, and yellow coloured matter used in textile and leather industries (**Ambasta, 1986**; **Watt, 1972**).

Rhus semialata Murr.+ (Pl. 58D).

Syn. : *R. chinensis* Mill., *R. javanica* L., *R. buckiamela* Roxb., *R. amela* Don.

Family : Anacardiaceae.

Vern. : Titri.

Regional Names

Ass. : Naga-Tenga;
H. : Tatri;
P. : Arkhar, Hulashing, Kakkaram, Totri, Vrash;
U.P. : Dampela, Dasmila.

Distribution : Kashmir-S. W. China: 1200-2400m.

Description : Trees. Leaves pinnate, rachis winged. Flowers pale-yellow. Drupe red-brown, tomentose.

Reproductive Cycle : April-August.

Habitat Ecology : Forests, shrubberies.

Material Examined : EBH- 886, 26-07-2001.

Parts Used : Leaves. Fruits.

Folk Uses : Fruits **edible** and a chutney prepared from its leaves and fruits useful for **gastro-intestinal** disorders. Also, ink made from its galls used for **writing**.

Uses in Literature : Known to be used as an appetizer, edible (fruits), fuel, gunpowder, purgative; and for amoebic dysentery, bronchitis, candles, cholera, colic, diarrhoea, gastric complaints, heart diseases, pulmonary tuberculosis (phthisis), skin disease, stomachache, and tanning (**Ambasta, 1986**; **Jain, 1991**; **Kapur & Srivastava, 1996**; **Lalramnghinglova, 1996**; **Manandhar, 1987b**; **Mao, 1993**; **Negi & Gaur, 1991**; **Sinha, 1996**; **Watt, 1972**).

Rhus succedanea L. (Pl. 59A).

Syn. : *R. acuminata* DC

Family : Anacardiaceae.

Vern : Rikhal.

English and Regional Names

Eng. : Wild Varnish Tree, Crab's Claw, Japan Wax Tree, Red Lac Sumach;

B. : Kakrasinghi;

H. : Kakrasingh, Kakrasinghi;

Kan. : Karkata-Shringhi;

Mar. : Kakada-Shinghi (galls);

P. : Arkhar, Arkhol, Choklu, Hala, Halai, Holashi, Kakkrim, Lakhar, Rikhul, Tatri, Titar, Titri;

Tam. : Karkkadagachinhi;

Tel. : Karkkarasinghi.

Distribution : Temperate Himalaya, from Kashmir-Sikkim:1200-2400m.

Description : A tree with glabrous, odd pinnate leaves. Flowers yellowish-green. Drupe globose, glabrous, pale-yellow.

Reproductive Cycle : June-September.

Habitat Ecology : Shrubberies.

Material Examined : EBH-629, 05-06-2001.

Parts Used : Stem-bark. Wood. Fruits.

Folk Uses : Wood suitable for **agricultural implements**. Ash of stem-bark given for **tuberculosis** (2-3g, twice a day with honey for 15-25 days). Fruits used in **phthisis**.

Uses in Literature : So far recorded in India as an expectorant, lubricant, stimulant, vesicant; and for agricultural implements, anticancer and antiviral action, bees-wax, candles, cordage, diarrhoea, dysentery, gunstock and tobacco pipes, lacquer-work, laundry glazer, leather and furniture polishes, lipsticks, metal drawing, pencils, phthisis, pulmonary tuberculosis, special soaps, textile and leather finisher, varnish, vulcanization of rubber, wax crayons, wax-matches, and pomads (**Kapur *et al.*, 1996; Lalramnhinglova, 1996; Watt, 1972**).

Ricinus communis L.[+] (Pl. 59B).

Syn. : *R. inermis* Jacq., *R. lividus* Jacq., *R. speciosus* Burm., *R. spectabilis* Blume, *R. viridis* Willd., *Croton spinosu* Linn.

Family : Euphorbiaceae.

Vern. : Arand, Erand.

English, Sanskrit & Regional Names

Eng. : Castor Oil-Plant, Palma-Christi;

Sans. : Dirghadantaka, Triputiphala, Vyadatvaka;

Ass.	:	Eri;
B.	:	Bherenda;
Bo.	:	Erendi;
G.	:	Diveligo;
H.	:	Arandi, Erand, Ind;
Kan.	:	Haralu;
Mal.	:	Avanakka, Chittavanaku;
Mar.	:	Yorandicha;
Oriya	:	Bheronta, Tindoko;
P.	:	Aneru, Bedanjir, Harnauli;
Tam.	:	Amanakku, Muttukkottlai;
Tel.	:	Chittamudamu;
U.P.	:	Rend.

Distribution : Planted and possibly naturalised in Himalayan villages upto 2400 m.

Description : Shrubs. Leaves peltate, 2-10 palmately-lobed. Flowers yellowish-red. Capsule prickly.

Reproductive Cycle : March-May (Intermittent flowering).

Habitat Ecology : Common in the hills in cultivated areas and wasteland.

Material Examined : EBH-919, 05-4-2001.

Parts Used : Leaves. Stem. Flowers. Roots. Seeds.

Folk Uses : Stem used for **scouring teeth**. Flowers hung on the main entrance of house for warding off **evil spirits**, and also offered to **deities**. Roots used as an **abortifacient**, and extr. of leaves applied in anus for expelling **intestinal worms**. Seed oil beneficial in checking **dermatitis**, **constipation** and also massaged for relief from **joint pains**. Mustard oil boiled with its seeds considered as a **hair tonic**.

Active : Chief constituents are hyperoside, rutin, ricinine

Constituents (flowers), apigenin, chlorogenic (endosperms), b-amyrin, stigmasterol, b-sitosterol, ricinine, quercetin, rutin, hyperoside, n-dimethylricinine, 3-0-b-d-xylopyranosides, 3-0-b-d-glucopyranosides and 3-0-b-D-rutinosides of kaempferol and quercetin (leaves).

Biological Activity : Anticholestatic and hepatoprotective activities confirmed.

Uses in Literature : Known earlier as an anthelmintic (vet), antidote to food poison, antipyretic (vet), anti-rheumatic, cathartic, contraceptive jellies, edible, emmanagogue, fodder, fuel, insulating materials, lactagogue, laxative, lubricant, lumbago, luminant, manure, psychoactive plant, skin and hair dressing, silkworm host, soothing agent used in ointments, symbolic (wards off evil), vermicide; and for abortion, amenorrhoea, anasarca, asthma, backache, body swelling, boils, bone fracture (vet), burns, candles, cholera, chronic ulcer, constipation (vet), culinary purposes, cutaneous affections, dental caries, dermatitis, dropsy, dyeing, dysentery, dysentery (vet), eczemas, facilitating child birth, fever, flatulence, fly-paper, gum trouble, hard and dry wax in ears, headache, heat stroke, heat-sealing adhesives, hydrocele, imitation leather, inflammation of breast during lactation, inflammation of eyes, injury, jaundice, loosening of limbs, muscular pain, nervous disorders, night blindness, plastic moulding composition, pneumonia, poultice to boils and sores, preservation of fruits and vegetables, preserving leather and hair oil, producing sterility for one month, rheumatism (vet), scorpion and snake-bites, soap making, sores, sprain and muscle cramps, swelling, synthetic nylon, transparent soaps, typewriter inks, ulcers, warts, wound (vet), wounds, and also in bone sprain of horses and cough (**Ambasta, 1986; Bhandary *et al.*, 1996; Borthakur, 1993; Hembrom, 1996; Huidrom, 1996; Jain, 1968, 1991; Jain & Saklani, 1992; Jain *et al.*, 1994; Jha *et al.*, 1991; Katewa & Arora, 1997; Khanna *et al.*, 1996a; Kharkonger & Joseph, 1981; Lalramnghinglova, 1996; Mahato & Mahato, 1996; Manandhar, 1996; Mao, 1993; Masih, 2003;**

Mishra *et al.*, 1996a, 1997; Panda, 1996; Rama Rao & Henry, 1996; Saini, 1996a; Satapathy & Brahmam, 1996; Sharma & Borthakur, 2003; Sharma *et al.*, 1979; Shome *et al.*, 1996; Singh *et al.*, 1996b, 1997; Siwakoti & Varma, 1996; Srivastava *et al.*, 2003; Tirkey *et al.*, 2003b; Tiwari & Tiwari, 1996; Vihari, 1995; Vishwanathan & Singh, 1996; Watt, 1972).

Major Ayurvedic Formulations : Eranda-pak, Eranda-saptaka-kvatha, Eranda-taila. Gangharvahastadi-kvatha-churna, Erandimubdi-kwath, Gandharvahasta-taila, Gan-gadhara-haritaki, Laghuvisgrava-taila, Rasna-saptak-kwath, Sadanga-guggulu, Simhanada-guggulu, Vatari-guggulu.

Trade : Erandi, Eranda Bija, 'Manjeeth' (roots).

Robinia pseudoacacia L.+ (Pl. 59C).

Family : Fabaceae.

Vern. : Robinia.

English Name

Eng. : Black-Locust, False-Acacia, Robinia;

Distribution : Himalaya upto 2400m.

Description : Trees. Leaves odd pinnate, leaflets entire. Flowers white, fragrant. Pod brown, flat, hairless.

Reproductive Cycle : March-May.

Habitat Ecology : Commonly planted along roadsides in W. Himalaya.

Material Examined : EBH-375, 05-04-2001.

Parts Used : Twigs. Wood.

Folk Uses : Planted for **shade, fodder, fuel** and **ornamental importance**. Wood used for **house building, tool-handles, agricultural implements** and low grade **furniture**. Twigs used as **toothbrush** for **scouring teeth.**

Active Constituents : The plant yields coumarin, scopoletin (leaves), a wax containing lauric (8.21), myristic (14.34), palmitic (19.43), stearic (24.73), and arachidic (33.29%), acids (leaves) and 2,4-Dihydroxybenzoic acid, its methyl, 1-(4-hydroxycinnamoyl) and 1-(3,4,5-trihydroxycinnamoyl) esters, liquiritigenin, butein, butin, robinetin and dihydrorobinetin (heartwood).

Biological Activity : Robinin possesses potent anti-inflammatory activity in rats and rabbits.

Uses in Literature : Known to be used earlier as an antispasmodic, axle-trees, diuretic, edible (seeds), emetic, fuel, laxative; and for agricultural implements, boast ribs, brackets, cart-wheels, digestive disorders, fence posts, flowers, furniture, gates, hard and heavy wood for construction, insulators or telegraph and telephone wires, paper making, renovating exhausted soil and instructing poor land, ship-building, shoe-lasts, sleepers, sports goods, tool-handles, tree nails, turnery, and wagon-hubs (**Ambasta, 1986**; **Singh *et al.*, 1996a**; **Watt, 1972**).

Rosa alba L.+ (Pl. 59D).

Syn. : *R. glandulifera Roxb.*

Family : Rosaceae.

Vern. : Gulab.

English, Sanskrit & Regional Names

Eng.	:	Common White Dog Rose, White Cottage Rose;
Sans.	:	Bhrigavallabha, Shivaballabha;
B.	:	Shvetigulab;
Bo.	:	Gul;
H.	:	Gulab;
Kan.	:	Mullusevantige;
Oriya	:	Sonuti, Seboti;
P.	:	Sulsoti.

Distribution : Cultivated.

Description : Shrubs with 5-7 leaflets. Flowers large-sized pale. Sepals often pinnatifid.

Reproductive Cycle : June-August.

Habitat Ecology : Cultivated.

Material Examined : EBH-140, 12-08-2001.

Parts Used : Leaves. Flowers.

Folk Uses : Grown as an **ornamental** and for **perfumery**. Flowers considered **sacred** for offerings, and its extr. used as an **eye drops** for relief from **burning sensation** and **skin ailments**, and also its sweetened infusion prescibed with milk before going to bed for **constipation**. 10-20 ml decoction of leaves given twice daily for 3-5 days to cure **urinary infection**. Also, a dose comprising 3-5g mixture of its flowers and those of *Tinospora cordifolia* and 'soa' (*Anethum sowa* Roxb.) and a pinch of sugar, given for **spermatorrhoea** (one teaspoonful in the morning with cow's milk).

Active Constituents : Essential oil affords geranial, nerol, rodinol, phenylethanol and linalool.

Uses in Literature : Reported earlier as refrigerant (flowers); and for palpitation of the heart, and perfumery (**Ambasta, 1986; Watt, 1972**).

Rosa moschata Mill.+ (Pl. 60A).

Syn. : R. brunonii *Lindl. Monogr.*, R. lindleyi *Harb.* Wall., R. pubescens *Roxb.*, R. glandulifera *Herb. Roxb.*, R.. recurva *Roxb.*

Family : Rosaceae.

Vern.: Kuja.

English, Sanskrit & Regional Names

Eng. : Himalayan Musk Rose;

Sans. : Sewati;

H. : Karer, Kuji, Kunja, Kwiala;

Kash. : Phulwari, Chal;

P. : Tarni.

Distribution : Temperate Himalaya: 1350-2650m.

Description : Climbing prickly shrubs. Flowers white. Fruits globose, dark brown.

Reproductive Cycle : April-October.

Habitat Ecology : Frequent on slopes, waste places.

Material Examined : EBH-461, 28-11-2001.

Parts Used : Roots. Flowers.

Folk Uses : Flowers used in **perfumery** and held **sacred** for **offerings**, and its extr. considered **antidiarrhoeal** and **emetic** and for **eye troubles** and also its squash serves as useful **refrigerant**. Poultice of roots useful against **joint pains.**

Uses in Literature : Considered useful in eye troubles, perfumery; and for making walking sticks. (**Ambasta, 1986**).

Roylea cinerea (D. Don) Baill.[+] (Pl. 60B).

Syn. : *R. elegans* Wall., *R. calycina* (Roxb.) Briq., *Ballota cinerea* Don, *Phlomis calycina* Roxb.

Family : Lamiaceae.

Vern. : Karkoi.

Regional Names

- **H.** : Patkarru;
- **P.** : Kaur, Kauri;

Distribution : Sub-Tropical W. Himalaya: 600-2400m.

Description : Shrubs with shortly-stalked, coarsely toothed leaves. Flowers white, tinged with pink. Nutlets smooth.

Reproductive Cycle : March-November.

Habitat Ecology : Frequent on open, rocky slopes.

Material Examined : EBH-352, 26-11-2002.

Parts Used : Leaves. Flowers.

Folk Uses : Paste of leaves useful for **boils,** and in **dermatitis**. In veterinary, juice of leaves applied as an **emollient** for healing open **wounds** and **mouth sores**, and also a few drops of its extr. added to **kill lice**, if present.

Active Constituents : Leaves yield betulin, b-amyrin, b-sitosterol, stigmasterol, cetyl alcohol, glucose, fructose, arabinose and palmitic, stearic, oleic, oxalic and tartaric acids.

Biological Activity : Plant extr. depresses CNS and relaxed skeletal muscle activities.

Uses in Literature : Reported in India as bitter tonic, blood purifier, febrifuge, snuff in tonsil; and for contusion produced by winds, fever, and pimples (**Ambasta, 1986**; **Jain, 1991**; **Watt, 1972**).

Rubia cordifolia L.[+] (Pl. 60C).

Syn. : *R. manjith* Roxb., *R. munjistha* Roxb., *R. mungisth* Desv., *R. javana* DC., *R. secunda* Moon, *R. alata* Wall., *R. purpurea* Dcne., *R. scadens* Zoll. & Moss., R. chinensis *Regel & Maack,* R. mitis *Miq.*

Family : Rubiaceae.

Vern. : Manjith.

English, Sanskrit & Regional Names

Eng. : Indian Madder;

Sans. : Aruna, Bhandi, Bhandiralatika, Bhandiri, Bhanditaki, Chhatra, Chitrangi, Gandhira, Harini, Hemapushpi, Janani, Javarihantri, Kalabhandika, Kandhira, Latayastni, Lohitalata, Manjistha, Nagakumarika, Raktangi, Raktayashtika, Tamramuli, Vashrabhusana, Vitasa, Yojnaparnika;

Ass. : Majathi, Majelti;

B. : Majith, Manjistha, Manjit;

Bo. : Madar, Manjit;

H.	:	Majit, Manjit;
Kan.	:	Manjustha, Siomalate, Siragatti;
Kash.	:	Dandu;
Mal.	:	Manjeti;
Mar.	:	Manjestha, Masai;
Oriya	:	Barheipoeni, Manjistha;
P.	:	Khuri, Kukarphali, Majit, Mitu, Munzat, Runang, Sheuhi, Tuara;
Tam.	:	Manjitti, Shevelli;
Tel.	:	Chiranji, Manjistha Tige, Tamravalli.

Distribution : Throughout Hilly regions of India, ascending to 2700m.

Description : A climbing herb with elongated, 4-angled stems and branches. Leaves in whorls of 4. Flowers small, reddish brown. Fruit succulent; black.

Reproductive Cycle : July-October.

Habitat Ecology : A frequent climber on shrubs in forests.

Material Examined : EBH-571, 20-07-2001.

Parts Used : Roots. Stem. Leaves. Bark.

Folk Uses : 1-3g powdered roots given with cow's milk twice daily for 5-7 days as **tonic** and for **regulating menstruation** and **colouring** medicinal oils. Poultice of leaves, bark and roots useful against **swellings, ringworm** and **skin ailments**.

Active Constituents : Roots contain purpurin, munjisthin, xanthopurein or purpuroxanthin, psuedo purpurin, alizarin and anthraquinones.

Biological Activity : Methanol and chloroform extrs. exhibit significant anticancer activity against P-388 lymphocytic leukaemia. Ethanolic extr. of the aerial parts of the plants showed hypoglycaemic activity in albino rats.

Uses in Literature : So far known as an alexipharmic, antidote to scorpion sting, antiseptic, astringent, deobstruent, diuretic, dye, nasal drops, tonic, vermifuge; and for amenorrhoea, bone fracture to reduce inflammation and swelling, chest complaint, colouring medicinal oils, cosmetics, dyeing, expelling placenta (vet), giddiness, headache, hepatic obstructions, leucoderma, menstrual disorders, obstructions in the urinary passages, paralysis, poison stings of insects, cobras and scorpion, postnatal problem (vet), removal of feckles, rheumatism, rhinosinal infections, skin discolouration, stomachache, ulcers and urinary complaints (**Ambasta, 1986; Aminuddin & Girach, 1991; Chauhan, 1999; Henry *et al.*, 1996; Hosagoudar & Henry, 1996a, b; Jain, 1991; Masih, 2003; Mukherjee *et al.*, 1986; Rama Rao & Henry, 1996; Rana *et al.*, 1996; Sharma & Sood, 1997; Tiwari & Tiwari, 1996; Uniyal, 1968; Watt, 1972**).

Major Ayurvedic Formulations : Devadararista, Karañjadi, Laghiuvisgarva-taila, Mahamanjasthadi-kvatha, Manjistha-arka, Manjisthadi-arak, Manjisthadi-kwath, Manjisthadi-talsar, Manjistha-malahara, Manjistha-phanta, Medipurine, Panchgun-taila, Pinda-taila, Septilin.

Trade : Manjit.

Rubus ellipticus Sm.+ (Pl. 60D).

Syn. : *R. rotundifolius* Wall., *R. gowry-phul* Roxb., *R. gowree-phul* Roxb., *R. flavus* Ham., *R. sessilifolius* Miq., *R. hirsutus* Roxb., *R. wallichianus* W. & A., *R. affinis* Madden.

Family : Rosaceae.

Vern. : Aakhe, Kharyadi.

English & Regional Names

Eng. : Himalayan Yellow-Raspberry;

Ass. : Jotelupoka;

H. : Anehhu, Hinsalu;

Kash. : Gouriphal;

P. : Akhi.

Distribution : S. E. Asia: 600-2300m.

Description : Shrubs with bristly branches. Leaves compound. Leaflets-3. Flowers white. Fruits globose, yellow, succulent.

Reproductive Cycle : January-May.

Habitat Ecology : Shrubberies; cultivated areas.

Material Examined : EBH-599, 14-02-2000.

Parts Used : Roots. Stem. Leaves. Fruits.

Folk Uses : Plant used as **fuelwood** and also relished by sheep and goats. Fruits **edible**, and its decoction prescribed as a mild **laxative**. A gargle of its crushed roots along with a pinch of 'black pepper' useful against **tonsilitis**. Wood suitable for **gun-powder**.

Active Constituents : Chief principles are octacosanol, b-sitosterol, its glucoside, octacosanoic, ursolic and rubitic acids (roots) and deoxysericoside (leaves).

Biological Activity : Effect on CNS and antifertility activity found positive; spasmolytic activity confirmed (Jain *et al.*, 1991).

Uses in Literature : Known earlier for cracked heals, dyeing, dysentery, malaria, stomachache, worms; and as ripe fruit edible (**Jain, 1991; Lal *et al.*, 1996; Mao 1993; Megoneitso & Rao, 1983; Panda, 1996; Watt, 1972**).

Rubus lasiocarpus Sm.+ (Pl. 61A).

Syn. : *R. distans* Don, *R. horsefieldii* Miq., *R. albescens* Roxb., *R. rosaeflrus* Roxb., *R. rotundifolius* Royle, *R. mysorensis* Heyne, *R. indicus* Heyne, *R. furfuracens* Wall., *R. bijugus* Focke Batogr.

Family : Rosaceae.

Vern. : Kala-Akha.

English and Regional Names

- **Eng.** : Raspberry, Mysore Raspberry, Mahabaleshwar raspberry;
- **H.** : Kala hinsalu;
- **Mar.** : Gauri phal;
- **P.** : Gunachal;

Distribution : Throughout the Himalaya. 4000-10000 ft.

Description : Prickly shrubs. Leaves compound. Leaflets 5-11, ovate-lanceolate. Flowers dark pink. Drupe succulent, black.

Reproductive Cycle : March-May.

Habitat Ecology : Forests, shrubberies; common.

Material Examined : EBH-791, 07-04-2001.

Parts Used : Fruits. Roots.

Folk Uses : Plant grown as **hedges**. Ripe fruits **edible**. 2-3g of its powdered roots in combination with 'mulathi' (*Glycyrrhiza glabra* L.), given on empty stomach for 8-15 days to cure **diabetes** and **spermatorrhoea**, and as **refrigerant**.

Uses in Literature : Known earlier to be useful as edible, hedge plant, stimulant; and for diarrhoea, facilitating delivery and minimising pain, fevers, heart, jams, mouth ulcers, nocturnal micturition in children, skin rashes, sore throat, ulcer, and wounds (**Ambasta, 1986; Jamir & Rao, 1999; Kapur, 1996; Kurian, 1995; Watt, 1972**;).

Rubus niveus Wall.+ (Pl. 61B).

Syn. : *R. pedunculosus* D. Don, *R. gracilis* Roxb., *R. albexens* Roxb.

Family : Rosaceae.

Vern. : Kala-Akha.

English & Hindi Name

- **Eng.** : Mysore Raspberry;
- **H.** : Kala Hinsalu, Kalianchii;

Distribution : Temperate Himalaya: 1000-2350m.

Description : Prickly shrubs. Leaves compound. Leaflets 3 rarely 5. Flowers pink. Drupelets succulent, black.

Reproductive Cycle : April-June.

Habitat Ecology : Forests, shrubberies; common.

Material Examined : EBH-781, 08-04-2000.

Parts Used : Fruits. Leaves.

Folk Uses : Fruits **edible**, and its decoction useful as **laxative** and **diuretic**. Leaves used as **fodder**.

Uses in Literature : Considered useful as an antidote to snake bite, diuretic, edible (berries), laxative, vegetable; and for stomachache (**Ambasta, 1986; Hosagoudar & Henry, 1996c; Jain, 1991; Kapur & Srivastava, 1996; Mao, 1993; Negi & Gaur, 1991; Rana *et al.*, 1996; Watt, 1972**).

Saccharum officinarum L.+ (Pl. 61C).

Syn. : *S. sinense* Roxb.

Family : Poaceae.

Vern. : Ganna.

English, Sanskrit & Regional Names

Eng.	:	Sugar Cane, Noble Cane;
Sans.	:	Adhipatra, Gudakastha, Rasola;
Ass.	:	Pura-kuhiyar;
B.	:	Ak, Kushiar, Puri;
Bo.	:	Gol, Serdi;
G.	:	Naisakar, Sheradi, Uns;
H.	:	Ganna, Ikh, Kumad;
Mal.	:	Kantarakam, Madhutrianam;
Mar.	:	Aos, Kabbo;
Oriya	:	Aku, Gudodaru;

P.	:	Ganna, Khaud, Paunda;
Tam.	:	Angarigai, Madudirunam;
Tel.	:	Arukanupulakranugu, Tellacheraku;

Distribution : Cultivated throughout India.

Description : Stems solid. Upper leaves long, scabrid. Spikelets lanceolate, surrounded by dense, silky white hairs.

Reproductive Cycle : January-April.

Habitat Ecology : Cultivated.

Material Examined : EBH-46, 12-01-2000.

Parts Used : Cane Juice.

Folk Uses : Stem constituent of local wine, '**sur**'. Dried canes used as **torches** and **thatching** material for huts. Its juice has a **cooling effect** and effectively prescribed for **jaundice**, and also as **diuretic**.

Active Constituents : Six glycans-saccharans A,B,C,D,E & F, neocar-linoside, neoisoschaftoside, orientin, swerti-ajaponin, vicenin-2, p-coumaric, ferulic, caffeic and 3,4-di-0-methylcaffeic acids isolated from stalk juice.

Biological Activity : Saccharans A,B,C,D,E & F exhibited hypoglycaemic activity in normal and alloxan produced hyperglycaemic rats.

Uses in Literature : Reported to be useful in India as an anthelmintic, antidote to arsenical poisoning, antiseptic, aphrodisiac, apierent, condiment, detergent, diuretic, emollient, fodder, thatching material for huts, laxative, manure, nutritious, pectoral, refrigerant, sweetener, symbolic in sculptures (bow of Kamdev-god of love made from canes), tonic, torches (dried canes); and for bagasse or fibre, beverages, bronchitis, calculus complaints, carbon paper, catarrhal affections, disorders of bile and wind, fever (Postnatal), gall bladder, incipient consumption, jaundice (juice as drink), insanity, lactation (vet), logs in to rafts, preservation of meat and fish, press-mud, production of lime, rough cordage, shoe and other polishes, soothing and protecting the alimentary tract inflammation, strengthening the gum, sugarcane wax, tying, ulcers, wax paper, and wounds (vet) (**Ambasta, 1986; Borthakur, 1996a; Dagar & Dagar, 1996; Gogoi *et al.*, 2003; Jain, 1991; Kurian, 1995; Lalramghonglova, 1996; Sahu, 1996; Siwakoti & Varma, 1996; Tirkey *et al.*, 2003b; Watt, 1972**).

Major Ayurvedic Formulations : Bala-taila, Navaratnaralamrganka-rasa.

Sageretia oppositifolia Brongn (Pl. 61D).

Syn. : *S. parviflora* (Roem & Schult.) G. Don, *S. filiformis* G. Don, *S. oppositifolia* (Wall.) Brogn., *Rhamnus parvifolius* Roem & Schult., *R. filiformis* Roth., *R. riggynus* Don, *Zizyphus oppositifolia* Wall.

Family : Rhamnaceae.

Vern. : Baakar ghaa.

Regional Names

- **Kash.** : Gidarak, Kanak;
- **P.** : Drange, Girthan;
- **Tel.** : Surabhi;

Distribution : Throughout India: 650-2000m.

Description : Rambling shrubs. Leaves ovate-lanceolate, aromatic long pointed. Flowers yellow. Fruit obovoid, black when ripe.

Reproductive Cycle : June-September.

Habitat Ecology : Forests, shrubberies; common.

Material Examined : EBH-881, 20-07-2001.

Parts Used : Leaves. Fruit. Wood.

Folk Uses : Ripe fruits **edible** and leaves used as **fodder**. Wood used as **fuel** and for **toolhandles**.

Uses in Literature : Used in India for axe handles, edible purposes and as fuelwood (**Ambasta, 1986**; **Watt, 1972**).

Salix babylonica L.⁺ (Pl. 62A).

Family : Salicaceae.

Vern. : Majnu, Weeping Willow.

English & Regional Names

Eng. : The Weeping Willow;
H. : Majnun;
Kash. : Bisa, Giur;
P. : Bedmaju, Bitsuber;
Tel. : Attuppalai;

Distribution : Himalaya upto 3650m.

Description : Trees. Leaves narrowly-lanceolate, finely toothed. Flowers appearing with leaves. Capsule glabrous, sessile.

Reproductive Cycle : February-May.

Habitat Ecology : Along water courses.

Material Examined : EBH-160, 12-02-2000.

Parts Used : Twigs. Wood. Roots.

Folk Uses : Plant grown as ornamental. Its wood considered suitable for **basketry**, **building works** and **charcoal**. Twigs used as **fodder**. Paste of roots applied as an **emollient** for **wrinkles** and **cracked heels**.

Active Constituents : Plant contains salicin.

Uses in Literature : Reported to be used as an anthelmintic, antipyretic, astringent, cattle feed, firewood, ornamental, tonic; and for basketry, brake blocks, building work, charcoal, cooling, dams, fences, furniture, gate fixtures, intermittent and remittent fevers, rheumatism, sports goods, weirs (**Ambasta, 1986, Jain, 1991, Singh** ***et al.*****, 1996a; Watt, 1972**).

Salix karelinii Turcz.* (Pl. 62B).

Syn. : *S. hastata* (non L.) Hook. f.

Family : Salicaceae.

Vern. : Pahadi-Buins.

Distribution : Afghanistan-C.Nepal, C. Asia; 2100-4500m.

Description : A shrub. Leaves elleptical-ovate, finely toothed. Young leaves woolly haired. Catkins borne on leaflets stems.

Reproductive Cycle : May-July.

Habitat Ecology : Open slopes, streamsides; common.

Material Examined : EBH-391, 14-05-2000.

Parts Used : Leaves. Wood.

Folk Uses : Plant used as **fodder** and **fuelwood.** Wood suitable for **tool handles.**

Salix tetrasperma Roxb.⁺ (Pl. 62C).

Syn. : *S. disperma* Don, *S, horsefieldiana* Miq., *S. nilagirica* Miq., *S. tetrasperma â-nilagirica* Andress

Family : Salicaceae.

Vern. : Buins.

English, Sanskrit & Regional Names

Eng. : Indian Willow;

Sans. : Varuna;

Ass.	:	Bhi, Veh;
B.	:	Boishaki, Punijami;
Bo.	:	Bacha, Wallunj;
H.	:	Bains, Bainsh, Bilsa;
Kan.	:	Bariche, Niravanji;
Kash.	:	Bins, Yir;
M.P.	:	Dhjanie;
Mal.	:	Arali, Attuphala, Vanti;
Mar.	:	Bitaso, Walunj;
Oriya	:	Baisi, Panijamo;
P.	:	Bakshel, Magsher;
Tam.	:	Niruvanji, Suvedam;
Tel.	:	Etipisinika.

Distribution : Throughout Tropical and Sub-Tropical India, below 1650m.

Description : Tree. Leaves ovate-lanceolate. Flowers in scattered clusters. Capsule long-stalked, glabrous.

Reproductive Cycle : April-July.

Habitat Ecology : Open slopes, roadsides.

Material Examined : EBH-432, 24-07-2001.

Parts Used : Leaves. Bark. Latex. Wood.

Folk Uses : 2-3g dose of mixture containing 500g of its bark, 10g 'cardamom', 10g 'black pepper', 10g 'mulathi' (*Glycyrrhiza glabra* L.) given with milk twice daily for 5-7 days to cure **hernia**. Wood suitable for **agricultural implements, furniture** and as **fuel**. Twigs suitable for **scouring teeth,** and also as **fodder**. Latex applied for promoting **hair growth**.

Active Constituents : Leaves and bark yield rhamnose and salicin.

Uses in Literature : Reported in India for agricultural implements, baskets, bent, brush-backs, cabinet and fancy work, epilepsy, gunpowder, charcoal, house construction, match boxes, packing cases, pencil and cricket stumps, pen holders, piles and swellings, ploughs, rheumatism, stone in the bladder, tanning, venereal diseases, wood for posts and planks; and also as cattle feed and febrifuge (**Ambasta, 1986, Watt, 1972**).

Trade : Indian Willow.

Sapindus mukorosii Gaertn.+ (Pl. 62D).

Syn. : *S. abruptus* Lour.

Family : Sapindaceae.

Vern. : Dodan, Ritha.

English, Sanskrit & Regional Names

Eng.	:	Soap Nut Tree of North India.
Sans.	:	Arishta, Phenila;
Ass.	:	Haithaguti;
B.	:	Chhota-Ritha, Ritha;
Bo.	:	Kanmar, Ritha;
H.	:	Aritha, Ritha;
Oriya	:	Ita;
P.	:	Dodan.

Distribution : Cultivated throughout N.India, ascending to 1650m.

Description : Trees. Leaves alternate, glabrous, paripinnate. Leaflets around 15. Flowers purple. Drupe yellow.

Reproductive Cycle : April-September.

Habitat Ecology : Common along edges of fields, roadsides.

Material Examined : EBH-1265, 14-08-2003.

Parts Used : Fruit. Seeds.

Folk Uses : Fruits used as **detergent** for woollen and silken clothes, and also as **emetic** and **hair wash**. Poultice of seeds useful against **bleeding piles.**

Active Constituents : These mainly contains sapindoside V, sapindosides A & B, hederagenin as common aglyconel, apindoside V, mukorosside (fruits), and protein (31.87), carbohydrate (2.21) and starch (11.94%) (seeds kernel).

Biological Activity : Saponins show high spermicidal activity in both spot and IPPF tests with human semen, with MED of 0.05 & 0.51, respectively.

Uses in Literature : Reported in India as an anthelmintic, cattle feed, contraceptive, detergent, emetic, expectorant, fish poison, hair wash, nauseant, oil seeds, purgative; and for charcoal making, chlorosis, epilepsy, polishing jewellery pneumonia, salivation, washing and bleaching cardamoms (**Ambasta, 1986; Banerjee & Pal, 1996; Borthakur, 1993; Chauhan, 1999; Jain, 1991; Rao, 1996; Rao *et al.*, 1996; Singh, 1991, 1996b; Siwakoti & Varma, 1996; Vihari Vishwa, 1995**).

Major Ayurvedic Formulations : Mahalaxadi-tel.

Trade : Dried fruits are sold in the local and national markets.

Sapium insigne (Royle) Benth.+ (Pl. 63A).

Syn. : *Falconeria insignis* Royle, *F. wallichiana* Royle, *Carumbium insigne* Kurz.

Family : Euphorbiaceae.

Vern. : Balodar.

Regional Names

Ass. : Mah-Kola;

Bo. : Duella;

H. : Khina, Khindra;

Kan. : Kanuupade, Kurda, Kurudanandi, Nanaka;

Mar. : Hura, Ura;

P. : Bilodar, Bilodja;
Tam. : Garpashola, Tillai;
Tel. : Garbhasula, Kadura.

Distribution : Kashmir-Bhutan: 500-1800m.

Description : Trees. Leaves alternate, toothed, oblong-lanceolate. Flowers yellow-green. Capsule long, fleshy when young.

Reproductive Cycle : May-August.

Habitat Ecology : Open slopes, drier areas.

Material Examined : EBH-120, 03-09-2001.

Parts Used : Latex. Wood.

Folk Uses : Wood used for making **drums**. Latex applied to check **itching** and for removing **warts**.

Uses in Literature : Known to be used for drums, floats, furniture, match-boxes, packing cases, sandals, toys, wounds; and also as cattle feed and germicide (**Ambasta, 1986; Jain, 1991; Kulkarni & Kumbhojkar, 1996; Watt, 1972**).

Sapium sebiferum (L.) Roxb.+ (Pl. 63B).

Syn. : *Exooecaria sebifera* Muell., *Stillangia sebifera* Michaux., *S. sinense* Baill., *Stillingfleetia sebifera* Bojer, *Carumbium sebiferum* Kurz., *Croton sebiferus* L.

Family : Euphorbiaceae.
Vern. : Makhan, Pahadi-Taali.
English, Sanskrit & Regional Names

Eng. : Chinese Tallow Tree;
Sans. : Toyapippali;
B. : China;
Bo. : Pipalyank;

H. : Vilayati-Shisham;
Kan. : Meenadabattimara;
Oriya : Ranojita.

Distribution : Cultivated in various parts of India.

Description : Trees. Leaves long-pointed, petioled, ovate. Capsule coriaceous.

Reproductive Cycle : June-August.

Habitat Ecology : Forests, shrubberies.

Material Examined : EBH-1048, 16-07-2003.

Parts Used : Leaves. Inner-bark.

Folk Uses : Planted for **ornamental** purposes, **fodder** and as **fuelwood**. Oil oozed out during burning of inner-bark, considered useful for massaging **painful joints**. 5-7g dose of powdered bark given with cow's milk twice a day for 10-15 days for **spermatorrhoea**.

Active Constituents : Active principles are 6,7,8-trimethoxy coumarin (bark), gallic acid, astragallin, (-) loliolide, kaempferol, quercetin and b-sitosterol glycoside (leaves).

Biological Activity : Antihypertensive activity positive.

Uses in Literature : Reported to be used as black dye, block-mount, feed, fertilizer, flour for human consumption purgative; and for bedsteads, candles, cosmetics, cricket bats, cutaneous troubles, dressing cloth, dyeing, dyspepsia, furniture, inferior quality pencils, paints, plastics, printing blocks, soaps, stillingia oil, tables, toys, varnishes; and also for rearing silk worms (**Ambasta, 1986**; **Jain, 1991**; **Watt, 1972**).

Saraca asoca (Roxb.) De Wilde[+] (Pl. 63C).

Syn. : *S. indica* auct. Non L., *S. arborescens* Burm., *S. minor* Miq., *S. zollingeriana* Miq., *Jonesia asoca* Roxb., *J. pinnata* Willd.

Family : Fabaceae.

Vern. : Asoka.

English, Sanskrit & Regional Names

Eng.	:	Asoka Tree;
Sans.	:	Karnapuraka;
B.	:	Asok, Asoka;
Bo.	:	Ashok, Jasundi;
G.	:	Ashopalava;
H.	:	Ashok;
Kan.	:	Aksunkar, Asokadamara; Kenkali;
Mal.	:	Hemapushpam, Vanjulam;
Mar.	:	Asoka, Jasundi;
Oriya	:	Osoka;
Tam.	:	Asogam, Malaikkarunai;
Tel.	:	Asokamu, Vanjalamu.

Distribution : W. Ghats; Often cultivated.

Description : Trees. Leaves sessile. Leaflets oblong-lanceolate. Pod 10-25cm. Seeds oblong, compressed.

Reproductive Cycle : March-September.

Habitat Ecology : Planted.

Material Examined : EBH-10, 14-04-2000.

Parts Used : Flowers. Bark.

Folk Uses : Plant considered **sacred**. Decoction of bark (10-20 ml) useful against **chest infections, leucorrhoea** and as **uterine tonic** (twice daily for 7-10 days). Powdered flowers mixed with honey good for **urinary discharges**.

Active Constituents : Chief constituents are leucopelarogonidin-3-0-b-D-glucoside, leucopelargonodin, leucocyanidin, b-sitosterol (stembark), b-sitosterol, quercetin, leucocyanidin, kaempferol-3-O-b-D-glucoside, apigenin-7-O-b-D-glucoside and gallic acid (flowers).

Biological Activity : Flower bud extr. inhibits *Escherichia coli, Salmonella viballerup, Vibrio cholerae, Shigella dysenteriae* and *Shigella boydii.*

Uses in Literature : Known in India as abode of God (sacred), astringent in treatment of excessive menstruation, diuretic, masticatory, psychoactive plant, substitute for milk in tea, tonic to prevent miscarriage, uterine sedative; and also for diabetes, easy delivery, edible purposes, haemorrhagic dysentery, heart diseases, internal haemorrhoids, leucorrhoea, scorpion sting, urinogenital diseases, ploughs and shafts (**Ambasta, 1986; Banerjee, 1996; Bhandary** ***et al.*****, 1996; Jain, 1968, 1991; Jain** ***et al.*****, 1994; Lalramnghinglova, 1996; Siwakoti & Varma 1996; Srivastava** ***et al.*****, 2003; Watt, 1972**).

Major Ayurvedic Formulations : Ashoka-dikvatha, Ashoka-vasi, Asoka-ghrita, Asoka-rishta.

Sarcococca saligna (D. Don) Muell. Arg.[+] (Pl. 63D).

Syn. : *S. pruniformis* Lindl., *S. trinervia* Wight., *S. sumatrana* BC., *S. salicifolia* Baill., *Tricera nepalensis* Wall., *Buxus saligna* Don.

Family : Buxaceae.

Vern. : Raabal.

English & Regional Names

Eng. : Christmas Box, Sarcocca, Sweet Box.

U.P. : Paliala, Sukat.

Distribution : Aghanistan-W. Nepal: 1200-2400m.

Description : Shrubs. Leaves entire, leathery, lanceolate. Flowers white. Fruit purple, 3-horned.

Reproductive Cycle : May-October.

Habitat Ecology : Moist shady places, forests; common.

Material Examined : EBH-1230, 18-09-2002.

Parts Used : Root. Seeds.

Folk Uses : 2-3g dose of powdered roots prescribed with honey (twice daily for 10-15 days) for **gonorrhoea**, and its poultice applied for **suppuration of boils.** Heated leaves applied on **joints** for relief from **pains**. Seeds eaten by animals.

Uses in Literature : Reported to be used for rheumatism, fever making brushes (**Ambasta, 1996**; **Kapur & Nanda, 1996**).

Saurauia nepaulensis DC (Pl. 64A).

Syn. : *S. paniculata* Wall., *Ternstroemia racemosa* Don, *Zanthoxylum serra* Turcz.

Family : Saurauiaceae.

Vern. : Gogna.

Hindi Names

H. : Gogandha, Gogina, Pangara;

Distribution : H.P.-S.W. China. Burma: 600-2100m.

Description : A small tree. Leaves large-sized, rusty beneath. Flower pink, in clusters. Fruit green.

Reproductive Cycle : March-October.

Habitat Ecology : Forests; planted in cultivated areas as a fodder tree.

Material Examined : EBH-1264, 15-04-2003.

Parts Used : Leaves. Fruit.

Folk Uses : Leaves used as cattle **fodder**. Fruits **edible**, and its 5-10 ml decoction given on empty stomach as **mild laxative.**

Uses in Literature : Reported to be used for edible purposes, stomach pain, packing cases, interior construction; and as fodder and purgative (**Ambasta, 1986; Jain, 1991; Jamir & Rao, 1990; Negi & Gaur, 1991; Sinha, 1996;** Watt, 1972).

Sechium edule Sw.+ (Pl. 64B).

Family : Cucurbitaceae.

Vern. : Lanku.

English & Regional Names

Eng. : Chow-Chow, Chayote;

Kan. : Seema Badane;

Tam. : Kattirikkai, Seeme.

Distribution : Cultivated.

Description : Climber.

Reproductive Cycle : June-September.

Habitat Ecology : Grown locally for its fruits.

Material Examined : EBH-1252, 18-07-2003.

Parts Used : Fruit. Seeds.

Folk Uses : Fruits cooked as **vegetable**, and considered good for **gastro-intestinal** disorders. Roasted seeds also **eaten**.

Active Constituents : These are a-amino-g-ureidobutyric acid (seeds), and glucosides of sitosterol and stigmasterol (fruits).

Biological Activity : b-sitosterol and stigmasterol-b-D-glucosides exhibit anti-inflammatory activity in rat paw oedema.

Uses in Literature : Known to be used in India for its edible fruits, fibre, and as fodder (**Ambasta, 1996; Negi & Gaur, 1991; Watt, 1972**).

Senecio rufinervis DC* (Pl. 64C).

Family : Asteraceae.

Vern. Name : Dhanpatra.

Distribution : W. Himalaya. 2000-2650m.

Description : Shrubs with tomentose stems and inflorescence. Leaves sharpely toothed, red-nerved and lower surface white tomentose. Heads numerous, yellow. Achenes glabrous.

Reproductive Cycle : August-October.

Habitat Ecology : Hilly slopes, moist higher forests.

Material Examined : EBH-1234, 19-07-2003.

Parts Used : Leaves. Inflorescence.

Folk Uses : Rolled dried leaves smoked as '**beedi**', and also used as cattle **fodder**. Inflorescence constituent of local wine, '**sur**'.

Sida cordifolia L. (Pl. 64D).

Syn.: *S. herbacea* Cav. Diss., *S. mican* Cav. Diss., *S. rotundifolia* Cav. Diss., S. athaeifolia Swartz.

Family : Malvaceae.

Vern. : Dredain.

English, Sanskrit & Regional Names

Eng.	:	Country-Mallow;
Sans.	:	Kathorayashtika, Raktatandalu;
Ass	:	Saru-Sonborial;
B.	:	Bala, Svetabarela;
G.	:	Baladana, Khareti;
H.	:	Barial, Kharenti;
Kan.	:	Hettugigada, Hettuthi, Kisangi;

Mal.	:	Katturam;
Mar.	:	Chikana, Khiranti;
Oriya	:	Beudia Naula, Bisvokapari;
P.	:	Kharent;
Tam.	:	Arivalamaippundu;
Tel.	:	Muttuvapulagamu, Suvarnamu.

Distribution : Throughout Sub-Tropical parts of India.

Description : Climbers. Leaves cordate-oblong, obtuse, crenate. Flowers yellow. Fruit reticulated.

Reproductive Cycle : August-September.

Habitat Ecology : Rocky wet slopes, moist places.

Material Examined : EBH-42, 23-09-2000.

Parts Used : Roots. Seeds. Aerial Plant Parts.

Folk Uses : 2-3g powdered seeds mixed with honey (twice a day for 10-15 days) useful against **sexual weakness**. Also, seeds ingredient of '**sund**'- a preparation considered good for **flatulence, joint pains, body tone** and **vitality**. 10 ml decoction of the aerial plant parts (thrice daily for 3-5 days) useful for excessive **menstrual flow** and **urinary disorders**.

Active Constituents : These contain sterculic, malvalic, coronaric acids (seed oil), ephedrine (stems and roots), palmitic stearic, hexacosanoic acids and b-sitosterol (aerial parts) and 0.45 rutin (plants).

Biological Activity : Antiprotozoal (*Entamoeba histolytica* strain STA) and CNS depressant.

Uses in Literature : Used in India as an antirheumatic, aphrodisiac, astringent, febrifuge, cardiac stimulant, colic, demulcent, diuretic, ingredient of 'chywanprash', tonic;

and for boils, bloody flux, cooling, cystitis, dysentery, excessive menstrual flow, eye diseases, facial paralysis, fibres, fits, dog and snake bites, general debility, gonorrhoea, haematuria, heart irregularity, hemiplagia, improving sexual strength, infective viral hepatitis, leucorrhoea, nervous disorders, opthalmia, pregnant micturition, sciatica, shivering (vet), shivering fits, spermatorrhoea, stomachache, urinary troubles, venereal diseases, wounds, and in cordage (**Ambasta, 1986; Borthakur, 1993; Dey, 1980; Gogoi *et al.*, 2003; Jain, 1968, 1991; Jha *et al.*, 1991; Khanna *et al.*, 1996b; Mehrotra, 1995; Saini, 1996b; Watt, 1972**).

Major Ayurvedic Formulations : Balaamool, Bala-churna, Baladi-kwath, Baladya-ghrita, Baladyarista, Bala-svarasa, Bala-taila, Bala-yadi-kvatha, Balayadirishta, Candana balalaksadi-taila, Kakuvadi-churna, Satapaki-ksiabala, Suda-rshan-taila.

Trade : Beejbanda.

Skimmia laureola Sieb. & Zucc. *ex* Walp.[+] (Pl. 65A).

Syn.: *S. arborescens* T. anders *ex* Gamble, S. anquetilia Taylor & Airy Shaw, *Limonia laureola* DC, *L. laureola* Wall., *Anquetilia laureola* Done., *Laureola fragrans* Roem.

Family : Rutaceae.
Vern. : Ner, Patti Dhup.
Regional Names

- **Kash.** : Nera, Patar;
- **P.** : Barru, Ner, Shalanghi.

Distribution : Temperate Himalaya: 2000-3500m.

Description : Shrubs having simple, thick leathery oblanceolate leaves. Flowers yellowish-white. Fruit red, ellipsoid, fleshy.

Reproductive Cycle : April-May.

Habitat Ecology : Forests, shrubberies, shady places, gregarious.

Material Examined : EBH-790, 26-04-2000.

Parts Used : Leaves. Inflorescence.Wood.

Folk Uses : Wood used for **agricultural implements** and **tool handles**. Inflorescence ingredient of local wine, '**sur**'. Leaves burnt as **incense** for purifying air.

Active Constituents : Main ingredients are skimmim, scopoletin (bark), lupenone, lupeol b-sitosterol, skimmianine, scopoletin, umbelliferone, beyapten, isopimpinellin linalyl acetate, furanocouniarin, bergapten, laureoline, b-phellandrene, geraneol, myrcene, nerol, terpenes (d-a-pinene and d-b-phellendene), azulene and traces of menthene (leaves).

Biological Activity : Skimmianine selectively inhibits 5-hydroxytrypt-amine-induced vasopressor responses in rats as well as in isolated atria.

Uses in Literature : Reported so far to be used in India as air purifier, antiseptic, base material for high grade perfumes and cosmetics, flavouring perfumery, incense; and for boils, gastric pain, rheumatic pains, scabies, small pox, making hoe and axe handles (**Ambasta, 1986; Jain, 1991; Jain *et al.*, 1991; Kapur & Srivastava, 1996; Rana *et al.*, 1996; Singh, 1991; Watt, 1972**).

Smilax aspera L.$^{+}$ (Pl. 65B).

Syn. : *S. maculata* Roxb., *S. fulgens* Wall., *S. nilgarensis* Steud.

Family : Liliaceae.
Vern. : Bragh Bel.

Distribution : Pakistan-Bhutan. India. Europe. E. Africa: 1200-2500m.

Description : Prickly climbers. Leaves linear-lanceolate. Flowers small, white, fragrant. Berry bluish-black on ripening.

Reproductive Cycle : September-November.

Habitat Ecology : A usual climber on shrubberies, moist rocky situations.

Material Examined : EBH-619, 11-10-2002.

Parts Used : Roots. Aerial Plant Parts.

Folk Uses : Plant ingredient of '**chyawanprash**'; cooked as **vegetable** and also **pickled**. Its root powder (1-2g) useful against **fever, joint pains** and **gastric ailments** (twice daily for 3-5 days).

Active Constituents : Chief constituents are sarasapogenin, its pseudogenin, a new sapogenin-asperagenin, 31-norcyloartanol, b-sitosterol (roots), a new bisdermosine 22-hydroxyflarostanol saponin asperoside (leaves) and rutin (plant).

Uses in Literature : Used in India as a sheaf binder in crop fields with the belief of good yield; and for cordage, skin eruptions, sores, tanning and wounds (**Jain 1991**; **Hosagoudar & Henry, 1996c**).

Solanum erianthum Don+ (Pl. 65C).

Syn.: *S. verbascifolium* L., *S. pubescens* Roxb.

Family : Solanaceae.

Vern. : Radka.

English, Sanskrit & Regional Names

Eng. : Potato Tree;
Sans. : Gandhira, Pathi,
Kan. : Kadusinde, Sowdangigida;
M.P. : Chichora;
Mal. : Chunta, Erichunta, Melanjunta;
P. : Kalamewa, Kharawine, Ola, Tiari;
Tam. : Anaichundai, Malaichundai, Sundai;
Tel. : Budamu, Rasagadi.

Distribution : Throughout India.

Description : A shrub. Leaves alternate at both ends or base sub-rhomboid. Berry globose-yellow.

Reproductive Cycle : July-October.

Habitat Ecology : Wastelands, roadsides, rare among boulders.

Material Examined : EBH-1095, 12-10-2003.

Parts Used : Leaves.

Folk Uses : Poultice of leaves heals open **wounds**.

Active Constituents : Leaves yield solasodine (57.0, 71.0), solasodiene (3.0, 3.5), diosgenin (8.5, 25.0%), solafloridine (8.0%, trt.), tomatidenol (23.0%), vespertilin, 5,16-dehydropregnenolone, solanine, solasonine, solamargine and solanidine.

Uses in Literature : So far reported to be used for body pains, burns, diarrhoea, dysentery, fever, flavouring in curries, horse afflicted with glanders, killing worms in stomach, urinary troubles, vaginal discharges, vertigo; and as vegetable (**Ambasta, 1986**; **Henry *et al.*, 1996**; **Hosagoudar & Henry, 1996b**; **Jain, 1991**; **Siwakoti & Varma, 1996**; **Watt, 1972**).

Solanum melongena L. (Pl. 65D).

Syn. : *S. incarnum* L., *S. insanum* L., *S. zeilanicum* Scop., *S. undulatum* Lamk., *S. trongum* Lamk., *S. longum* Roxb., *S. pseudo-undulatum* Blume, *S. ovigerum* Blume, *S, esculentum* Dunal, *S. heteracanthum* Dunal, *S. melanocarpum* Dunal, *S. sativum* Dunal, *S. pressum* Dunal, *S, cumingii* Dunal, S. torrum var. *inerme* Dalz. & Gibs, *S. ferox* var. *â*-Kurz.

Family : Solanaceae.

Vern. : Baingan.

English, Sanskrit & Regional Names

Eng. : Brinjal, Egg Plant;

Sans. : Dirghavartaki, Nilakantaka;

Ass. : Jatibengani;

B.	:	Bartaku, Kulibegun;
Bo.	:	Waingi;
G.	:	Rihgni;
H.	:	Baigun, Baingan;
Kan.	:	Badanekayi, Doddabadane;
Kash.	:	Tukmibadanjanurmi, Vangun;
Mal.	:	Hinguli, Nilavalutina;
Mar.	:	Vangi;
Oriya	:	Bartakki;
P.	:	Bengan, Vataun;
Tam.	:	Kalliyam, Mullukkattari;
Tel.	:	Mettavanga, Nidupuvanga.

Distribution : Widely cultivated for its edible fruits.

Description : Prickly shrubs. Leaves wooly lobed. Flowers blue. Berry glabrous, whitish-yellow.

Reproductive Cycle : June-September.

Habitat Ecology : Cultivated, near houses.

Material Examined : EBH-236, 23-07-2001.

Parts Used : Fruit. Stem.

Folk Uses : Fruit cooked as **vegetable**. Ash of its burnt stem checks **tongue** and **mouth sores**.

Active Constituents : Fruit yields solasodine and diosegin.

Uses in Literature : Known to be used in India as alternative, antiasthmatic, cathartic, diuretic, hydrogogue, laxative, stimulant, sialagogue, tonic and vegetable; and also for asthma, bronchitis, fever, diarrhoea, disorders of the sight, dropsy, dysuria, dyspepsia, enlargement of liver, epistaxis (vet), heart diseases, hydrophobia, inflammation of kidneys and bladder, otitis, producing uncomfortable pregnancy sleep, psoriasis, skin diseases, snake bite, stomatitis, toothache, ulcers, virulent gonorrhoea and in anasarca (**Ambasta, 1986; Gill *et al.*, 1993; Jain, 1991; Rama Rao & Henry, 1996; Watt, 1972**).

Solanum pseudocapsicum L.* (Pl. 66A).

Family : Solanaceae.

Vern. : Charoti Mirch, Gol Mirch.

English Name

Eng. : Jerusalem Cherry.

Distribution : Temperate Himalaya, Kashmir - Sikkim.

Description : A shrub Flowers white, deeply lobed. Fruit orange-yellow.

Reproductive Cycle : June-October.

Habitat Ecology : Planted near houses.

Material Examined : EBH-464, 16-07-2001.

Parts Used : Fruits.

Folk Uses : Fruit used as **spice**.

Active Constituents : Solanocapsine, solacapine, isosolacapine, episolacapine and 0-methylsolanocapsine isolated from aerial parts.

Solanum torvum Swartz[+] (Pl. 66B).

Syn.: *S.indicum* L., *S. stramoniflium* Roxb., *S. ferrugineum* Jacq., *S. pseudosaponaceum* Blume, *S. wightii* Miq.

Family : Solanaceae.

Vern. : Chhota-Baingan.

Regional Names

Ass.	:	Hathibhekuri, Bhit-Tita;
B.	:	Gotabegun, Titbaigun;
H.	:	Bhurat, Ranabaigan;
Kan.	:	Kadusunde, Sondaegida, Sundakkayi;
Mal.	:	Chunta, Kattuchunta;
Tam.	:	Kotukattarai, Malaichundai, Sundai;
Tel.	:	Chundakai, Kondavuste, Kottuvastu, Ustekaya.

Distribution : Naturalised in Himalaya: 250-750m.

Description : Prickly shrubs. Leaves ovate, with sinuate margins. Flowers white. Berry globose, yellow when ripe.

Reproductive Cycle : April-August.

Habitat Ecology : Wastelands.

Material Examined : EBH-1246, 13-06-2003.

Parts Used : Root. Fruit.

Folk Uses : Ripe fruits edible and unripe ones cooked as **vegetable**. Poultice of roots considered good for **itching** and **skin eruptions**. Also, roots chewed with honey to check **conception**.

Active Constituents : Sapogenins sisalagenone and torvogenin isolated after acid hydrolysis of fruit extr.

Uses in Literature : Reported to be used as an abortifacient, analgesic, antidote to poison, digestive, diuretic, oral contraceptive, sedative, psychoactive plant, vegetable (leaves, fruit); and also for aches, bleeding, chapped hands and feet, cough, diarrhoea, dog bite, ear pain, itching and enlargement of spleen, malaria, respiration (vet), skin diseases, snake bite, spinal cord, stomachache, toothache, tooth decay, and wounds (**Ambasta, 1986;**

Borthakur, 1996b; Chowdhery, 1996; Dwarakan & Ansari, 1996; Gill *et al.*, 1993, 1997; Jain, 1991; Jain *et al.*, 1994; Mao, 1993; Negi & Gaur, 1991; Negi *et al.*, 1993; Rama Rao & Henry, 1996; Sinha, 1996; Srivastava *et al.*, 2003; Vishwanathan & Singh, 1996).

Solidago canadensis L. (Pl. 66C).

Family : Asteraceae.

Vern. : Sone-ki-chhadi.

English Name

Eng. : Golden Rod.

Distribution : Temperate Himalaya: cultivated.

Description : A woody shrub. Leaves lanceolate, 3-veined. Heads bright yellow in dense unilateral, axillary racemes forming a pyramidal panicle.

Reproductive Cycle : October-December.

Habitat Ecology : Planted.

Material Examined : EBH-1244, 19-11-2002.

Plant/Parts Used : The Whole Plant.

Folk Uses : Grown for its **ornamental** value.

Active Constituents : Bayogenin, isorhamnetin, quercetin and isorhamnetin3-0-b-D-glucopyramoside isolated from aerial parts.

Biological Activity : Extr. of aerial parts exhibits hypoazotemic and diuretic activities in rabbits.

Uses in Literature : Used in India for hemorrhagic nephritis (**Ambasta, 1986**).

Sorbaria tomentosa (Lindl.) Rehder* (Pl. 66D).

Syn. : *S. sorbifolia* Hook. f., *S. lindleyana* Wall. *ex* Lind., *Schizonotus tomentus* Lindl.

Family : Rosaceae.

Vern. : Khusti, Kusth, Khusk.

English Name

Eng. : Hardhack Steeple Brush.

Distribution : Afghanistan-C. Nepal; C. Asia; N. China. 1800-2900m.

Description : Shrubs. Leaves pinnate, stipulate. Leaflets 13-19. Flowers white. Follicles glabrous.

Reproductive Cycle : June-August.

Habitat Ecology : Open slopes, riversides, cultivated areas; gregarious and common.

Material Examined : EBH-380, 29-06-2000.

Parts Used : Wood. Stem. Branches. Leaves.

Folk Uses : Wood suitable for making **walking sticks**, smoking rod of **hubble-bubbles** and as **fuel**. Stem yields fibres for **ropes**. Poultice of leaves and young branches considered suitable for healing **wounds** and **sprains**. Also, its leaves mixed with barley flour, and given to milching cattle for better yield of milk.

Active Constituents : These are sorbifolin (leaves), astralgin, arbutin, isorhamnetin-3-glucoside, kaempferol-3-arabinofuranoside, scutellarein and its 7-O-a-L-rhamnopyranoside and chlorogenic acid (flowers and leaves).

Spirarea canescens D. Don⁺ (Pl. 67A).

Syn.: *S. cuniefolia* Wall.

Family : Rosaceae.

Vern. : Chakhu.

Regional Names

Kash. : Chaku, Taku.

Distribution : Temperate Himalaya: 200-4000m.

Description : Deciduous shrubs. Twigs ribbed, reddish-brown. Leaves ovate, toothed towards the tip. Flowers pale-pink. Follicles hairy.

Reproductive Cycle : May-October.

Habitat Ecology : Shrubberies, cultivated areas.

Material Examined : EBH-1032, 23-09-2002.

Parts Used : Seeds. Wood.

Folk Uses : Planted for **ornamental purposes**. Wood suitable for **walking sticks**, and seeds used for **flavouring curries**.

Active Constituents : Seeds rich in saponins.

Uses in Literature : Used so far as sedative and for walking sticks (**Ambasta, 1986; Kapur & Srivastava, 1996**).

Spondias mangifera Willd. (Pl. 67B).

Syn.: *S. pinnata* Kurz., *S. acuminata* Roxb., *S. macrophylla* Wall *ex* Hook f., *S. amara* Lamk., *Evia amara* Comm., *Mangifera pinnata* L. f.

Family : Anacardiaceae.

Vern. : Muara.

English, Sanskrit & Regional Names

Eng. : Bile Tree, Indian Hog Plum. Traveller's Delight, Wild Mango;

Sans. : Advagabhjogya, Madhuramlaka;

Ass. : Amara;

B. : Ambra, Amna;

Bo. : Amarah, Jangliam;

G. : Ranamba;
H. : Amara, Ambodha;
Kan. : Amatekaya, Ambattemara;
Mal. : Ampalam, Kattampalm, Pulimavu;
Mar. : Amb, Ambada, Ranamba;
Oriya : Ambula, Amrotoko;
P. : Ambara, Bahamb;
Tam. : Ambiram, Kattumagirangai;
Tel. : Adavimamena, Irurumamidi.

Distribution : Temperate Himalaya; upto 900m.

Description : A medium-sized tree. Leaflets 9-13, elliptic-oblong. Flowers greenish-white. Drupes fleshy-yellow.

Reproductive Cycle : April-June.

Habitat Ecology : Cultivated.

Material Examined : EBH-679, 14-04-2000.

Parts Used : Fruit. Inflorescence. Leaves.

Folk Uses : Ripe fruits **edible**, and unripe ones cooked as vegetable, and **pickled**. Leaves used as **cattle feed**, and flowers for **flavouring**.

Active Constituents : Principle constituents are b-amyrin, oleanolic acid (fruits), lignoceric acid, 24-methylene cycloartanone, stigmast-4-en-3-one, b-sitosterol and its glucoside (aerial parts), glycine, cystine, serine, alanine, and leucine (fruits).

Uses in Literature : So far used for bilious dyspepsia, boats, cholera, chutney, diarrhoea, dysentery, earcahe, edible purposes, floats, fumigation, horses in pneumonia, itching, jams, flavouring, match splints, nigth fever, non-ornamental plywood, pickles, regularising menstruation, reviving taste, rheumatism, ringworm, stomachache, tanning, tea-chests, unbleachable, wood-pulp; and also as an aromatic, anticancerous, astringent, tonic after delivery and vegetable (**Ambasta, 1986; Borthakur, 1996b; Jain 1991, 1996; Kulkarni & Kumbhojkar, 1993; Lalramnghinglova, 1996; Mohanty & Padhy, 1996; Nazarudeen** ***et al.*****, 1996; Negi & Gaur, 1991; Sharma & Borthakur, 2003; Singh** ***et al.*****, 1996b; Watt, 1972**).

Major Ayurvedic Formulations : Dadhika ghrta.

Stephania elegans Hook. f. & Thoms.+ (Pl. 67C).

Family : Menispermaceae.

Vern. : Dhartu Bel.

Distribution : Sikkim-Shimla: 1000-2100m.

Description : Climbing shrubs. Leaves varying much in size, triangular-ovate, base rounded, lower surface pale. Flowers red-purple. Drupe red.

Reproductive Cycle : August -October.

Habitat Ecology : Moist situations.

Material Examined : EBH-562, 03-08-2001.

Parts Used : Aerial Plant Parts. Leaves.

Folk Uses : Poultice of the aerial plant parts used as an antidote to **snake bite**, and leaves used as **cattle feed**.

Active Constituents : Chief principles are epinernandolinol, N-methylcorydalmine, hasubanonine, aknadinine, cyclandine, magnoflorine, isotetranelrine, isochondrodendrine and cyeleanine (leaves).

Uses in Literature : Used in India for removing worms from calf stomach (**Manandhar, 1990**).

Stephania glabra (Roxb.) Miers. (Pl. 67D).

Syn.: *S. rotuda* Hook. f., *Cissampelos glabra* Roxb., *C. longa* Miers., *C. japonica* Miers., *C. roxburghiana* Miers., *C. glabra* Miers., *C. glandulifera* Miers., *Cocculus glabra* Roxb., *C. roxburhianus* Wall. Cat., *C. finlaysoniansus* Wall., *Clypea wightii* Arn.

Family : Menispermaceae.

Vern. : Bish-Khapar.

Regional Names

U.P. : Gindaru, Purha.

Distribution : Hilly districts throughout India, ascending to 2000m.

Description : Climbers. Leaves broadly-ovate. Flowers greenish-yellow. Fruit red, pisiform.

Reproductive Cycle : June-August.

Habitat Ecology : Common in moist slopes.

Material Examined : EBH-1001, 22-07-2003.

Parts Used : Root. Tuber.

Folk Uses : Poultice of tubers considered useful for **skin diseases**. Pieces of boiled roots given to livestock for **gastro-intestinal** problems.

Active Constituents : Tubers yield palmatine, dehydrocorydalmine, palmatrubine stepharanine, tetrahydropalmatine, crydalmine and stepholidine.

Biological Activity : Anticholinesterase, antiinflammatory and spasmo-lytic activities found positive.

Uses in Literature : Reported to be used for asthma, body and headache, eye disease, intestinal complaints, pulmonary tuberculosis, skin diseases, tumour in stomach and as psychoactive plant (**Ambasta, 1986; Jain, 1991; Jain *et al.*, 1994**).

Sterculia villosa Roxb. (Pl. 68A).

Family : Sterculiaceae.

Vern. : Leandrein.

Regional Names

Ass.	:	Odol, Udal;
H.	:	Udal, Udar;
Kan.	:	Bilidale, Chauri, Savaya;
Mal.	:	Vakka;
Mar.	:	Kuthada, Sarda, Sardal;
Oriya	:	Kodalo;
P.	:	Gulbodla, Gulkandar;
Tam.	:	Arnivakenar, Murattham;
Tel.	:	Kummari-Poliki.

Distribution : N.W. India.

Description : A tree. Leaves glabrescent, deeply palmate. Flowers pinkish. Seeds black.

Reproductive Cycle : June-August.

Habitat Ecology : Along hilly slopes; wastelands.

Material Examined : EBH-1084, 12-06-2003.

Parts Used : Leaves. Wood. Seeds. Bark.

Folk Uses : Leaves as **fodder**. Wood used for making **boxes** and as **fuel**. Seeds **edible**. 1-3g powdered bark (twice a day for 3 days) prescribed with lukewarm water for **amoebic dysentery** and also for making **ropes**.

Uses in Literature : Reported to be used for blood dysentery, constipation, cordage, edible purposes (cotyledons), diarrhoea, dyeing, dysentery, dyspepsia, fistula, hydrocele, light packing cases, match boxes, night fever, ply work, rough bags, ship building, sores, splints, swelling, tonsillitis and as fuelwood and cattle halters (**Ambasta, 1986**; **Jain, 1991**; **Jain & Borthakur, 1980**; **Kothari & Moorthy,**

1996; Lalramnghinglova, 1996; Molla & Roy, 1985; Mukherjee & Namhata, 1990; Rama Rao & Henry, 1996; Sinha, 1996; Upadhye & Kumbhojkar, 1996).

Trade : Udal (Wood), Udal Lakri.

Strobilanthes glutinosis Nees* (Pl. 68B).

Syn.: *Ruellia glutinosa* Wall., *R. jacquemontiana* Nees

Family : Acanthaceae.
Vern. : Neela Phool.

Distribution : Temperate Himalaya, 1000-2000m.

Description : Shrubs. Leaves-stalked, sharpely toothed. Flowers pale blue. Capsule viscous, pubescent, 4-seeded.

Reproductive Cycle : October-November.

Habitat Ecology : Usually in forest undergrowth.

Material Examined : EBH-743, 23-10-2000.

Parts Used : Leaves.

Folk Uses : Leaves serve as an excellent **fodder** for sheep and goats.

Symplocos racemosa Roxb.[+] (Pl. 68C).

Syn.: *S. hamiltoniana* Wall., *S. rigida* Wall., *S. nervosa* A. DC, *S. propinqua* Hance

Family : Symplocaceae.
Vern. : Lodh, Lodhar.
English, Sanskrit & Regional Names

- **Eng.** : Californian Cinchona, Chinanora, Lodh Tree;
- **Sans.** : Bhillatary, Kandakilaka, Marjana, Shahara lodhra;
- **Ass.** : Bhomroti, Kaviang;
- **B.** : Lodh;

Bo.	:	Hura, Lodhra;
G.	:	Lodar;
H.	:	Lodh;
Kan.	:	Balalodduginamaru, Pachettu;
Mal.	:	Pachotti;
Mar.	:	Lodhra;
Oriya	:	Lodho;
Tam.	:	Velli-Lothi;
Tel.	:	Lodduga, Sabaramu.

Distribution : Throughout N. E. India. At 0-2500 ft.

Description : A small-sized tree. Leaves obtuse, coriaceous with distant nerves.

Reproductive Cycle : October-May.

Habitat Ecology : Forests, shrubberies.

Material Examined : EBH-632, 06-02-2000.

Parts Used : Wood. Branches. Leaves.

Folk Uses : Wood suitable for **agricultural implements** and for **cheap furniture**. 2-4g powdered young branches given with honey for irregular menstruation and **digestive disorders**. A gargle of decoction prepared from leaves good for **bleeding piles**.

Uses in Literature : Used in India as an astringent, blood purifier and gargle; and also for bleeding of gums, bowel complaints, bronchitis, conjuctivitis, cooling, cough, diarrhoea, digestive disorders, dropsy, dyesentery, elephantiasis, eye diseases, fat in urine, furniture, hardening adherent teeth, liver complaints, loose motions, menorrhagia, opthalmia, plastering, preventing miscarriage, promoting maturation of wounds, tumours, ulcers and uterine disorders (**Ambasta, 1986**; **Brahmam** *et al.*,

1996; Jain, 1968, 1991; Jha & Varma, 1996; Mukherjee & Namhata, 1990; Rao *et al.*, 1996; Upadhye *et al.*, 1994; Watt, 1972).

Major Ayurvedic Formulations : Bhrat-gangadhara-churna, Lodhraasav, Lodhradi-kwath, Pippalayasav, Pushanuga-churna.

Symplocos spicata Roxb.+ (Pl. 68D).

Syn. : *S. laurina* (Retz.) Wall. *ex* Rehd., *S. polycarpa* Wall., *S. loha* Don, *S. racemosa* Wall., *S. cochinchinensis* (Lour.) Moore *ssp. Laurina* (Retz.) Nooteboom var. *Laurina*

Family : Symplocaceae.

Vern. : Sodgar, Sodhar.

English, Sanskrit & Regional Names

Eng.	:	Chunga;
Sans.	:	Lodra;
B.	:	Bholica, Buri, Lodh;
H.	:	Bholia, Sodh;
Mal.	:	Pachotti;
Oriya	:	Bhaunni;
Tam.	:	Kambli Vetti.

Distribution : N. & E. India.

Description : Small trees. Leaves broadly-lanceolate, acute serrulate. Spikes long, branched, rusty-pubescent.

Reproductive Cycle : May-July.

Habitat Ecology : Forests; shady ravines.

Material Examined : EBH-745, 14-06-2000.

Parts Used : Leaves. Bark.

Folk Uses : Bark used as a source of **yellow dye**. Leaves used as **fodder**.

Active Constituents : Constituents isolated are a new glycoside of leucoanthocyanidin (I), a-spinasterol (stem bark) and a new flavanol glycoside rhamnetin-3-0-b-D-galactosyl-(4→1) and 0-b-D-galactopyranoside (II) (leaves).

Biological Activity : a-spinasterol at a dose of 25.0 mg/kg exhibits significant anti-inflammatory activity.

Uses in Literature : Reported to be used in biliousness, diarrhoea, diseases of eye, dyeing, gonorrhoea, hemorrhages; and also amulet of seeds hung round the necks of children to avert evil spirits (**Ambasta, 1986; Watt, 1972**).

Tabernaemontana divaricata R. Br. *ex* Roem. & Sch.⁺ (Pl. 69A).

Syn. : *T. divaricata* Blume, *T. coronaria* R. Br., *Nerium Divaricatum* L., *N. coronarium* Jacq., *Jasminum zeylanicum* Burm., *Ervatamia coronaria* Stapf.

Family : Apocynaceae.

Vern. : Chandni.

English, Sanskrit & Regional Names

Eng. : East India Rosebay.
Sans. : Nandivriksha, Tagara;
Ass. : Tagar;
G. : Sagar, Tagar;
H. : Chameli, Chandni, Tagar;
Kan. : Kotuhalle, Nandebatlu;
Mal. : Kuttampale, Nantiyavattam, Takaram;
Mar. : Ananta, Gondetagara, Sagar, Tagar;
Tam. : Adukkunandiyavattai, Kuruduoolai, Nandiyavattam;
Tel. : Gandhitagarpu, Nandivardhanamu.

Distribution : Cultivated throughout India.

Description : An evergreen, dichotomously branched, glabrous shrub. Leaves oblong-lanceolate. Flowers white, fragrant. Seeds embedded in a red, pulpy aril.

Reproductive Cycle : July-October.

Habitat Ecology : Cultivated; often met as an escape.

Material Examined : EBH-1249, 16-07-2003.

Parts Used : Flowers. Wood.

Folk Uses : The plant is known for **ornamental purposes**. Wood used as an **incense**. 10-20 ml decoction of flowers prescribed twice a day for 5-7 days as **refrigerant**, **anthelmintic** and **febrifuge**.

Active Constituents : Flowers yield kaempferol, ibogamine, janetine, voaphyllene, voacangine, olivaline, dregamine, tabernaemontanine, voacamine and loronaridine.

Biological Activity : Indole alkaloid (I) inhibits HCl induced ulcer in mice by 48.8%.

Uses in Literature : Used in India for boils, dental caries and pyrrhoea, decorative purposes, dyeing, eye complaints, fever, headache, jaundice, sores, toothache and perfumery, and also as anodyne, anthelmintic, antidote to scorpion sting, incense, red dye, refrigerant and vermicide (**Ambasta, 1986; Gogoi & Borthakur, 1991; Jain 1991; Khanna *et al.*, 1996a; Pandey *et al.*, 1996; Rao *et al.*, 1996**).

Taxus baccata L.[+] (Pl. 69B).

Syn. : *T. baccata* Hook. F., *T. baccata* L. *subsp. Wallichiana* (Zucc.) Pilger, *T. wallichiana* L., *T. nucefera* Wall., *T. virginata* Wall., *T. orientalis* Bertoloni, *T. nepalensis* Jacq.,*T. contorta* Griff.

Family : Taxaceae.

Vern. : Rakhaal.

English, Sanskrit & Regional Names

Eng. : Yew, Common Yew;

Sans. : Manduparni;

B. : Bhirmik, Sugandhu;

Bo. : Barmi;

H.	:	Thuno, Zirnubbirmi;
Kash.	:	Chatung, Postil, Sungal;
P.	:	Barmi, Chogu, Rikhai;
U.P.	:	Gallu, Thuner.

Distribution : S. E. Asia: 2100-3400m.

Description : Trees. Leaves flattened, aromatic when dry. Fruits red, fleshy.

Reproductive Cycle : Fruiting: September-November

Habitat Ecology : Forests, shady ravines; usually in the understory.

Material Examined : EBH-1021, 24-07-2003.

Parts Used : Leaves. Fruit. Wood.

Folk Uses : Wood used as **incense**, and also suitable for low grade **furniture**. Fruits **edible**. Leaves and bark used as substitute for **tea**. 1-3g powdered leaves and bark given with cow's milk till cure for **cough, cold, bronchitis, breast cancers** and **spasm**. Bark also used as **hair dye**.

Active Constituents : Taxine A, taxagifin, tetrol (I), amentoflavone, mono and di-0-methufamentoflavones, sciadopitysin, ginkgetin, sequoiaflavone, taxinine, hydrocyanic acid, formic acid, reducing sugars, resins, tannins, ephedrine, a glucoside, taxiphyllin, ferredoxin, acidyrterone and b-sitosterol (leaves).

Biological Activity : Aq. extr. of leaves exhibits a depressant effect on CNS of rats.

Uses in Literature : Reported to be used as an anticancerous, antispasmodic, aphrodisiac, aromatic, beauty and shaving creams, beverage, carminative, dentrifices, diuretic, emmenagogue, expectorant, fish poison, fodder, hair colour, incense, lithotropic, ornamental, sacred tree, sedative, stomachic; and also for abortions, asthma, back of combs, bronchitis, bone fracture (vet), carrying poles, carvings and layings, cheap grade pencils, cold and cough, convulsions, diarrhoea, dyeing, edible purposes, epilepsy, flooring, furniture, gates, giddiness, handles of knives, headache, hysteria, indigestion, irregular and depressal action of heart, nervousness, ploughs, severe biliousness, strength and vigour, tumours, turnery, veneers and water-wheel (**Ambasta, 1986; Aswal, 1996; Chauhan, 1999; Jain 1991; Lal *et al.*, 1996; Negi & Gaur 1991; Negi & Pant, 1990; Rana *et al.*, 1996, Sharma & Sood, 1997; Singh, 1991; Watt, 1972**).

Major Ayurvedic Formulations : Kankasava, Lavan-bhaskar-churna, Sudarshan-churna, Talisadi-churna, Talisadi-vati.

Trade : The taxol prepared from leaves and bark fetches a price of 1.25 lakh US$ per kg in world market.

Tecoma grandiflora Loisel.[+] (Pl. 69C).

Syn. : *Campsis grandiflora* (Thunb.) K. Schum., *Bignonia grandifora* Thunb.

Family : Bignoniaceae.

Vern. : Shehnai Phool.

English Name

Eng. : Chinese Trumpet Creeper.

Distribution : Planted.

Description : A handsome shrub. Leaflets 7-9. Flowers large orange cloured in drooping clusters.

Reproductive Cycle : Summer season.

Habitat Ecology : Cultivated.

Material Examined : EBH-998, 14-05-2002.

Parts Used : Flowers.

Folk Uses : The plant is grown for **ornamental purposes**, and its flowers considered sacred for **offerings**.

Active Constituents : These are furfural, 5-methylfural, furfuryl alcohol and 2-acetylfuran (essential oil of flowers).

Uses in Literature : So far used in India for diabetes (**Ambasta, 1986**).

Terminalia arjuna Wt. & Arn.⁺ (Pl. 69D).

Syn. : *T. berryi* W. & A., *T. glabra* W. & A., *T. ovalifolia* Rottl., *Pentaptera glabra* Roxb., *P. angustifolia* Roxb.

Family : Combretaceae.

Vern. : Arjun.

English, Sanskrit & Regional Names

Eng. : White Murdah;
Sans. : Dhananjaya, Karaviraka, Savya Sachi;
Ass. : Orjun;
B. : Arjun, Kahu;
Bo. : Arjunasadadra, Jamla;
G. : Arjunasadada, Dhaulasadr;
H. : Arjun, Kangach, Kowa;
Kan. : Maddi;
Mal. : Vellamarutu;
Mar. : Arjunladada, Saumadat;
Oriya : Orjuno, Pornoyosahajo;
P. : Arjan, Jumla;
Tam. : Attumarudu, Kulamatrudu;
Tel. : Kakubhamu, Tittumaddi;
U.P. : Kauha, Arjun.

Distribution : Commonly along riverian areas in the sub-Himalayan tracts.

Description : Trees. Leaves narrowed at the base, often cordate. Spikes usually panicled. Fruits glabrous, ovoid.

Reproductive Cycle : May-September.

Habitat Ecology : Planted along agricultural fields, roadsides.

Material Examined : EBH-741, 14-04-2000.

Parts Used : Bark. Leaves.

Folk Uses : 10-20 ml decoction of bark useful for the cure of **heart problems**, **sciatica**, **leucorrhoea** and as **blood purifier**, **purgative** and **cardiac tonic** (twice daily for 30-45 days). Leaves kept in grain storage containers for repelling insects. Wood used for **agricultural implements**, **building works** and **cart-wheels**.

Active Constituents : Bark contains cevasidin, b-sitosterol, friedelin, methyl deanolate and gallic, ellegic & arjunic acids, b-sitosterol, (+) catechol, (+) gallocateenol, epicateenol, epigalco catechol, arjunolone, arjunglucoside III, arjunoside I and II, arjunflucoside I, arjunetin, leucoganidin, terminoic acid and arjunin.

Uses in Literature : So far reported to be used as an antidote to poison, antidysenteric, astringent, detergent, deobstruent, expectorant, febrifuge, fish poison, hair wash, psychoactive plant, purgative, sacred, sex stimulant, styptic; and for agricultural implements, anaemia, apthae, asthma, bilious affections, blood diseases, blood dysentery, blood pressure, boats and canoes, boils, chest pain and heart palpitation, cirrhosis of liver, cuts and wounds, dysuria, earache, electric poles, epilepsy,fatigues, fever, fractures, house construction, hypertension, leprosy, leucoderma, leucorrhoea, masts, menstrual disorders, musical instruments, painful urination, pleurisy, plywood manufacture, pneumonia, sores, tanning, tumours, ulcers, water traps, water trough, wheels of country carts, wounds (**Ambasta, 1986; Chauhan, 1999; Datt, 1996; Hembrom, 1991; Henry *et al.*, 1996; Hosagoudar & Henry, 1996a; Jain, 1968, 1991; Jain *et al.*, 1994; Kothari & Moorthy, 1996; Kurian, 1995; Lalramnghinglova, 1996; Masih, 2003; Rama Rao & Henry 1996; Satapathy & Brahmam,**

1996; Sharma & Sood, 1997; Singh & Prakash, 1996; Singh *et al.*, 1997; Srivastava *et al.*, 2003; Watt, 1972).

Major Ayurvedic Formulations : Arjunaghrita, Arjunaksirapalla, Arjunarishta, Arjunatwal-churna, Arjun-kwath, Arvindaasava, Devadarista, Hridayarnava-rasa, Kakubhaadi-churna, Nagarjunabhra-rasa, Parthadyarista

Trade : Arjun.

Terminalia bellirica (Gaertn.) Roxb.⁺ (Pl. 70A).

Syn. : *Myrobalamus bellirica* Gaertn.

Family : Combretaceae.

Vern. : Bahera.

English, Sanskrit & Regional Names

Eng.	:	Bastard Myrobalan, Beddanuts, Belleric Myrobalan;
Sans.	:	Anilaghnaka, kaliyugalaya, Karshaphala;
Ass.	:	Bauri, Hulluch;
B.	:	Bahera, Buhura;
Bo.	:	Buhudda, Bherda, Yella;
G.	:	Bahedamunjad, Bahedo;
H.	:	Bahera, Bhairach, Sagona;
Kan.	:	Tare Kai, Shanti Kayi;
Kash.	:	Babelo, Balali;
Mal.	:	Tanni, Tusham;
Mar.	:	Beharda, Bhirda, Goting, Vavara;
Oriya	:	Horinda, Jonghihorida;
P.	:	Bahira, Bayrah;
Tam.	:	Akkam, Kalitturuman;
Tel.	:	Karakkaya, Nallakaraka.

Distribution : Throughout the plains, Sub-Himalyan tracts.

Description : Trees. Leaves alternate. Drupe hard, ovovoid, more often pentagonal.

Reproductive Cycle : May-September.

Habitat Ecology : Planted near houses, temples, agricultural fields.

Material Examined : EBH-921, 13-07-2002.

Parts Used : Fruits. Seeds.

Folk Uses : The plant commonly grown as **ornamental** in courtyard of temples. Wood employed for **agricultural tools,** and its gum said to check **constipation**. Fruits **edible,** and important constituent of **'triphala'** (prepared by mixing dried, powdered fruits of *Terminalia chebula, T. bellerica* and *Emblica officinalis* in the ratio of 1:2:3) and prescribed on empty stomach with lukewarm water for **gastro-intestinal** disorders, **liver complaints, bronchitis, piles,** and as **refrigerant, tonic** and **laxative.** Besides, 1-3g powdered fruits taken with honey for curing **cough, bronchitis, epilepsy, stomachache, spleen enlargement, purgative, body tone** and **vitality.** Seeds considered **nutritious,** and its oil also used for **cooking** purposes.

Active Constituents : Fruits afford b-sitosterol, gallic acid, ellagic acid, ethyl gallate, galloyl glucose and chebulegic acid.

Uses in Literature : Known in India as an anthelmintic, antiseptic, aperient, astringent, attenuant, brain tonic, colic, demulcent, diuretic, expectorant, fish-poison, fodder best for milching cows, hallucinogenic, ornamental around templeyards, important constituent of triphala, laxative, narcotic, psychoactive plant, vegetable and writing inks; and for acidity and stomach pain, ageing, agricultural tools, asthma, black hairs, boards, boats, bronchitis, carts, and in coal mines as pit-props, cholera, coffee boxes, cold, constipation, cooling, cooper's work, cough, cramps, diabetes, diarrhoea (vet), dropsy, dugouts, dyeing, dysentery (vet), dysuria, enlargement of spleen, fever, gastric complaints, headache, hoarseness, horses in dysentery, inflammation, inflammatory diseases of eye, intestinal complaints, itching by ticks, leprosy, liver complaint, loose motions, manufacture of molasses, muscular pain, opthalmia, packing cases, piles, ploughs, rafters, respiratory disease, rheumatic, soothing lotion, swellings, skin diseases, slate frames, small crafts, snake bite, soap manufacture, tannin, throat troubles, turning, ulcer (vet), vigour and vitality, vomiting, and weakness (**Ambasta, 1986; Aminuddin & Girach 1993, Borthakur & Gogoi, 1994; Borthakur, 1996b; Chauhan, 1999; Gogoi & Borthakur, 1991; Goud & Pullaiah, 1996b; Jain, 1968, 1991; Jain *et al.*, 1994; Kothari & Moorthy, 1996; Kumar & Pullaiah, 2003; Lalramnghinglova, 1996; Masih, 2003; Mishra *et al.*, 1997; Mohanty & Padhy, 1996; Nazarudeen *et al.*, 1996; Prasad *et al.*, 1996; Pullaiah & Kumar, 1996; Rama Rao & Henry, 1996; Rao, 1996; Sharma & Borthakur, 2003; Sharma & Sood, 1997; Singh, 1991; Siwakoti & Varma, 1996; Tirkey *et al.*, 2003a; Verma *et al.*, 1995; Watt, 1972**).

Major Ayurvedic Formulations : Agnitundi-vati, Aragyavardhini-vati, Bibhitak-taila, Chandraphabha-vati, Goksh-uvadi-guggulu, Herbo-tone, Karanjadi-yoga, Khadirarista, Kutajavaleha, Laghuvisgarva-taila, Lavangadi-vati, Navayas-loaha, Punarnava-mandur, Sanjivani-vati, Sringadi-churna, Sudarshan-churna, Talisadi-churna, Trip-hadi-taila, Triphala-churna, Triphala-kwath, Vidan-gadi-louha.

Trade : Bahera, Belleric Myrobalan.

Terminalia chebula Retz.[+] (Pl. 70B).

Syn.: *T. aruta* Ham., *T. reticulata* Roth., *Myrobalanus chebula* Gaertn., *Embryogonia arborea* Teys & Binn.

Family : Combretaceae.

Vern. : Harad.

English, Sanskrit & Regional Names

Eng. : Chebulic Myrobalan;

Sans. : Harra;

Ass. : Silikha;

B. : Haritaki;

G. : Hardo;

H.	:	Har, Harar, Harra;
Mal.	:	Kadukka;
Mar.	:	Hirda;
Oriya	:	Haridra;
P.	:	Har, Harar;
Tam.	:	Kadukkai;
Tel.	:	Karaka;
U.P.	:	Harr.

Distribution : Throughout India.

Description : Trees. Leaves deciduous. Flowers pubescent, deciduous.

Reproductive Cycle : April-June.

Habitat Ecology : Barren lands, open slopes, along agricultural fields.

Material Examined : EBH-740, 14-06-2000.

Parts Used : Bark. Fruit. Seeds.

Folk Uses : Fruits important constituent of '**triphala**' and seeds and bark that of local wine, '**sur**'. Roasted fruits good for **constipation, indigestion, cough, tuberculosis, kidney disorders** and **flatulence**. Also, paste of fruits made with butter milk by rubbing on bronze plate, considered good for **itching, skin diseases** and as a **cooling wash** to eyes.

Active Constituents : Fruits possess ellagitannin, terchebulin, punicalagin, terflavins A, B, C and D, shilimic, gallic, triacontanoic and palmitic acids, b-sitosterol, daucosterol, triethyl ester of chebulic acid and ethyl ester of gallic acid arjungenin, terminoic acid and arjunolic acid.

Biological Activity : Antibacterial and spasmolytic activities positive.

Uses in Literature : Used in India as an alterative, antidote to poisons, astringent, cardiac tonic, colic, cooling wash to eyes, dentrifice, dye for hairs, gargle in inflammation of mucous membrane of mouth, fish poison, fodder, important constituent of 'triphala', ingredient of 'chywanprash', laxative, ornamental, psychoactive plant, sacred for making writing ink, stomachic, tonic, blood purifier; and also for acidity, ageing, agricultural implements, anaemia, anthrax (vet), apthae, ascites, asthma, bile troubles, bleeding, blood pressure, burn, chronic diarrhoea and ulcer, cold, colouration of houses, conjunctivitis, cough, cuts, diabetes, diarrhoea, dysentery (vet), dysuria, eczema, edible purposes, enlarged spleen and liver, eye diseases, fever, flatulence, gastropathies (vet), haematuria, headache, hiccup, horses in dysentery, house construction, intestinal worms, liver (vet), measles, menstrual complaints, muscular pain, musical instruments, ornamental rosaries, piles, pneumonia, preventing the effects of ageing, rheumatic pain, skin diseases, sore throat, sores, stomach complaint, stomachache (vet), tannin, teeth bleeding and ulceration of gums, urinary diseases, vaginal discharge, vigour and enhances vitality, vomiting, worms and wounds (**Ambasta, 1986; Aminuddin & Girach, 1993; Borthakur, 1996a, b; Borthakur & Gogoi, 1994; Chauhan, 1999; Chowdhery, 1996; Chhetri, 1994; Girach, 1992; Gogoi & Borthakur, 1991; Hosagoudar & Henry, 1996a; Jain, 1968, 1991; Jain *et al.*, 1994; Kumar, 1996; Lalramnghinglova, 1996; Mohanty & Padhy, 1996; Masih, 2003; Mehrotra *et al.*, 1995; Mishra *et al.*, 1997; Rama Rao & Henry, 1996; Saini, 1996a; Sharma & Borthakur, 2003; Sharma & Sood, 1997; Singh, 1991; Siwakoti & Varma, 1996; Srivastava *et al.*, 2003; Tirkey *et al.*, 2003a; Verma *et al.*, 1995; Watt 1972**).

Major Ayurvedic Formulations : Abhaya-lavana, Abhaya-modak, Abhyarista, Agastya-haritaki-rasayana, Amlakyadi-churna, Amrita-haritaki, Arogyavardhini, Avleha, Brahma-rasayana, Chandra-prabha-vati, Chitrak-haritaki, Dantiharitaki, Dasmula-haritaki, Gokshuraadi-gugglu, Haritaki-churna, Karanjaadi-yoga, Medigatone, Panchaskar-churna, Pathyadi-lepa, Sanjivani-vati, Sudarshan-churna,

Triphala, Triphala-churna, Triphaladi-taila, Triphala-ghrit.

Trade : Myrobalan, Chebulic Myrobalan (tree & fruit), Harada, Jangi Harad (for small variety).

Terminalia tomentosa (DC) Wt. & Arn.+ (Pl. 70C).

Syn. : *T. alata* Heyne *ex* Roth., *T. crenulata* W. & A.

Family : Combretaceae.

Vern. : Alsan, Dhamni.

English, Sanskrit & Regional Names

Eng. : Black Murdah;
Sans. : Krishnatvaka, Nissaraphalaka, Shyamasa-Raka;
Ass. : Amari;
B. : Ashan, Piashel;
Bo. : Ain, Marthi, Sadri;
G. : Hadri, Sadada, Sajadio;
H. : Asluna, Kauha, Sein;
Kan. : Sadada;
Mal. : Karimarutu, Temparu;
Mar. : Madat, Sadada, Yeu;
Oriya : Kolashahko, Pondosahajo;
P. : Aisan, Sein;
Tam. : Aruchananam, Karuppumarudu, Puddavem;
Tel. : Inumaddi, Nalamaddi.

Distribution : Sub-Himalayan tracts, Deccan peninsula.

Description : Trees. Leaves sub-opposite, elliptic. Spikes panicled. Flowers dull yellow. Fruit brown, glabrous.

Reproductive Cycle : August-March.

Habitat Ecology : Along countryside roads and agricultural fields.

Material Examined : EBH-700, 21-01-2000.

Parts Used : Leaves. Wood. Bark.

Folk Uses : Wood suitable for **agricultural implements, building works, tool handles** and **charcoal**. 1-3g powdered bark given twice daily for 3-5 days to check **diarrhoea** and for **stomachache**. Leaves considered good for rearing silk worms, and also as **fodder**.

Active Constituents : Chief constituents are b-sitosterol, oleanolic acid, arjunolic acid, maslinic lactone and termonolic (heartwood), arjunic, arjunolic, betulinic and ellagic acids and b-sitosterol (bark).

Uses in Literature : Reported to be used in India for agricultural implements, anaemia, beams, burns, charcoal, cholera, construction of carts, cuts, diarrhoea, doors, dyeing, dysentery, edible purposes (ash of bark), electric casing, engine, brake blocks, fever, fire-proof building, furniture, haematuria, loose motions, musical instruments, oil plywood manufacturing, rafters, rearing silk worms, rough carpentry, sores, stomachache, tannin, tool handles, toys, ulcers, windows, wounds; and as adhesive, antiemetic, cardio-tonic, fish poison, fodder, purgative, rice pounders, stimulant, symbolic (sacred for idols) and tonic (**Ambasta, 1986**; **Jain, 1991**; **Manandhar, 1990**; **Watt, 1972**).

Trade : Alaurel.

Thevetia neriifolia Juss. *ex* Steud.$^{+}$ (Pl. 70D).

Syn. : *T. peruviana* (Pers.) K. Schum, *Cerbera peruviana* Pers., *Cascabela thevetia* (L.) Lipp.

Family : Apocynaceae.

Vern. : Kaner.

English, Sanskrit & Regional Names

Eng. : Bastard Oleander, Exile Oleander, Lucky Nut Tree, Yellow Oleander;

Sans. : Ashvamaraka, Nakharatlia, Sidhapushpa;

B. : China-Karab, Kokil-Phul;

Bo. : Pilakaner, Zardakunel;

G. : Pilakanir;

H. : Pila, Pila Kananir;

Kan.	:	Kadukasi, Kanogalu;
Mal.	:	Pachchaarali;
Mar.	:	Pivalakanhera;
Oriya	:	Konyar-phul;
Tam.	:	Pachaiyalari, Tiruvachippu;
Tel.	:	Pachchaganeru;
U.P.	:	Pila-Kaner.

Distribution : Throughout India.

Description : Shrubs. Leaves linear, lanceolate. Flowers yellow. Drupe angular, poisonous.

Reproductive Cycle : July-September.

Habitat Ecology : Forests, shrubberies.

Material Examined : EBH-961, 14-09-2001.

Parts Used : Flowers. Fruit. Bark. Leaves.

Folk Uses : Fruits **edible**, and its decoction considered as a useful **cardiotonic**. Oil obtained by boiling its bark with mustard oil is used for massaging **painful joints**. Flowers used as **pot herb**, and also its decoction (2-5 ml) used as **anthelmintic**, **purgative**, and for **skin diseases** (twice daily for 3-7 days). Powdered leaves given with honey as a **blood purifier**.

Active Constituents : Main ingredients are thevetolin, querectin-3-galactoside, 3-digalastoside, tamarixetin, theretolin characterised as tamarixetin-3-0-digalautoside (flower), arabinose galactose, glucose, galacturonic acid and xylose (fruits).

Biological Activity : Cardiotonic and isotropic effects and musculotropic activity confirmed.

Uses in Literature : Used in India as an abortifacient, acro-narcotic poison, alexetric, antiperiodic, bitter cathartic, bitter tonic, cardio tonic against pentobarbital induced heart failure, emetic, febrifuge, laxative, ornamental (beads), poison (vet), vegetable; and also useful for carotene diseases, children diseases, criminal poisoning of cattle, dropsy, heart diseases, impotence, insecticide, intermittent fevers, menstrual disorder, rheumatism, skin diseases, snake bite, swelling (vet), swelling of foot, tumours and worms (**Ambasta, 1986**; **Bhandary *et al.*, 1996**; **Chauhan, 1999**; **Huidrom, 1996**; **Jain, 1991**; **Jha *et al.*, 1996**; **Kurian, 1995**; **Mishra *et al.*, 1996a**; **Saini, 1996a**; **Sharma & Sood, 1997**; **Singh & Pandey, 1996**; **Siwakoti & Varma, 1996**; **Watt, 1972**).

Thuja orientalis L.+ (Pl. 71A).

Syn.: *Biota orientalis* (L.) Endl.

Family : Cupressaceae.

Vern. : Sarua, Morpankh.

English & Hindi Names

Eng. : Oriental Arbor-Vitae;

H. : Mayur Pankh, Morpankhi.

Distribution : Cultivated.

Description : Ornamental shrubs. Globose.

Reproductive Cycle : July-October.

Habitat Ecology : Cultivated often met as an escape.

Material Examined : EBH-15, 14-07-2000.

Parts Used : Leaves. Seeds.

Folk Uses : Planted for **ornamental purposes**. Leaves used as substitute for **tea** during shivering, and its poultice or that of seeds useful for healing **open wounds**. Also, mixture of powdered seeds in equal combination with yellow sandal wood, 'triphala', and 'trikuta' [Saunf

(*Foeniculum vulgare*) + maggan (*Piper longum*) + black pepper (*Piper nigrum*)] considered good for **bleeding gums** and **loosening of teeth**. Besides, seeds are prescribed for **paralytic problems** in veterinary.

Active Constituents : These are pimaric and isopimaric acids (leaves), and palmitic (5.28), stearic (7.30), C-18 acids (36.29), C-20 acids (6.10) and linoleic acid (44.6%) (seed oil).

Biological : Quercitin exhibits haemostatic activity in mice. Deoxypodophyllotoxin and ethyl p-methoxy-trans-cinnamate showed cytotoxic activity against HeLa cells.

Uses in Literature : So far known for barrels and casks, dysentery, fencing posts, furniture, house building and tannin; and also as antipyretic, tonic, diuretic, and psychoactive plant (**Ambasta, 1986; Jain *et al.*, 1994; Singh & Prakash, 1996**).

Thunbergia coccinea Wall. (Pl. 71B).

Syn.: *T. pendula* Hassk., *Hexacentris acuminata* Nees, *H. coccinea* Nees, *H. dentata* Nees, *H. acuminata* Nees

Family : Acanthaceae.

Vern. : Ghantion-vbali-bel.

Regional Name

Ass. : Chongalota.

Distribution : U.P.-S.W. China: 300-2100m.

Description : Climbers with pendent branches. Leaves ovate, heart-shaped. Corolla tube orange-pink.

Reproductive Cycle : October-March.

Habitat Ecology : Found growing on shrubs.

Material Examined : EBH-1188, 26-11-2002.

Plant/Parts Used : The Whole Plant.

Folk Uses : Grown as **ornamental**.

Uses in Literature : Used in India as ornamental (**Ambasta, 1986**).

Tinospora cordifolia Miers.+ (Pl. 71C).

Syn. : *T. palminervis* Miers., *Coccculus cordifloius* DC, *C. convolvulaceus* DC, *C. verrucosus* Wall. Cat., *Menispermum cordifloium* Willd.

Family : Menispermaceae.

Vern. : Gloa, Gulja.

English, Sanskrit & Regional Names

- **Eng.** : Gulancha Tinospora;
- **Sans.** : Amritalata, Jwaranashini, Madhuparnika;
- **Ass.** : Amarlata, Siddhilata;
- **B.** : Gadancha, Gulancha;
- **Bo.** : Gharol, Giroli;
- **G.** : Gado, Gulo;
- **H.** : Gilol, Gurach;
- **Kan.** : Amrutaballi;
- **Kash.** : Amrita, Gilo;
- **Mal.** : Amrytu, Peyamrytum;
- **Mar.** : Gharol, Gulavella;
- **Oriya** : Gulancha, Gulochi;
- **P.** : Batindu, Gilogularich;
- **Tam.** : Amridavalli, Kaippuchindil;
- **Tel.** : Duyuntige, Jivantika;
- **U.P.** : Gurch.

Distribution : Throughout Tropical India.

Description : Climber. Leaves cordate, glabrous, acuminate. Flowers yellow. Drupe red, pea-sized.

Reproductive Cycle : June-July.

Habitat Ecology : Common in hedges.

Material Examined : EBH-547, 29-06-2000.

Parts Used : Stem. Leaves.

Folk Uses : Plant one of the important constituents of '**havan**' for performing relegious ceremonies, and as **cattle feed** for enhancing lactation. Its decoction (5-20 ml) considered useful for **piles**, **diabetes**, **leucorrhoea**, **jaundice**, **spermatorrhoea**; and as **aphrodisiac**, **refrigerant**, **febrifuge** and **anthelmintic** (twice daily for 5-10 days); and also its paste useful for healing **cuts** and **wounds**.

Biological Activity : Anti-inflammatory, analgesic, antipyretic, CNS depressant and hypoglycaemic activities found to be postive.

Uses in Literature : So far reported as an analgesic, anthelmintic, antiarthritic, antiperiodic, antipyretic, aphrodisiac, appetiser, bitter tonic, blood purifier, cardiac stimulant, carminative, colic (vet), diuretic (vet), emetic, expectorant, fibre for tying bundles, fodder, ingredient of 'chywanprash', psychoactive plant, sacred, stimulant, tonic (vet), vermifuge (vet); and for acidity of urine, anthrax (vet), asthma, biliousness, bone fracture, bone fracture (vet), bronchitis, chronic gonorrhoea, cough, cough (vet), debility, diarrhoea, diptheria, dysentery, dyspepsia, fever (vet), gastric trouble, general debility, gout, haematuria (vet), headache, impaction (vet), improving the health of a lean and thin horse (vet), indigestion (vet), insect bite (vet), intermittent fevers, jaundice, jaundice (vet), leprosy, leucorrhoea, malaria, mastitis (vet), nocturnal emission, pthisis, piles, pulmonary tubercular, reducing bile production, rheumatism, scorpion sting, skin disease, snake bite, spermatorrhoea, splenic affections, stomach disorder, stomach ulcers, syphilis, ulcer (vet), urinary diseases, vein injuries, venereal complaints and visceral obstruction (**Ambasta, 1986; Chopra *et al.*, 1956; Goud & Pullaiah, 1996b; Jain & Saklani, 1992; Jain, 1968, 1991; Jain *et al.*, 1994; Jha *et al.*, 1991;**

Katewa & Arora, 1997; Lalramnghinglova, 1996; Mahato & Mahato, 1996; Mehrotra *et al.*, 1995; Pandey & Verma, 2003; Prasad *et al.*, 1996; Rama Rao & Henry, 1996; Reddy *et al.*, 1996; Sahoo & Mudgal, 1995; Saini, 1996a; Sensarma, 1991; Sharma, 1996; Singh & Pandey, 1996; Singh, 1991; Siwakoti & Varma, 1996; Srivastava *et al.*, 2003; Verma *et al.*, 1995; Watt, 1972).

Major Ayurvedic Formulations : Amritaastaka-kwaatha, Amritasava, Amrtarista, Amrotottara-kvatha-churna, Ashwagandha-churn, Chandra-prabha-vati, Chinnodbhavadi-kvatha-churna, Giloya-satva, Guduchayadi-churna, Gudu-chi-ksaya-churna, Guduchi-ksaya-taila, Guduchi-taila, Guduchi-tel, Guduchyadi-churna, Guduch-yadi-kwath, Guduci-sattva, Kaishore-guggulu, Rasnapanchak-kwath, Sangivani-vati, Sudarshan-churna, Yograja-guggulu

Trade : 'Giloy' or 'Gulancha' (dried stems).

Trachelospermum lucidum (D. Don) Schuamann+ (Pl. 71D).

Syn.: *T. fragrans* (Wall. *ex* G. Don) Hook. f., *Alstonia lucida* D. Don, *Ichnocarpus fragrans* Wall., *I. inflata* Herb., *Blaberopus lucidus* A. DC

Family : Apocynaceae.

Vern. : Naraun.

Regional Names

Ass. : Akhahilata;

U.P. : Dudhi.

Distribution : Sub-Tropical & Temperate Himalaya: 600-2100m.

Description : Climbing, glabrous shrubs with pedunculous branches. Leaves shortly-stalked, ovate-lanceolate and long-pointed. Flowers white, fragrant.

Reproductive Cycle : May-September.

Habitat Ecology : Climber in forests and shrubberies.

Material Examined : EBH-802, 27-05-2001.

Parts Used : Leaves. Inflorescence.

Folk Uses : Inflorescence constituent of local wine, '**sur**'. Leaves used as **fodder**, and its poultice useful against **swellings** and **rheumatism**, and its decoction given to animals for checking **constipation**.

Active Constituents : Leaves yield b-amyrin, b-sitosterol its glucoside ursolic acid, quercetin, hyperin, quercetin and astilbin.

Uses in Literature : Reported to be used as an anthelmintic, bitter tonic, depressant on the CNS, febrifuge, galactagogue, haemostatic; and for asthma, beri-beri, cardiac troubles, chronic diarrhoea, congested liver, malarial fever (vet), sores, rheumatic swellings, tumours and ulcers (**Ambasta, 1986; Jain, 1991**).

Trichosanthes tricuspidata Lour.+ (Pl. 72A).

Syn.: *T. palmata* Roxb., *T. bracteata* (Lour.) Cogn., *T. lacionosa* Wall., *T. aspera* Heyne., *T. tricuspis* Miq., *T. cordata* Wall., *T. anguina* Wall., *T. lepiniana* (Naudin) Cogn, *Cucurbita melopepo* Wall., *Involucraria wallichii* Seringe, *Bryonia palmata* Wall.

Family : Cucurbitaceae.

Vern. : Bim Bel.

Sanskrit & Regional Names

Sans. : Mahakala, Svetapushpi;

B. : Makal;

Bo. : Kaundal;

G. : Ratanindravanam;

H. : Indrayan, Mahakal;

Kan. : Avaguda-Hannu;

Mal. : Kakatonti;

Mar. : Kavandala, Mukal;

Tam. : Ankorattai, Shavaripalam;

Tel. : Abuba, Avaguda, Avagudapanda.

Distribution : W. Nepal-Bhutan, India, China. S.E. Asia; 600 -2100m.

Description : A climbing undershrub with branched tendrils. Leaves 3-7 lobed, toothed. Flowers white. Fruit smooth, globose, red stripped with orange.

Reproductive Cycle : May-October.

Habitat Ecology : Climber on shrubberies; common.

Material Examined : EBH-589, 23-05-2000.

Parts Used : Seed. Bark. Fruit.

Folk Uses : Fruits **edible**, and said to be useful in **asthma**, **constipation**, **stomachache** and **enlargement of spleen**. Poultice of its powdered seeds helpful in **suppuration of boils**, and also its powder mixed with black pepper, given to check **dysentery** in animals. Also, roots are fed to animals for **inflammation of lungs** and **cough**. 1-3g powdered bark given once daily for 2-5 days for **gastric disorders**.

Uses in Literature : Used in India for acidity (vet), arthritis, asthma, burns, carbuncles, cough (vet), diarrhoea, dysentery (vet), ear infections, edible, fever, fever (vet), gastric (vet), giddiness, gonorrhoea, headache, inflammation of lungs (vet), mouth (vet), pneumonia, regulates fertilty (vet), rheumatism, snake bite, sores in eczema, stomach (vet), throwing out poison, dysentery (vet), destroying crows, vomiting, worshipping; and as purgative and tonic (**Ambasta, 1986; Bhandary *et al.*, 1996; Jain, 1991; Kapur & Nanda, 1996; Kothari & Moorthy, 1996; Rama Rao & Henry, 1996; Satapathy & Brahmam, 1996; Watt 1972**).

Tropaeolum mazus L. (Pl. 72B).

Family : Tropaeolaceae.

Vern. : Tindu Phool.

English Name

Eng. : Garden Nastrutium.

Distribution : Commonly cultivated in Gardens.

Description : Tuberous rooted, climber. Leaves small, divided into 5-narrow, pointed lobes.

Reproductive Cycle : May-November.

Habitat Ecology : Hilly slopes.

Material Examined : EBH-457, 24-10-2000.

Parts Used : Flowers.

Folk Uses : Flowers used for **offerings**.

Ulmus laevigata Royle[+] (Pl. 72C).

Syn. : *U. villosa* Brandis *ex* Gamble Man., *U. erosa* Wall., *U. pedunculata* Herb., *U. wallichiana* Planch.

Family : Ulmaceae.

Vern. : Marinu.

Hindi Name

H. : Bhamri.

Distribution : Afghanistan-West Nepal: 1800-3200m.

Description : Trees. Leaves smooth, slightly oblique at the base. Young fruit villous.

Reproductive Cycle : May-August.

Habitat Ecology : Roadsides.

Material Examined : EBH-149, 11-05-2002.

Parts Used : Wood. Branches. Leaves.

Folk Uses : Young branches and leaves lopped for **fodder**. Wood used for **construction works**, and **furniture**.

Uses in Literature : Reported to be used for ark poles, bed-strings, bone fractures (vet), endeye, panels for dog-carts, sandals, furniture, and as fodder for sheep and goats (**Ambasta, 1986; Watt, 1972**).

Viburnum cotonifolium Don+ (Pl. 72D).

Syn.: *V. polycarpum* Wall.

Family : Caprifoliaceae.

Vern. : Ricchnabu.

Regional Names

Kash. : Guch, Kilmich, Richabi;

P. : Bankunch, Bathor, Guch, Kilmach, Richh, Richanabi, Uklu.

Distribution : Afghanistan-C. Nepal, Bhutan; 1800-3600m.

Description : Shrubs. Leavse crenate, orbicular. Flowers white, tinged with pink. Drupe red turning black when ripe.

Reproductive Cycle : April-September.

Habitat Ecology : Forests, shrubberies.

Material Examined : EBH-442, 12-05-2000.

Parts Used : Fruit. Branches.

Folk Uses : Ripe fruits **edible**, and considered to be an **appetiser**. Dried branches used as **fuelwood**.

Active Constituents : Ursolic acid and uraol isolated from aerial parts.

Uses in Literature : So far known in India for edible fruits, haemorrhage, menorrhagia and meterrhagia (**Ambasta, 1986; Jain, 1991; Kapur, 1996, Negi & Gaur, 1991; Watt, 1972**).

Viburnum stellulatnm Wall.[+] (Pl. 73A).

Syn.: *V. mullaha* Buch.-Ham. *ex* D. Don

Family : Caprifoliaceae.
Vern. : Dhaab.

Distribution : Temperate Himalaya: 1500-3000m.

Description : Shrubs with ovate-lanceolate, sharpely toothed leaves, having upper surface glabrous and lower ones stellately hairy. Flowers white. Drupe bright red.

Reproductive Cycle : May-July.

Habitat Ecology : Forests, shrubberies.

Material Examined : EBH-1212, 18-07-2003.

Parts Used : Leaves. Fruits. Branches.

Folk Uses : Branches suitable for making **walking sticks**. Fruits **edible** and consumed as **tonic**. Leaves lopped for **fodder**.

Uses in Literature : So far known in India for edible fruits, walking sticks; and as uterine tonic (vet) (**Ambasta, 1986; Watt, 1972**).

Vigna unguiculata L. (Pl. 73B).

Syn.: *V. catiang* (Burm. f.) Walp., *V. sinense* (L.) Walp., *V. sinense* Endl., *V. cylindrica* (L.) Skeels, *V. sesquipedalis* Fruhw., *Dolichos tranquebarius* Jacq., *D. monachalis* Brot., *D. cajang* Burm. f., *D. sinensis* L., *Phaseolus cylindricus* L.,

Family : Fabaceae.

Vern. : Rajmah.

English, Sanskrit & Regional Names

Eng. : Black Eye Pea, Catiang Pea, Cow Pea, Yard Long Pea;

Sans. : Chapala, Dirghabijam, Dirghashimbi, Dvijasapta, Marutkara, Mahamasha, Kshudhabhijanaka, Nilamasha, Nripamasha, Nripochita, Rajamasha, Rishapara, Sitamasha, Varvata;

Ass. : Mahorpat, Urohi, Urhi-Mah;

B. : Barbati, Chim, Ramhikolai, Shim;

Bo. : Chola, Chowli, Gatval, Hurrea, Lobeh, Safed Lobeh;

G. : Chola, Chora;

H. : Bora, Chowli, Lobia, Rianish, Souta;

Kan. : Tadagunny;

M. : Alsando, Vervil;

Mal. : Alsaendi, Kottapayuru, Vellapayuru:

Mar. : Chaoli;

P. : Chota, Harwanh, Lobiya, Rawan, Rawangi, Souta;

Tam. : Caramunnipayari;

Tel. : Alusindi, Boberlu, Bobra, Duntuppesalu.

Distribution : Widely cultivated.

Description : Twinning herbs. Leaflets 3. Flowers yellowish-reddish. Pod linear, glabrous, pale at maturity.

Reproductive Cycle : June-October.

Habitat Ecology : Cultivated.

Material Examined : EBH-1017, 17-10-2002.

Parts Used : Pods. Seeds.

Folk Uses : Seeds consumed as '**dal**', and pods as **feed** for live-stock.

Active Constituents : Seeds contain trypsin inhibitors, A2-a, A2-b, A2-c, A2-d, A2-e and phytohemmagglutin.

Biological Activity : Diuretic activity confirmed.

Uses in Literature : Used in India as food, antibilious, diuretic, feed for livestock, green manure, thickening agent and for dyeing, hepatic troubles, smallpox, strengthening stomach and in textile-sizing (**Ambasta, 1986**; **Jain, 1991**; **Watt, 1972**).

Major Ayurvedic Formulations : Dhanvantra-taila, Saptasara-kvatha and -churna.

Trade : 'Rajmah'.

Vinca rosea L.⁺ (Pl. 73C).

Syn.: *Catharanthus roseus* G. Don, *Lochnera rosea* (L.) Reichb.

Family : Apocynaceae.

Vern. : Sadabahar.

English & Regional Names

Eng. : The Periwinkle;
B. : Nayantara;
H. : Sadabahar;
Mal. : Ushamalari;
Mar. : Sad-Phul;
Oriya : Ainskati;
Tel. : Billaganneru.

Distribution : Cultivated throughout India, also met as an escape.

Description : An undershrub. Leaves obovate. Flowers axillary, pink.

Reproductive Cycle : Throughout the year.

Habitat Ecology : Grown in gardens.

Material Examined : EBH-980, 23-06-2001.

Parts Used : Flowers. Leaves.

Folk Uses : Plant grown for **ornamental** purposes, and its 1-3g powder given with honey for 15-40 days, twice daily, to cure **cancer**. Poultice of leaves good for **boils, insect bites** and **swellings.**

Active Constituents : Active constituents of flowers are vinblastine, coronaridine, vindorosine, vindoline, catharanthine, mitraphlline, ajmalicine, tetrahydroalstonnie, 11-methoxyytabersonineb and reserpine.

Biological Activity : Antitumorous (catharanthamine), anticancerous (vinblastine-*i.p.* or *i.v.*), (yohimbine-1mg/kg of rats), cytotoxic, hypotensive and hypoglycaemic activities (aq. extr. of fruits and seeds) positive.

Uses in Literature : Known in India to cure anthrax (vet), boils, cold (vet), cough (vet), cut (vet), diabetes, diarrhoea, fever (vet), high blood pressure, madness (vet), mastitis (vet), menorrhagia, paralysis (vet), rabies (vet), swelling of body (vet), toothache, wasp stings, wounds; and as antileukemic, anticancerous, haemostat (vet), hypotensive, sedative, mosquito repellent (vet), purgative, stomachic (**Ambasta, 1986; Hosagoudar & Henry, 1996c; Jain, 1991; Kurian, 1995; Lalramnghinglova, 1996; Prasad** ***et al.***, **1996; Sharma & Sood, 1997; Siwakoti & Varma, 1996; Watt 1972**).

Vitex negundo L.[+] (Pl. 73D).

Syn.: *V. bicolor* Willd., *V. arborea* Desf., *V. panniculata* Lamk., *V. incisa* Lamk.

Family : Verbenaceae.

Vern. : Bana, Suro.

English, Sanskrit & Regional Names

Eng. : Common Chaste Tree, Indian Privet, Nigandi;
Sans. : Indrani, Nilapushpa, Shepali, Svetasuraso;
Ass. : Aggla-Chita, Pasutia;
B. : Nishinda, Samalu;
Bo. : Katri, Shiwari;
G. : Nagaol, Nirgari;
H. : Mewri, Nirgunda, Sinduari;
Kan. : Lakkigida, Nakkilu, Nekki;

M.P.	:	Nagad;
Mal.	:	Vellanochi;
Mar.	:	Nirgur;
Oriya	:	Begundia, Indrani;
P.	:	Bankahu, Moraun;
Tam.	:	Siduvavam, Tiributamerittan;
Tel.	:	Nellavavilli, Sinduvaramu;
U.P.	:	Tarvan, Kali-Nirgundi.

Distribution : Afghanistan-Bhutan;.India; China; S.E. Asia upto 2000m.

Description : Shrubs. Leaves digitate, 3-5 foliate. Flowers blue purple. Fruit black, ovoid, succulent.

Reproductive Cycle : March-August.

Habitat Ecology : Cultivated areas, wastelands; very common.

Material Examined : EBH-974, 02-08-2002.

Parts Uses : Root. Leaves. Stem.

Folk Uses : Stem prescribed for **scouring teeth**, and also chewed for curing **mouth sores**. Poultice of its heated leaves employed for **swelling of joints** by acute rheumatism, and for healing **internal wounds**, and also as an antidote to **snake bites**. Besides, leaves and young branches kept in storage containers for protecting grains from **infestation** by pests. In veterinary, crushed leaves mixed with wheat flour and a pinch of rock salt, given for **indigestion** and **enlarged spleen**. 10 ml decoction of roots, given twice a day for 15 days, considered **aphrodisiac** and for **body tone** and **vitality**.

Active Constituents : These are two new leueoanthoeyamidins determined as 6,8-di-0-methyllencodelphinidin and 3′,4-di-0-methyelencocyanidin-7-0-rhammogluco-side (stembark) and nishindaside, negrindoside, hydrocotylene, glucon561itol (essential oil).

Biological Activity : Extr. of leaves exhibits anticancerous activity against *Ehrlich ascites* tumour cells.

Uses in Literature : So far reported in India as an abortifacient, alterative, anasarca, anodyne, athelmintic, antifertility in man, antiparasitic, appetizer, astringent, attenuant, aromatic, colic, haemostatic, demulcent in dyesentery and piles, diaphoretic, discutient, diuretic, expectorant, febrifugal, fuelwood, insect reppellent, nervine tonic, nocturnal emission, psychoactive plant, sterilizer, and symbolic to keep evil spirits away; and for basketry, bladder complaints, blisters, boils, bone fracture, cardiac, catarrhal fever, cephalic, checking pain, cholera, chronic liver problem associated with loss of appetite, cold, conjuctivitis, cooling effect, cough, diarrhoea, dropsy, dullness of hearing, dyeing, dysentery and piles, dyspepsia, eczema, emmenagogue, epilepsy, eye complaints, gastric troubles, gout, gum troubles, haemorrhage of the digestive system, headache, resolving obstructions in kidney and spleen, indigestion, insect bites, intermittent fevers, irritable fevers, irritable bladder, itching, leprosy, lung diseases, making tooth sticks, mental disease, paralysis, paralysis of animals, pneumonia, post natal, reducing sexual desire and muscular pain of the body, rheumatoid arthritis, scabies, sciatica, scrofulous ulcers, sinuses, sores, stored grain pests, syphilis, testes disease, restoring fertility, tonsillitis, toothache, ulcers, urinary troubles, waist pain, warm bath to ladies after delivery, and wounds (**Ambasta, 1986; Aminuddin & Girach, 1996; Bhandary *et al.*, 1996; Borthakur & Sarma, 1996; Geetha *et al.*, 1996a, Girach & Aminuddin, 1995; Gogoi *et al.*, 2003; Hembrom, 1996; Jain, 1991; Jain & Sharma, 1996; Jain *et al.*, 1994; Katewa & Arora, 1997; Khanna *et al.*, 1996a; Kothari & Moorthy, 1996; Kulkarni & Kumbhojkar, 1996; Kurian, 1995; Lal *et al.*, 1996; Mahato & Mahato, 1996; Mishra *et al.*, 1997; Painuli and Maheswari, 1996; Ragupathy & Mahadevan, 1991; Rama Rao & Henry, 1996; Saini, 1996a; Satapathy & Brahmam, 1996; Sharma, 1996; Sharma & Sood, 1997; Singh, 1991; Singh *et al.*, 1996b;**

Siwakoti & Varma, 1996; Srivastava *et al.*, 2003; Watt, 1972).

Major Ayurvedic Formulations : Himocolin cream, Nirgundi-ghrit, Nirgundi-kalpa, Nirgundi-kwath, Nirgundi-taila, Pilex, Rumalaya (cream and tablets), Vishgarbh-tel, Vrana-sodhana-taila.

Trade : Nirgundi-phal, Samhaalu-phal.

Vitis himalayana Brand.* (Pl. 74A).

Syn.: *V. neilgherrensis* Wight, *Amplelopsis neilgherrensis* Wight, *A. himalayana* Royle, *Parthenocissus himalayana* Brandis, *Parthenocissus semicordata* Wall. var. Roylei (King) Raiz. & Saxe., *P.himalayana* Walp.

Family : Vitaceae

Vern. : Amru-Ri-Bel.

Punjabi Names

P. : Phlankur, Zemaro.

Distribution : W. Himalaya upto 3400m.

Description : A climber with digitately compound leaves. Leaflets three, long-pointed, terminal leaflet large and lateral ones unequal-sided. Flowers yellow-green, numerous. Fruits, black, pea sized.

Reproductive Cycle : April-November.

Habitat Ecology : A common climber in oak forests.

Material Examined : EBH-1008, 24-07-2003.

Parts Used : Leaves. Fruit.

Folk Uses : Fruit **edible**, and leaves used as **fodder**.

Vitis jacquemontii Parker* (Pl. 74B).

Syn. : *Ampelocissus jacquemontii* (Parker) Raizada

Family : Vitaceae.
Vern. : Jangli Dakh.

Distribution : Pakistan-E. Nepal: 1000-1800m.

Description : Woody climbers. Leaves large, ovate heart-shaped. Flowers greenish-yellow. Fruit globular.

Reproductive Cycle : April-May.

Habitat Ecology : Shrubberies, steep slopes.

Material Examined : EBH-239, 13-04-2000.

Parts Used : Fruit. Seeds.

Folk Uses : Decoction of seeds useful in **constipation, headache**, and **cold**, and fruits considered **carminative** and also good for **anaemia**.

Vitis rependa Wt. & Arn.* (Pl. 74C).

Syn.: *V. laeta* Wall., *V. rosea* Royle, *Cissus acuminata* Thwaites, *C. aquosa* Wall., *C. repanda* Vahl, *C. repens* Thwaites, *C. riparia* Wall., *C. vitiginea* Linn.

Family : Vitaceae.
Vern. : Maraingnu.
Regional Names

Ass. : Medmedia-Lop;
H. : Pani-Bel;

Distribution : N. W. Himalaya.

Description : A climber having stem and leaves apressed with woolly tomentum. Leaves broadly-cordate, coarsely crenate. Fruit pyriform.

Reproductive Cycle : April-June.

Habitat Ecology : Commonly growing on trees.

Material Examined : EBH-670, 16-05-2000.

Parts Used : Fruit. Stem.

Folk Uses : Fruits **edible**, and also used as **cattle feed.** Extr. of fruits considered **refrigerant**. Watery exudate oozed out from cut pieces of climber, given as **diuretic**.

Wendlandia heynei (Roem & Schult.) Sant. & Merch.⁺ (Pl. 74D).

Syn.: *W.* exserta DC, *R. cinerea* Wall., *Rondeletia exserta* Roxb., *R. cinerea* Wall., *R. thyrsiflora* Roth.,*R. orissensis* Roth., *R. heynei* Roem & Sch.

Family : Rubiaceae.
Vern. : Pansrai.
Regional Names

- **Ass.** : Kadam;
- **B.** : Mimri, Tilki;
- **H.** : Chaulai, Chila, Chilkiya, Tila;
- **Mar.** : Padgira;
- **P.** : Chilkaya, Pansara, Pudhara.

Distribution : Tropical Himalaya, ascending to 1350m.

Description : A small tree. Leaves opposite, stalked, ovate-lanceolate. Flowers white, numerous, fragrant. Capsule white, tomentose.

Reproductive Cycle : April-June.

Habitat Ecology : Dry forests, wasteplaces, edges of cultivated areas, slopes.

Material Examined : EBH-678, 14-04-2000.

Parts Used : Stem. Wood.

Folk Uses : Wood suitable for **agricultural implements**, and **building works**, and stem for **scouring teeth**. Leaves used as **cattle feed**.

Uses in Literature : Reported to be used for agricultural implements, building work, house posts, skin diseases, snake bite, urinary complaints, and as cattle fodder. Also, used for curing skin diseases in veterinary (**Ambasta, 1986**; **Jain, 1991**; **Watt, 1972**).

Woodfordia floribunda Salisb.+ (Pl. 75A).

Syn.: *W. fruticulosa* (L.) Kurz, *W. tomentosa* Bedd., *Lythrum fruticosum* L., *Grislea tomentosa* Roxb., *G. uniflora* Rich., *G. punctuata* Ham., *G. micropetala* Hocsht *ex* Steud.

Family : Lythraceae.

Vern. : Dhawain, Dhaun.

English, Sanskrit & Regional Names

Eng. : Fire Flame Bush, Red Bell Bush, Shirantitea, Woodfordia;
Sans. : Bahupushpika, Rodhrapushpini, Reyya-Manu;
Ass. : Eda, Era;
B. : Dawa, Dhadki, Dhawanji;
Bo. : Dhavri, Dowari;
G. : Dhawadina;
H. : Dhai, Dhaura, Thawi;
Kan. : Tamrapushpi, Dhataaki Huvvu;
Kash. : Thawi;
Mal. : Taativipu, Tatire;
Mar. : Dhaiphal, Phulsatti;
Oriya : Dhatubiphala, Horibadi;
P. : Dawi, Khurd;
Tam. : Thatarijargi, Velakkai;
Tel. : Gaddapisniki, Reyyamanu;
U.P. : Phoolchhari.

Distribution : Pakistan-S.W. China; India; Burma; W. Asia; Africa: 300-1800m.

Description : Shrubs. Leaves opposite, lanceolate, pointed. Flowers red. Capsule ovate.

Reproductive Cycle : March-June.

Habitat Ecology : Forests, open slopes; common.

Material Examined : EBH-696, 12-04-2000.

Parts Used : Flowers. Leaves. Wood.

Folk Uses : Plant grown for **ornamental** purposes. Flowers constituent of local wine, '**sur**', and its decoction (5-10 ml) considered useful in **leucorrhoea**, **dysentery**, **cough**, **cold** and **sexual weakness**. Also, flowers considered good for **bee keeping**. Poultice of leaves in combination with those of *Bauhinia variegata* useful against **vesicular eruptions**. Bark considered excellent for **tanning**.

Active Constituents : Various constituents are ellagic acid, polystachoside, myricetin-3-galactoside and pelar gonidin-3,5-dighnoside (leaves & flowers), cyaniding-3, 5-diglucoside, deta coganol, b-sitosterol, chrysophanol-8-D- b-D-glucopyranoside, quercetin and tannins (20%) (flowers), lupeol, (0.023), ursolic acid (0.22), b-sitosterol (0.008), betulin (0.004), betulinic acid (0.001%), and oleanolic acid (leaves).

Biological Activity : Extr. of flowers exhibits significant abortifacient activity in mice.

Uses in Literature : Reported to be used in India as an astringent, cool drink, dyeing and tanning material, fuelwood, ornamental mosquito and insect repellent, stimulant, styptic, supportive, tonic, vegetable, vermillion; and for axe handles, bleeding in pregnancy, body pains, bone fracture, bowel complaints, burns, cancer, carbuncle, cholera, complaints of liver, cough, diarrhoea, dropsy, dysentery, edible purposes, expelling worms from the body, fever, fermentations for most of 'aristas' and 'asvas' of ayurveda, foul ulcers, haemorrhages, haemorrhoids, injuries, leprosy, leucorrhoea, loss of appetite in pregnancy, mucous disorders, muscular pain, nausea, night blindness, pain, pimples, rheumatism, safe stimulant in pregnancy, sores, seminal weakness, skin diseases, smallpox, spleen complaints, sprain, tanning, vesicular eruptions and to toddy for enhancing taste and intoxication. Also, known in veterinary for rheumatsim, sore and wounds (**Ambasta, 1986; Aminuddin & Girach, 1996; Borthakur, 1993; Jain, 1991; Kothari & Moorthy, 1996; Kulkarni & Kumbhojkar, 1996; Kurian, 1995; Masih, 2003; Rama Rao & Henry, 1996; Saini, 1996a; Shah & Singh, 1990; Sharma & Sood, 1997; Singh, 1991; Siwakoti & Varma, 1996; Watt, 1972**).

Major Ayurvedic Formulations : Abhyayarista, Arvindasava, Asokarista, Brhat-gangadhara-churna, Dhatakyadi-churna, Dhata-kyadi-taila, Gangadhar-churna, Kanakasava, Kuta-jarista, Laghu-gangadhar-churna, Pippalayasava, Pushayanug-churn.

Trade : 'Dhaataki phul', 'Dhai phool'.

Xylosma longifolium Clos.[+] (Pl. 75B).

Family : Flacourtiaceae.

Vern. : Draindu.

Regional Names

Ass. : Mota Koli, Kata Holi,. Katahal;

P. : Chopra, Chiund, Thakola;

Distribution : Outer ranges of W. Himalaya upto 1650m.

Description : Spinous trees. Leaves thick, oblong-lanceolate, long-pointed. Flowers yellow. Berries black glabrous.

Reproductive Cycle : February-May.

Habitat Ecology : Shrubberies, forests; common.

Material Examined : EBH-893, 26-03-2002.

Parts Used : Wood. Roots.

Folk Uses : Wood used as **fuel** and good for making **charcoal**. Twigs used as **fodder**, and a dose of 2-3g mixture comprising 500g of its roots, 25g 'ashwagandha' (*Withania somnifera* Dunal) and 25g 'soa' (*Anethum sowa* Roxb.) considered useful for **body tone** and **vitality**.

Uses in Literature : Known in India for fencing, house building, opium like effect, stomachache; and as edible and fuel (**Ambasta, 1986; Jain, 1991; Watt, 1972**).

Zanthoxylum armatum DC+ (Pl. 75C).

Syn. : *Z. alatum* Roxb., *Z.planispinum* Sieb & Zucc., *Z.hostile* Wall.

Family : Rutaceae.

Vern. : Tirmir, Tumbar.

Sanskrit & Regional Names

- **Sans.** : Mahamuni, Sauravanaja, Tumburu,
- **B.** : Gaira, Nepali-Dhamia;
- **G.** : Tejabala, Tejbal;
- **H.** : Tejbal, Timur,
- **Kan.** : Jimmi, Tejovanti, Tejpatri, Tumbura, Tumburudra;
- **Mal.** : Thumboonal, Thumbooni, Valiyavaluzhavam;
- **Mar.** : Tejabal;
- **Oriya** : Tundapoda,
- **P.** : Tezbal, Timmal
- **Tam.** : Thejyovathi;
- **Tel.** : Tejovathi;

Distribution : Kashmir-S.W.China; S.E.Asia: 1100-2500m.

Description : Shrubs having stems and branches armed with long prickles. Leaves imparipinnate. Flowers yellow. Fruit pale-red. Drupe globose.

Reproductive Cycle : April-June.

Habitat Ecology : Shrubberies, cultivated areas.

Material Examined : EBH-218, 14-04-2000.

Parts Used : Roots. Stem. Branches. Fruits. Seeds. Bark.

Folk Uses : Wood suitable for building works and branches for making **walking sticks** and **pestles**. Stem used as **toothbrush**, and also these are hung on main door of the house **to ward off the evil spirits**. Besides, 'chutney' prepared from its young twigs considered **nutritious** and **carminative**. Fruits **edible**, and used for **flavouring curries;** and its powder (250g) in combination with black pepper (25g), and alum (25g) useful against **bleeding gums** and **toothache**. Roots form one of the constituents of local wine, '**sur**'. Poultice of crushed seeds checks **headache**, and that of bark useful for **suppuration of boils**. Also, a gargle of bark decoction said to cure **mouth ulcers**.

Active Constituents : Spilantho (2,6,8-decatrienoic isobutylamide) (roots); linalool, limonene, methyl cinnamate, lineole, trans-cinnamic acid, nevadensin, umbelliferone, b-sitosterol & its glucoside (fruits), b-sitosterol, pinoresionol dimethyl ether and a mixture of alkaloids (aerial parts), linalool (58.30), limonene (24.46), alloaroma dendrene (8.98), methyl linnamate (8.92) myrcene (3.55), a-thujene (1.65), a-fenchol (0.68), p-cymene (0.65), b-caryophyllane (0.50), a-terpineol (0.31), camphor (0.25), 1,8-lineole (0.25), carvone (0.20), tagetonol (0.16), lis-linnamic (0.12) and g-terpinene (0.08%) (fruit oil).

Biological Activity : Stem and bark exhibit hypoglycaemic and anthelmintic activities, respectively.

Uses in Literature : Used in India as an anthelmintic, antiseptic, aromatic, astringent, carminative, condiment, deodorant, disinfectant, febrifuge, flavouring agent in paan, stimulant, stomachic, tonic in fever, and vegetable; and also for boils, cholera, clubs, cough, dental diseases, diarrhoea, dyspepsia, eczema, fencing, gum troubles, hair lotions, indigestion, intoxicating fish, itching, leucoderma, mouth blisters, serving furs, paralysis, pestles, piles, purification of water, religious ceremonies, rheumatism, round worms, scabies, smallpox, snake bite, soap making, warding off house flies, and wood for house building (**Ambasta, 1986; Ansari, 1991; Borthakur & Sarma, 1996; Chauhan, 1999; Chowdhery, 1996; Jain, 1991; Kapur & Singh, 1996; Kirtikar & Basu, 1935; Lal *et al.*, 1996; Manandhar, 1996; Mao, 1993; Negi & Gaur, 1991; Rana *et al.*, 1996; Ravi Shankar & Henry, 1992; Singh, 1991; Siwakoti & Varma 1996; Watt, 1972**).

Major Ayurvedic Formulations : Brahmi-vati, Dantamajan, Hingodi-taila, Kalaka-churna (Lepa), Madhusnuhi-rasayan, Pancatikta-guggulu-ghrita, Tumbarwadi-churna.

Trade : Nepali 'dhaniyaa', 'timroo' or 'tejbal' (dried fruits), Tumbru.

Ziziphus jujuba Mill.+ (Pl. 75D).

Syn. : Z. *mauritiana* Herb., Z, *soraria* Schult., Z. *trinervis* Roth., Z. *sativa* Gaertn DC, Z. *vulgaris* Lamk., *Rhamnus jujuba* L.

Family : Rhamnaceae.

Vern. : Ber.

English, Sanskrit & Regional Names

Eng.	:	Chinese Date, Indian Cherry, Indian Jujube, Indian Plum;
Sans.	:	Dridhabija, Prithukoli, Svaccha;
B.	:	Bogri;
Bo.	:	Bhurmi, Bordi;
G.	:	Ber, Boreli, Boyedi;
H.	:	Baer, Ber, Beri;
Kan.	:	Elanji, Yalachi;
Mal.	:	Badaram, Badari, Perintulali;

Mar.	:	Baher, Bor;
Oriya	:	Barkoli, Bodokoli;
P.	:	Beri, Unab;
Tam.	:	Iratti, Kulvali, Sivagam;
Tel.	:	Gangaregu, Karkhanduvu.

Distribution : Throughout India.

Description : Shrubs. Flowers in axillary clusters. Drupe ellipsoid, red, shining.

Reproductive Cycle : July-August.

Habitat Ecology : Wastelands, rocky slopes.

Material Examined : EBH-597, 17-07-2001.

Parts Used : Leaves. Fruit. Roots.

Folk Uses : Ripe fruits **edible**, and its powder (2-3g) given with lukewarm water (twice a day for 3-5 days) for **opthalmia**, **dysentery**, **headache** and as **refrigerant**. Leaves used as **cattle feed**, and its poultice good for **skin** and **eye ailments** and **wounds**.

Uses in Literature : So far known in India as a blood purifier, colic, refrigerant, psychoactive plant, stomachic and sweet drink and for blood dysentery, cholera, chronic diarrhoea, edible fruits, eye ailments, falling hair, fever, headache, oedema, promoting conception, rheumatism, scorpion bites, sores, spleen complaints, stomach disorder, symbolic worshipping, ulcers, whooping cough and wounds (**Gogoi & Borthakur, 1991**; **Jain, 1991**; **Jain** ***et al.*****, 1994**; **Lalramghonglova, 1996**; **Mohanty & Padhy, 1996**; **Ragupathy & Mahedevan, 1991**; **Rama Rao & Henry, 1996**; **Sahu, 1996**; **Shome** ***et al.*****, 1996**; **Shukla & Verma, 1996**; **Sikarwar & Kaushik, 1992**; **Singh, 1996b**; **Srivastava** ***et al.*****,** 2003).

Major Ayurvedic Formulations : Madhusdndi-Kshay.

□□□

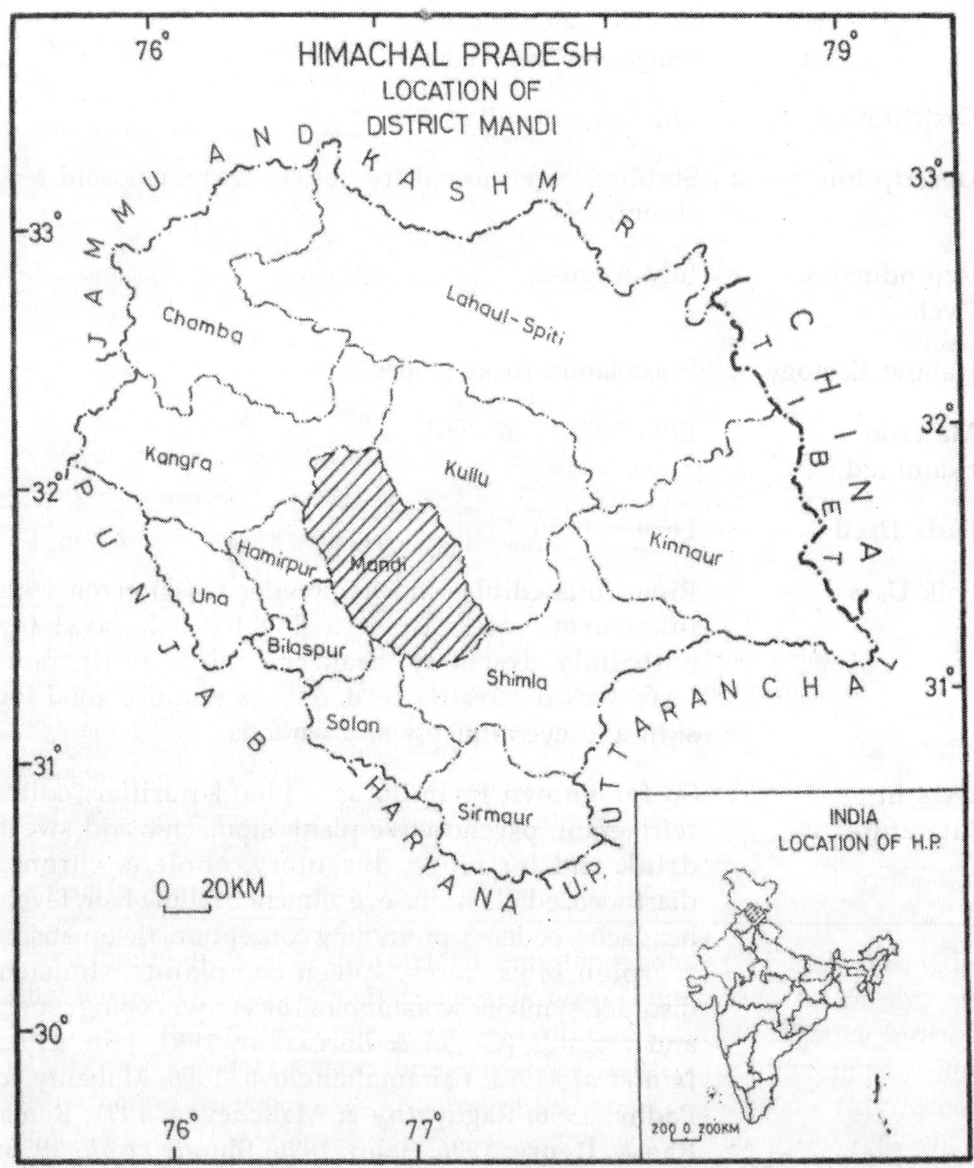

Fig. 1 : Map of Himachal Pradesh showing location of District Mandi.

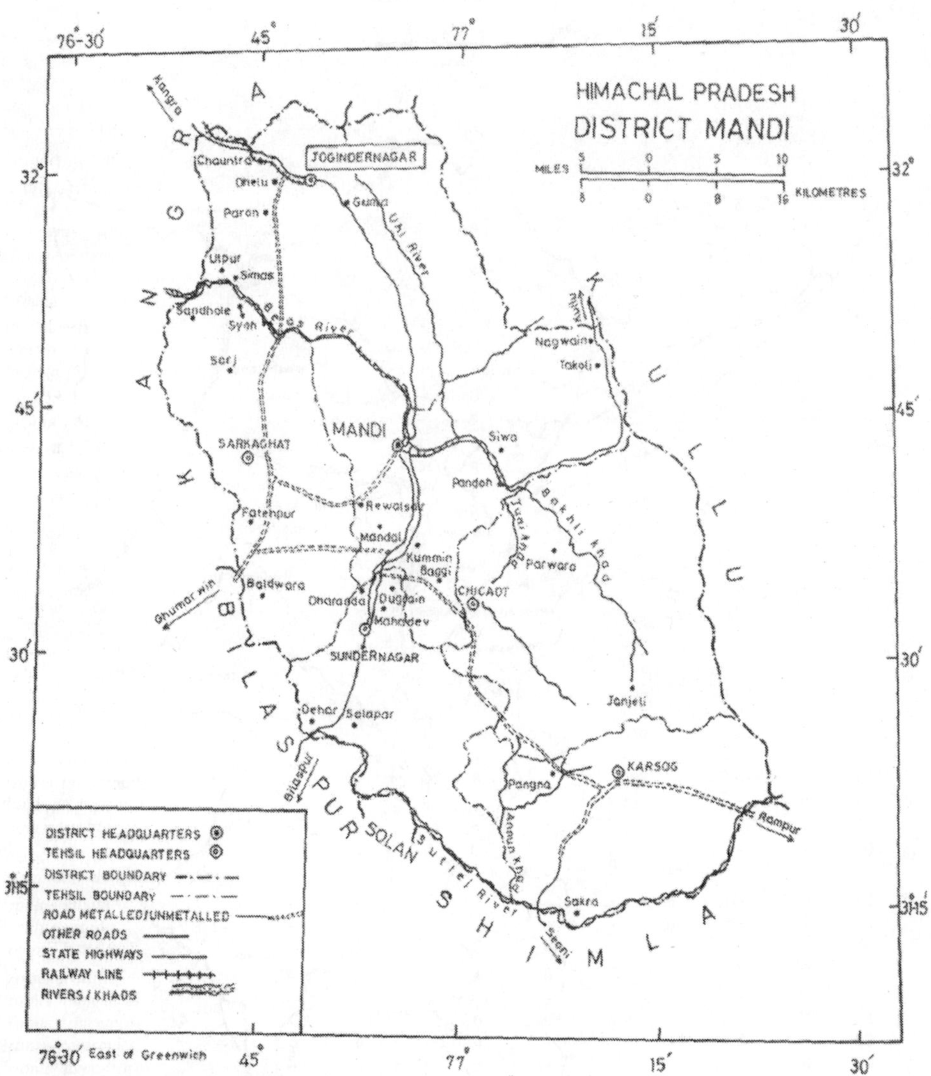

Fig. 2 : Enlarged view of outline of District Mandi showing location of the study area (Joginder nagar Himalaya).

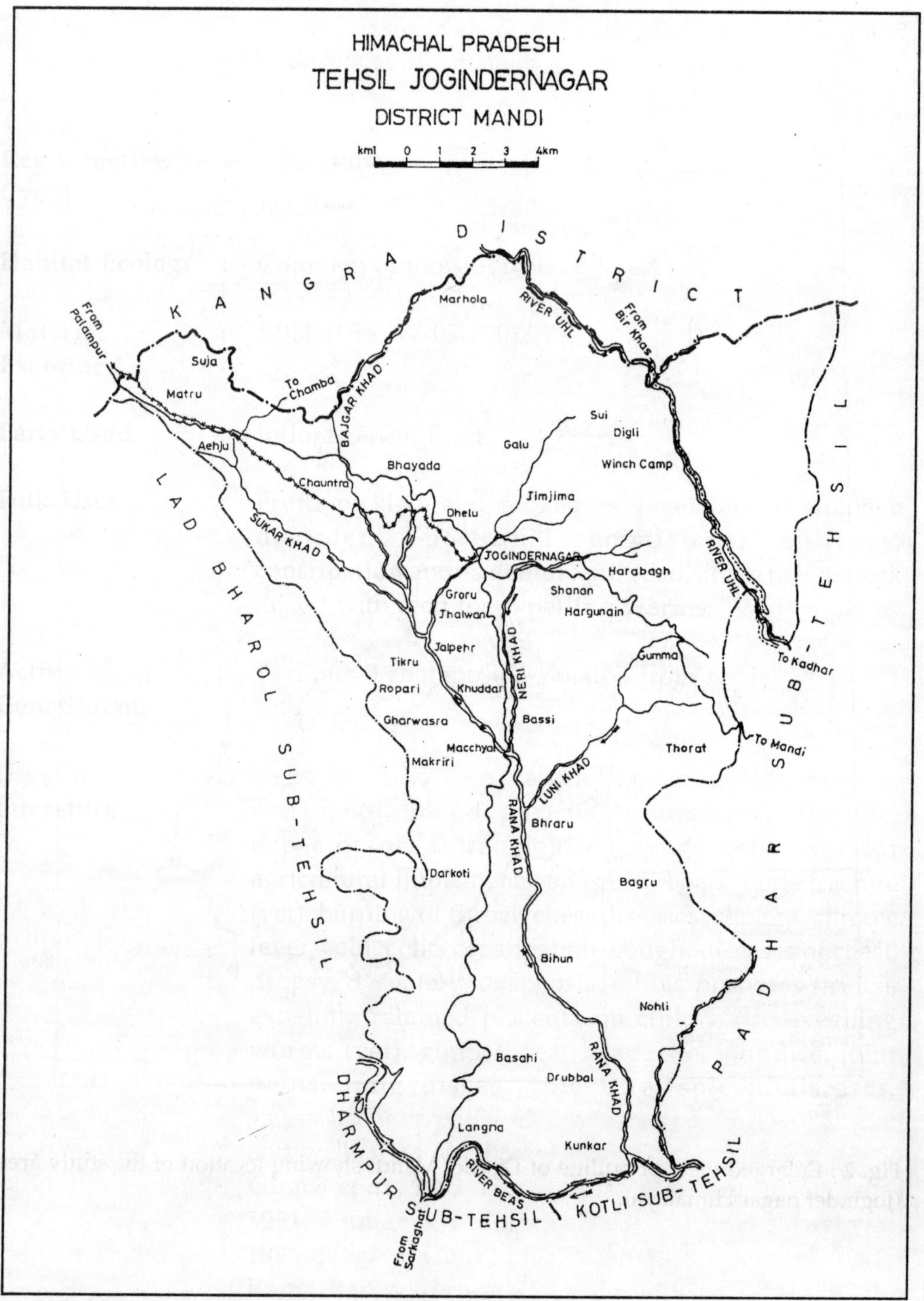

Fig. 3 : Map of Jogindernagar Tehsil showing location of the study area.

Pl. 1. **A.** *Abies pindrow* Royle x 0.025. **B.** *Abrus precatorius* L. x 0.36. **C.** *Acacia catechu* Willd. x 0.067. **D.** *Acacia concinna* DC x 0.31.

Pl. 2. **A.** *Achyranthes aspera* L. x 0.27. **B.** *A. bidentata* Blume x 1.26. **C.***Actinidia chinense* var. *deliocosa* (A. Chev.) A. Chev. x 0.36. **D.** *Adhatoda vasica* Nees x 0.17.

Pl. 3. **A.** *Aegle marmelos* (L.) Correa x 0.31. **B.** Same; a close up view of fruit x 0.33. **C.** *Aesculus indica* Colebr. x 0.18. **D.** *Agave americana* L. x 0.033. E. *Albizzia julibrissin* Durazz. x 0.23.

Pl. 4. **A.** *Albizzia lebbeck* Benth. x 0.025. **B.** *A. stipulata* Boiv. x 0.02. **C.** *Alnus nepalensis* Don x 0.3. **D.** *A. nitida* Endl. x 0.37.

Pl. 5. **A.** *Aloe vera* (L.) Webb & Berth x 0.06. **B.** *Andrachne cordifolia* (Decne.) Muell.-Arg. x 0.51. **C.** *Araucaria columnaris* (Forst.) Hook. x 0.27. **D.** *Arundinaria falcata* Nees x 0.17.

Pl. 6. **A.** *Asclepias currasavica* L. x 0.27. **B.** *A. physocarpa* Schlecht x 0.33. **C.** *Atyloshia platycarpa* Benth. x 0.43. **D.** *Bambusa nutans* Wall. x 0.034.

Pl. 7. **A.** *Bauhinia vahlii* Wight & Arn. x 0.052. **B.** *B. variegata* L. x 0.075. **C.** *Berberis aristata* DC x 0.22. **D.** *B. lycium* Royle x 0.025.

Pl. 8. **A.** *Boehmeria platyphylla* Don x 0.2. **B.** *Bombax ceiba* L. x 0.008. **C.** *Bougainvillea spectabilis* Willd. x 0.21. **D.** *Broussonetia papyrifera* Vent. x 0.034.

Pl. 9. **A.** *Buddleia crispa* Benth. x 0.18. **B.** *B. paniculata* Wall. x 1.12. **C.** *Butea monosperma* Kuntze x 0.035. **D.** *Buxus wallichiana* Baill. x 0.21.

Pl. 10. **A.** *Caesalpinia decapetala* (Roth.) Alston x 0.19. **B.** *Callicarpa macrophylla* Vahl. x 0.25. **C.** *Callistemon lanceolatus* DC x 0.38. **D.** *Calotropis procera* R. Br. x 0.35.

Pl. 11. **A.** *Camellia sinensis* (L.) Kuntze x 0.82. **B.** Same; a close up view of fruit x 0.5. **C.** *Canna indica* L. x 0.47. **D.** *Carica papaya* L. x 0.04.

Pl. 12. **A.** *Carissa congesta* Wight x 0.14. **B.** *Carpinus viminea* Wall. *ex* Lindl. x 0.48. **C.** *Carya illenoesis* (Wang.) Koch. x 0.083. **D.** *Caryopteris wallichiana* Schauer x 0.25.

Pl. 13. **A.** *Cassia fistula* L. x 0.076. **B.** *C. laevigata* Willd. x 0.46. **C.** *Cedrella serrata* Royle x 0.43. **D.** *C. toona* Roxb. *ex* Rottl. x 0.07.

Pl. 14. **A.** *Cedrus deodara* (Roxb.) Louden x 0.058. **B.** *Celtis australis* L. x 0.008. **C.** *Cestrum nocturnum* L. x 0.72. **D.** *Cinnamomum camphora* (L.) Nees & Eberm x 0.19.

Pl. 15. **A.** *Cinnamomum tamala* Nees & Eberm x 0.033. **B.** *Cissampelos pareira* L. x 0.32. **C.** *Citrus aurantium* L. x 0.024. **D.** *C. grandis* (L.) Osbeck x 0.046.

Pl. 16. **A.** *Citrus jambhiri* Lush. x 0.05. **B.** *C. limon* (L.) Burm. f. x 0.035. **C.** *C. medica* L. x 0.075. **D.** *Clematis buchananiana* DC x 0.33.

Pl. 17. **A.** *Clematis grata* Wall. x 0.48. **B.** *Clerodendron philippinum* Schauer x 0.20. **C.** *Colebrookia oppositifolia* Sm. x 0.35. **D.** *Cordia dichotoma* Forst. x 0.29.

Pl. 18. **A.** *Coriaria nepalensis* Wall. x 0.06. **B.** *Cornus macrophylla* Wall. x 0.41. **C.** *Costus speciosus* Smith x 0.22. **D.** *Cotoneaster bacillaris* Wall. *ex* Lindl. x 0.34.

Pl. 19. **A.** *Cotoneaster microphylla* Wall. *ex* Lindl. x 0.23. **B.** *Cryptolepis buchanani* Roem. & Sch. x 0.33. **C.** *Cucumis sativus* L. x 0.36. **D.** *Cucurbita hispida* Thunb. x 0.67.

Pl. 20. **A.** *Cucurbita melo* L. x 0.13. **B.** *C. pepo* L. x 0.12. **C.** *Cuscuta reflexa* Roxb. x 0.29. **D.** *Cyathula tomentosa* (Roth.) Moq. x 0.18.

Pl. 21. **A.** *Cycas circinalis* L. x 0.05. **B.** *Cyphomandra betacea* (Cav.) Sendt. x 0.71. **C.** *Dalbergia sissoo* Roxb. x 0.089. **D.** *Daphne cannabina* Wall. x 0.28.

Pl. 22. **A.** *Datura metel* L. x 0.20. **B.** *D. stramonium* L. x 0.18. **C.** *D. suaveolens* Humb. & Bonpl. *ex* Willd. x 0.15. **D.** *Debregeasia hypoleuca* Wedd. x 0.26.

Pl. 23. **A.** *Dendrocalamus hamiltonii* Nees & Arn. *ex* Munro x 0.0092. **B.** *Desmodium multiflorum* DC x 0.25. **C.** *Deutzia corymbosa* R. Brown x 0.30. **D.** *Dioscorea bellophylla* Voigt x 0.64.

Pl. 24. **A.** *Dioscorea bulbifera* L. x 0.19. **B.** *D. deltoidea* Wall. x 0.55. **C.** *Diospyros kaki* L. x 0.036. **D.** *Dodonea viscosa* Jacq. x 0.05.

Pl. 25. **A.** *Duranta repens* L. x 0.13. **B.** *Ehretia acuminata* R. Br. x 0.016. **C.** Same; a close up view of inflorescence x 0.27. **D.** *Ehretia laevis* Roxb. x 0.03.

Pl. 26. **A.** *Elaeagnus umbellata* Thunb. x 0.16. **B.** *Embellia tsjeriam-cottam* A. DC x 0.31. **C.** *Emblica officinalis* Gaertn. x 0.06. **D.** *Eriobotyrica japonica* Lindl. x 0.37.

Pl. 27. **A.** *Erythrina suberosa* Roxb. x 0.03. **B.** *Eucalyptus umbellata* Domin x 0.45. **C.** *Eugenia jambolana* Lam. x 0.057. **D.** *Euonymus echinatus* Wall. *ex* Roxb. x 0.45.

Pl. 28. **A.** *Euphorbia pulcherrima* Willd. x 0.13. **B.** *E. royleana* Boiss. x 0.042. **C.** *Ficus bengalensis* L. x 0.015. **D.** *F. clavata* Wall. x 0.2.

Pl. 29. **A.** *Ficus foveolata* Wall. x 0.05. **B.** *F. glomerata* Roxb. x 0.09. **C.** *F. hispida* L. x 0.17. **D.** *F. palmata* Forssk. x 0.032.

Pl. 30. **A.** *Ficus religiosa* L. x 0.01. **B.** Same; a close view of fruit x 0.48. **C.** *F. roxburghii* Wall. *ex* Miq. x 0.13. **D.** *F. rumphii* Bl. x 0.025.

Pl. 31. **A.** *Ficus semicordata* Buch.-Ham. *ex* Smith x 0.05. **B.** *Flacourtia ramontchi* L' Herit x 0.16. **C.** *Flemingia macrophylla* Blume *ex* Miq. x 0.83. **D.** *F. strobilifera* R. Br. x 0.15.

Pl. 32. **A.** *Fuchsia splendens* (Plum.) L. x 0.66. **B.** *Gardenia angusta* (L.) Merr. x 0.075. **C.** *Glochiodion velutinum* Wight x 0.016. **D.** Same; close up view x 0.50.

Pl. 33. **A.** *Gmelina arborea* Roxb. x 0.06. **B.** *Gossypium arboreum* L. x 0.53. **C.** *Grevillea robusta* A. Cunn *ex* R. Br. x 0.015. **D.** *Grewia oppositifolia* Roxb. x 0.008.

Pl. 34. **A.** *Hedera helix* auct (non L.) Clarke x 0.16 **B.** *Hibiscus rosa-sinensis* L. x 0.32. **C.** *Holarrhena antidysentrica* Wall. x 0.17. **D.** *Hydrangea grandiflora* Sieb. x 0.18.

Pl. 35. **A.** *Hypericum cernuum* Roxb. x 0.21. **B.** *H. patulum* Thunb. x 0.5. **C.** *Ilex dipyrena* Wall. x 0.52. **D.** *I. excelsa* (Wall.) Hook. f. x 0.36.

Pl. 36. **A.** *Indigofera gerardiana* Wall. x 0.22. **B.** *I. pulchella* Roxb. x 0.14. **C.** *Inula cappa* DC x 0.38. **D.** *Ipomoea cairica* (L.) Sweet x 0.16.

Pl. 37. **A.** *Ipomoea carnea* Jacq. x 0.09. **B.** *I. eriocarpa* R. Br. x 0.52. **C.** *I. muricata* Jacq. x 0.40. **D.** *I. quamoclit* L. x 0.75.

Pl. 38. **A.** *Jacaranda acutifolia* Humb. & Bonpl. x 0.023. **B.** *Jasminum dispersum* Wall. x 0.17. **C.** *J. grandiflorum* L. x 0.26. **D.** *J. humile* L. x 0.19.

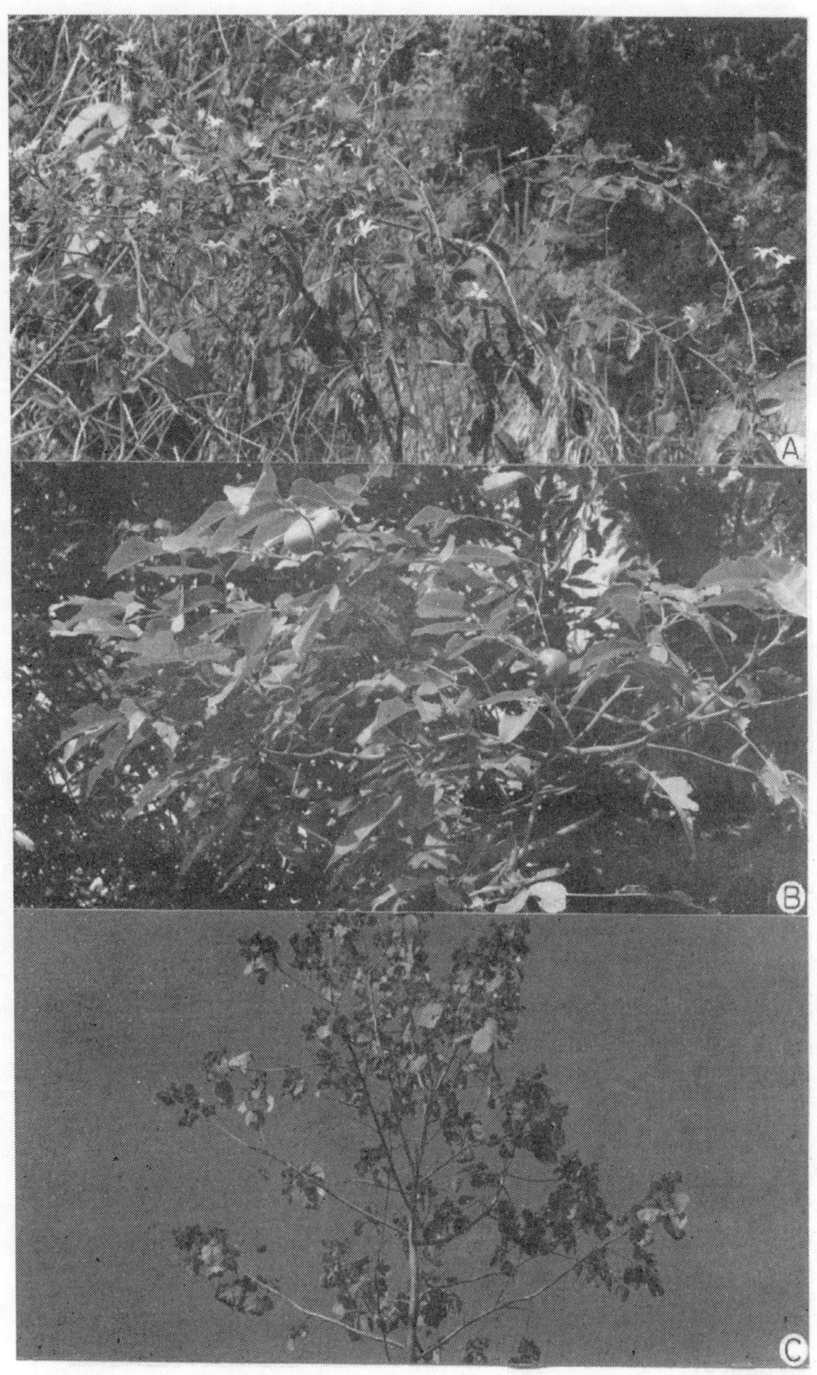

Pl. 39. **A.** *Jasminum multiflorum* (Burm. f.) Andrews x 0.20. **B.** *Juglans regia* L. x 0.066. **C.** *Kydia calycina* Roxb. x 0.022.

Pl. 40. **A.** *Lagerstroemia indica* L. x 0.028. **B.** *Lantana camara* L. x 0.53. **C.** *Leea aspera* Wall. *ex* Roxb. x 0.44. **D.** *Leptodermis lanceolata* Wall. x 0.15.

Pl. 41. **A.** *Lettsomia thomsonii* Clarke x 0.26. **B.** *Leucaena glauca* Benth. x 0.17. **C.** *Leycesteria formosa* Wall. x 0.25. **D.** *Litchi chinensis* Sonner. x 0.095.

Pl. 42. **A.** *Litsea polyantha* Juss. x 0.04. **B.** *L. sebifera* Pers. x 0.08. **C.** *L. umbrosa* Nees x 0.35. **D.** *Lonicera japonica* Wall. x 0.22.

Pl. 43. **A.** *Lonicera quinquelocularis* Hardw. x 0.25. **B.** *Loranthus ligustrinus* Wall. x 0.53. **C.** *Luffa cylindrica* (L.) Roem. x 0.094. **D.** *Lyonia ovalifolia* (Wall.) Drude x 0.37.

Pl. 44. **A.** *Machilus odoratissima* Nees x 0.33. **B.** *Mallotus phillipinensis* (Lam.) Muell.-Arg. x 0.18. **C.** *Malvaviscus conzatii* Greenm. x 0.10. **D.** *Mangifera indica* L. x 0.15.

Pl. 45. **A.** *Melia azedarach* L. x 0.21. **B.** *Melothria heterophylla* (Lour.) Cogn. x 0.36. **C.** *Michelia champaca* L. x 0.02.

Pl. 46. **A.** *Milletia extensa* Benth. *ex* Baker x 0.26. **B.** *Momordica charantia* L. x 0.67. **C.** *Moringa pterygosperma* Gaertn. x 0.08. **D.** *Morus alba* L. x 0.025.

Pl. 47. **A.** *Mucuna pruriens* (L.) DC x 0.36. **B.** *Murraya koenigii* Spreng. x 0.25. **C.** *Myrica esculenta* Buch.-Ham. *ex* **D.** Don x 0.02. D. *Myrsine africana* L. x 0.15

Pl. 48. **A.** *Nerium indicum* Mill. x 0.14. **B.** *Opuntia dillenii* Haw. x 0.17. **C.** *Oroxylum indicum* Vent. x 0.013.

Pl. 49. **A.** *Osbeckia stellata* Wall. x 0.21. **B.** *Osyris arborea* Wall. x 0.26. **C.** *Pavetta indica* L. x 0.18. **D.** *Phoenix sylvestris* Roxb. x 0.097.

Pl. 50. **A.** *Phyllanthus parvifolius* Buch.-Ham. *ex* D. Don x 0.40. **B.** *Picea smithiana* (Wall.) Boiss. x 0.012. **C.** *Pinus roxburghii* Sarg. x 0.02. **D.** *P. wallichiana* A. B. Jackson x 0.017.

Pl. 51. **A.** *Pistacia integrimma* J. L. Stewart *ex* Brand. x 0.023. **B.** *Plumbago zeylanica* L. x 0.19. **C.** *Plumeria rubra* L. x 0.05.

Pl. 52. **A.** *Pogostemon benghalensis* Burm. f. x 0.27. **B.** *Polygonum chinense* L. x 0.07. **C.** *Populus ciliata* Wall. x 0.09. **D.** *P. nigra* L. x 0.03

Pl. 53. **A.** *Premna latifolia* Roxb. x 0.13. **B.** *Prinsepia utilis* Royle x 0.13. **C.** *Prunus amygdalus* Baill. x 0.058. **D.** *P. armeniaca* L. x 0.072.

Pl. 54. **A.** *Prunus cerasoides* D. Don x 0.12. **B.** *P. domestica* L. x 0.038. **C.** Same; close up view of fruit x 0.46. **D.** *P. persica* Benth. & Hook. f. x 0.37.

Pl. 55. **A.** *Psidium guajava* L. x 0.07. **B.** *Punica granatum* L. x 0.16. **C.** *Pyrus communis* L. x 0.18.

Pl. 56. **A.** *Pyrus malus* L. x 0.25. **B.** *P. pashia* Ham. x 0.13. **C.** *Quercus glauca* Thunb. x 0.05. **D.** *Q. leucotrichophora* A. Camus x 0.016.

Pl. 57. **A.** *Quercus semicarpifolia* Sm. x 0.12. **B.** *Randia dumetorum* (Retz.) Lam. x 0.035. **C.** *Reinwardtia indica* Dumort. x 0.22. **D.** *Rhamnus persicus* Laws. x 0.05.

Pl. 58. **A.** *Rhamnus triquetra* Wall. x 0.26. **B.** *Rhododendron arboreum* Sm. x 0.028. **C.** *Rhus cotinus* L. x 0.078. **D.** *R. semialata* Murr. x 0.19.

Pl. 59. **A.** *Rhus succedanea* L. x 0.05. **B.** *Ricinus communis* L. x 0.18. **C.** *Robinia pseudoacacia* L. x 0.07. **D.** *Rosa alba* L. x 0.24.

Pl. 60. **A.** *Rosa moschata* Mill. x 0.34. **B.** *Roylea cinerea* (D. Don) Baill. x 0.16. **C.** *Rubia cordifolia* L. x 0.43 **D.** *Rubus ellipticus* Sm. x 0.17.

Pl. 61. **A.** *Rubus lasiocarpus* Sm. x 0.14. **B.** *R niveus* Wall. x 0.16. **C.** *Sachharum officinarum* L. x 0.081. **D.** *Sageretia oppositifolia* Brongn x 0.26.

Pl. 62. **A.** *Salix babylonica* L. x 0.024. **B.** *S. karelinii* Turcz. x 0.24. **C.** *S. tetrasperma* Roxb. x 0.018. **D.** *Sapindus mukorosii* Gaertn. x 0.021.

Pl. 63. **A.** *Sapium insigne* (Royle) Benth. x 0.12. **B.** *S. sebiferum* (L.) Roxb. x 0.26. **C.** *Saraca asoca* (Roxb.) De Wilde x 0.06. **D.** *Sarcococca saligna* (D. Don) Muell.-Arg. x 0.40.

Pl. 64. **A.** *Saurauia nepaulensis* DC x 0.09. **B.** *Sechium edule* Sw. x 0.34. **C.** *Senecio rufinervis* DC x 0.24. **D.** *Sida cordifolia* L. x 0.021.

Pl. 65. **A.** *Skimmia laureola* Sieb. & Zucc. *ex* Walp. x 0.29. **B.** *Smilax aspera* L. x 0.26. **C.** *Solanum erianthum* Don x 0.28. **D.** *S. melongena* L. x 0.17.

Pl. 66. **A.** *Solanum pseudocapsicum* L. x 0.32. **B.** *S. torvum* Swartz x 0.28. **C.** *Solidago canadensis* L. x 0.37. **D.** *Sorbaria tomentosa* (Lindl.) Rehder x 0.15.

Pl. 67. **A.** *Spirarea canescens* D. Don x 0.46. **B.** *Spondias mangifera* Willd. x 0.17. **C.** *Stephania elegans* Hook. f. & Thoms. x 0.39. **D.** *S. glabra* (Roxb.) Miers. x 0.45

Pl. 68. **A.** *Sterculia villosa* Roxb. x 0.83. **B.** *Strobilanthes glutinosis* Nees x 0.46. **C.** *Symplocos racemosa* Roxb. x 0.45. **D.** *S. spicata* Roxb. x 0.22

Pl. 69. **A.** *Tabernaemontana divaricata* R. Br. *ex* Roem. & Sch. x 0.25. **B.** *Taxus baccata* L. x 0.86. **C.** *Tecoma grandiflora* Loisel x 0.11. **D.** *Terminalia arjuna* Wt. & Arn. x 0.15.

Pl. 70. **A.** *Terminalia bellirica* (Gaertn.) Roxb. x 0.42. **B.** *T. chebula* Retz. x 0.047. **C.** *T. tomentosa* (DC) Wt. & Arn. x 0.035. **D.** *Thevetia neriifolia* Juss. *ex* Steud. x 0.33.

Pl. 71. **A.** *Thuja orientalis* L. x 0.13. **B.** *Thunbergia coccinea* Wall. x 0.20. **C.** *Tinospora cordifolia* Miers. x 0.06. **D.** *Trachelospermum lucidum* (D. Don) Schuamann x 0.30.

Pl. 72. **A.** *Trichosanthes tricuspidata* Lour. x 0.16. **B.** *Tropaeolum mazus* L. x 1.0. **C.** *Ulmus laevigata* Royle x 0.021. **D.** *Viburnum cotonifolium* Don x 0.16.

Pl. 73. **A.** *Viburnum stellulatnm* Wall. x 0.25. **B.** *Vigna unguiculata* L. x 0.43. **C.** *Vinca rosea* L. x 0.51. **D.** *Vitex negundo* L. x 0.40.

Pl. 74. **A.** *Vitis himalayana* Brand. x 0.27. **B.** *V. jacquemontii* Parker x 0.31. **C.** *V. rependa* Wt. & Arn. x 0.028. **D.** *Wendlandia heynei* (Roem. & Schult.) Sant. & Merch. x 0.041.

Pl. 75. **A.** *Woodfordia floribunda* Salisb. x 0.03. **B.** *Xylosma longifolium* Clos. x 0.04. **C.** *Zanthoxylum armatum* DC x 0.17. **D.** *Ziziphus jujuba* Mill. x 0.05.

Chapter 3

Epilogue

It is evident from the present study that the rural inhabitants of Jogindernagar tehsil are using 290 plants for their sustenance and healthcare needs (Adhesive-3 spp; Beverage- 3 spp; Brush- 1 sp; Detergent- 6 spp; Dye- 13 spp; Eyeliner- 1 spp; Fibre- 13 spp; Fish poison- 1 sp; Fodder- 67 spp; Fruits- 74 spp; Fuel- 30 spp; Human ailments- 211 spp; Illuminant-1; Incense- 3 spp; Ink- 2 spp; Local wine (sur)-15 spp; Magico-religious- 42 spp; Perfumery- 3 spp; Repellent-7 spp; Veterinary ailments- 13 spp; etc.). The over all diversity includes 126 trees, 106 shrubs, 11 undershrubs and 47 climbers (Figs. 4A, B) under 211 genera and 96 families of which 8 species belong to gymnosperms (*Abies pindrow, Cedrus deodara, Cycas circinalis, Picea smithiana, Pinus roxburghii, P. wallichiana, Taxus baccata, Thuja orientalis*), 13 spp (*Agave americana, Aloe vera, Arundinaria falcata, Bambusa nutans, Canna indica, Costus speciosus, Dendrocalamus hamiltonii, Dioscorea bellophylla, D. bulbifera, D. deltoidea, Phoenix sylvestris, Saccharaum officinarum, Smilax aspera*) to monocots; and the remaining 269 spp (193 genera) to dicots with predominance of Fabaceae (26 spp), followed by Rosaceae (19 spp), Moraceae (11 spp), Euphorbiaceae (10 spp), Cucurbitaceae (9 spp), Rutaceae (9 spp), Solanaceae (9 spp), Apocynaceae (8 spp), Verbenaceae (8 spp), Convolvulaceae (6 spp), Lauraceae (6 spp), Rubiaceae (6 spp), Anacardiaceae (5 spp), Caprifoliaceae (5 spp), Malvaceae (5 spp), Menispermaceae (5 spp), Salicaceae (5 spp), etc. (Table II). Amongst monocotyledons, Bambusaceae and Dioscoreaceae with 3 spp each predominate in uses, followed by Liliaceae (2 spp) and the rest of families, viz., Agavaceae, Arecaceae, Cannaceae, Poaceae and Zingiberaceae are represented by one species. Of all the plant parts, leaves predominate in usage (164), which is followed by fruits (104), wood (85), flowers (58), roots (56), bark (51), seeds (48), etc. (Figs. 5A, B); and the aerial plant parts are mostly used in comparison to underground ones (Figs. 6A, B). There could be no two opinions that the rural populace of this part of north-west Himalaya has vast knowledge of the local plant wealth and their dependence on a great diversity of plant species ensures a year-round supply of food and medicine, and at the same time this practise ensures less pressure on the surrounding plant diversity.

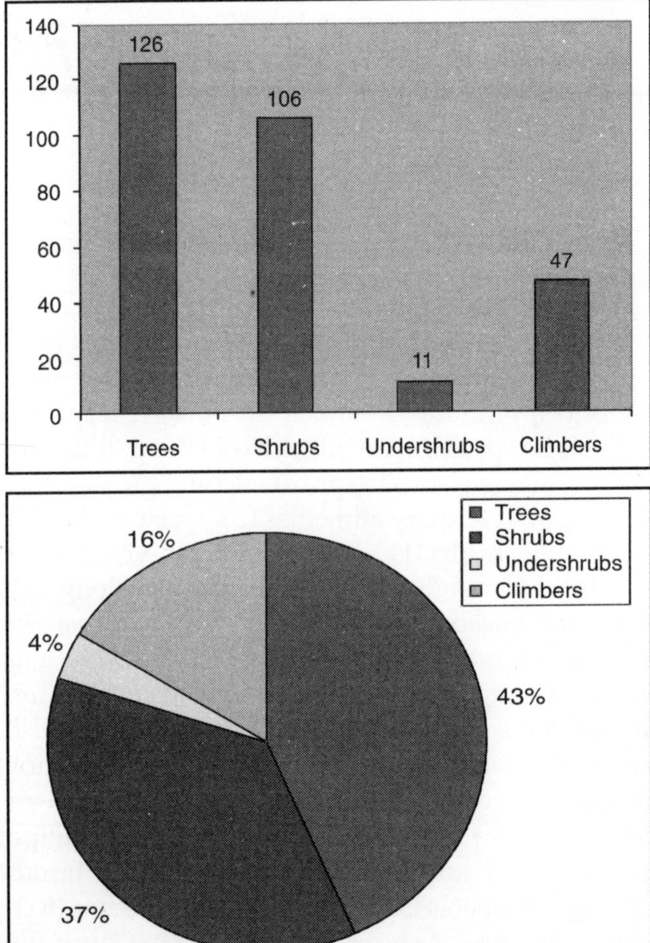

Fig. 4A, B. Habit-wise classification of ethnobotanical species of Jogindernagar tehsil indicating total number of trees, shrubs, undershrubs and climbers used by the rural populace, and their percentage respectively.

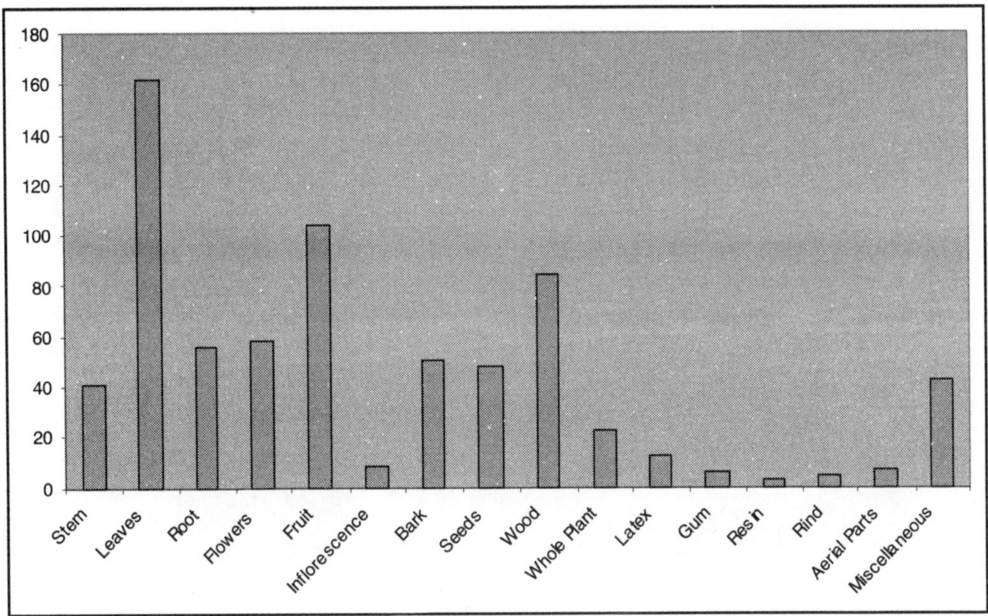

Fig. 5A. Quantitative reflection of uses of various plant parts by the rural populace of Jogindernagar tehsil.

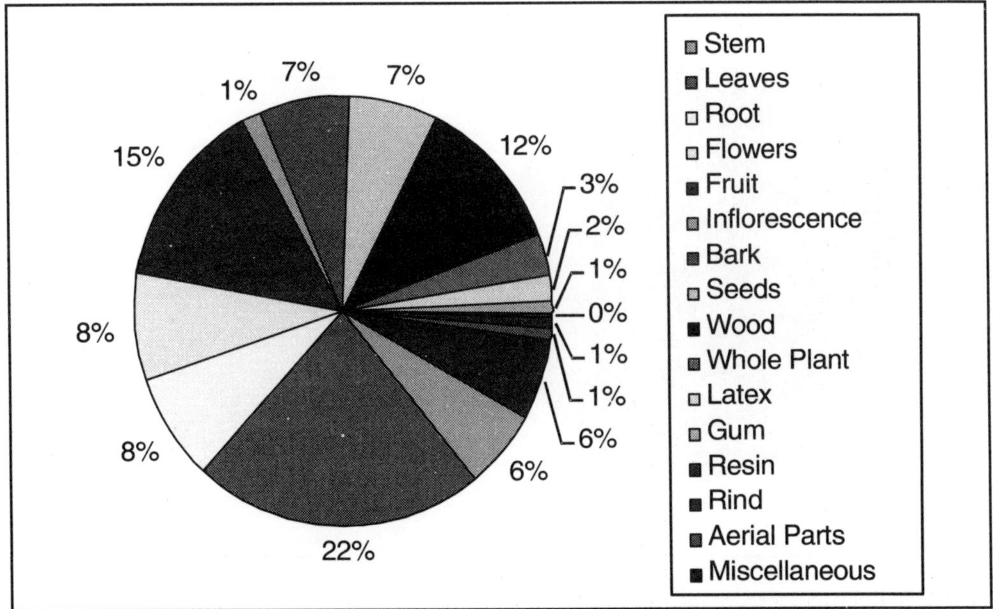

Fig. 5B. Per cent use of various plant parts by the rural populace of Jogindernagar tehsil.

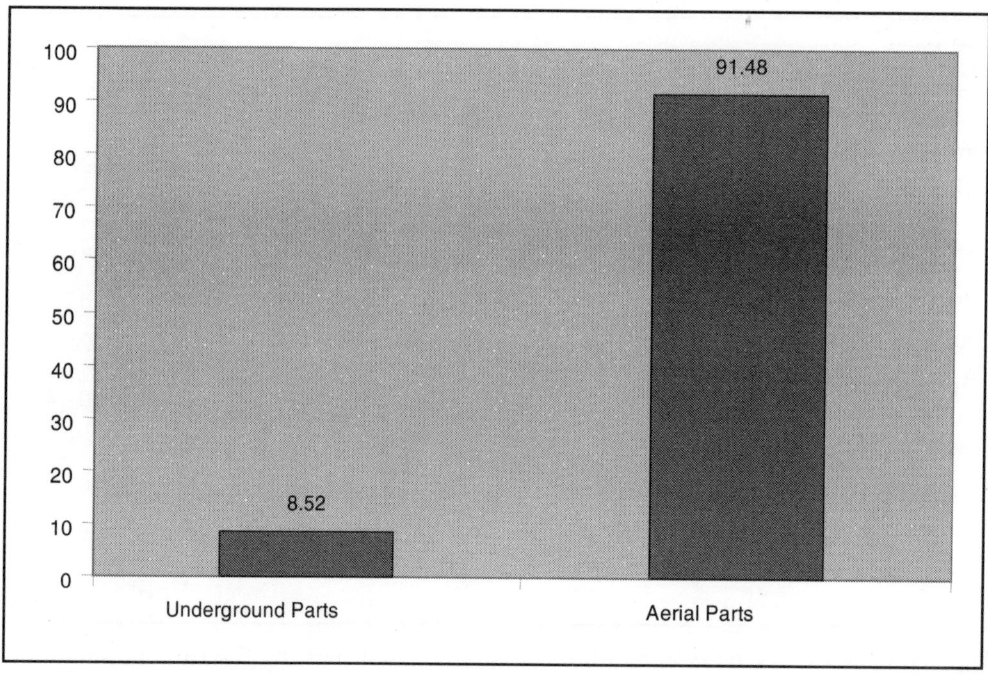

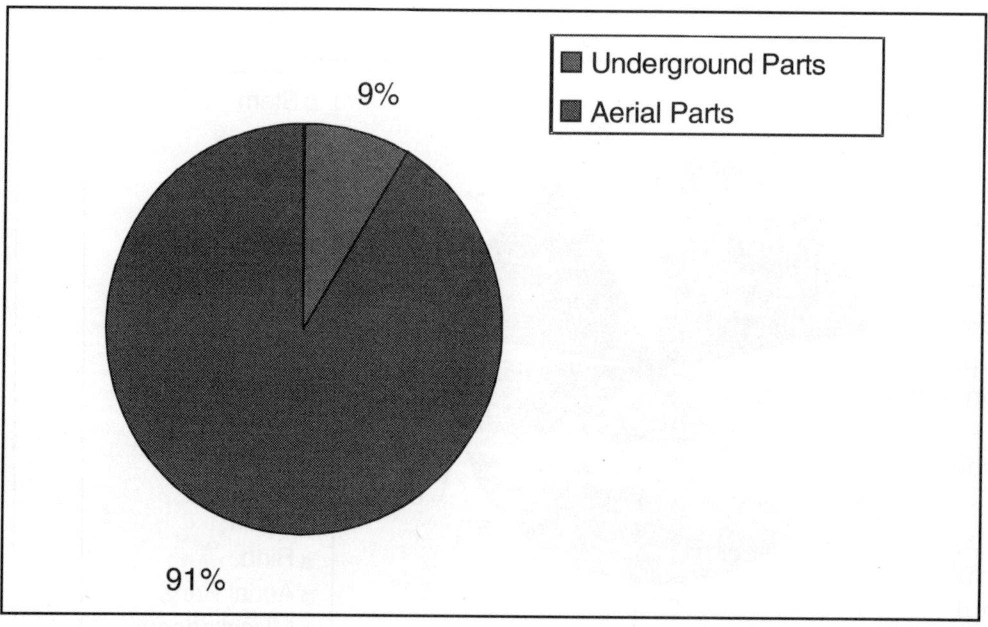

Fig. 6A, B. Relative percentage of underground and aerial parts used by the rural populace of Jogindernagar tehsil.

Table II. Predominant Ethnobotanically Important Families of Jogindernagar Tehsil

Family	Species of Ethnic Use	Percentage in the Taxa
Fabaceae	26	8.96
Rosaceae	19	6.55
Moraceae	11	3.79
Euphorbiaceae	10	3.44
Cucurbitaceae	9	3.10
Rutaceae	9	3.10
Solanaceae	9	3.10
Apocyanaceae	8	2.75
Verbenaceae	8	2.75
Convolvulaceae	6	2.06
Lauraceae	6	2.06
Rubiaceae	6	2.06
Anacardiaceae	5	1.72
Caprifoliaceae	5	1.72
Malvaceae	5	1.72
Menispermaceae	5	1.72
Salicaceae	5	1.72

Altogether 211 plant species belonging to 142 genera have been documented for their healing properties against 72 types of ailments prevailing in the region. Notably, the local inhabitants use the maximum number of plants for gastro-intestinal disorders (64 spp), followed by skin diseases (54 spp), joint diseases (33 spp), refrigerant (23 spp), anthelmintic (22 spp), piles (21 spp), lung diseases (19 spp), aphrodisiac (17 spp), blood purifier (17 spp), injuries (16 spp), liver complaints (16 spp), dental problems (15 spp), mouth sores/ ulcers (15 spp), pains/ aches (15 spp), diuretic (13 spp), menstruation complaints (12 spp), asthma (11 spp), blood sugar (11 spp), constipation (11 spp), febrifuge (10 spp), spermatorrhoea (10 spp), antidote (9 spp), throat complaints (9 spp), cardiac complaints (7 spp), eye diseases (7 spp), kidney disorders (7 spp), leucorrhoea (7 spp), cold (6 spp), epilepsy (6 spp), nasal problems (6 spp), abortifacient (5 spp), bone fracture (5 spp), internal swellings (5 spp), antiemetic (4 spp), hernia (4 spp), venereal diseases (4 spp), bronchitis (3 spp), internal injuries (3 spp), spleen complaints (3 spp), cancer (2 spp), fertility (2 spp), hair care (2 spp), hydrocele (2 spp), nervous disorders (2 spp), whooping cough (2 spp), etc. However, only one plant species has been recommended each for other ailments, namely, antiallergic, antiseptic, antispasmodic, bed wettings, bites, burns, chest infection, dehydration, demulcent, facial cosmetics, gall bladder disorders, giddiness, glandular adenytis, goitre, malaria, migraine, nail diseases, paralysis, pthsis, rubefacient, spitting, stimulant, suppressing hunger, typhoid, ulcer, and uric acid accumulation, thereby indicating less prevalence of these diseases in the area (Appendix II). It was specially observed that the local inhabitants are using all these species very carefully since time immemorial and with no harmful effects. Moreover, the documentation of most of these remedies are hitherto unreported and worthy of consideration as perspective health care plants for the welfare of humans. It is needless to say that their efficacies in the specific ailments require verification through clinical evaluation in order to ascertain and establish their utility.

For supplementing their diet, the rural populace of Jogindernagar Himalaya consume 101 species as food plants (Table III). They are represented by 55 families and 79 genera. Of these, buds and flowers of 5 species, tender stem and twigs of 6 species, leaves of 5 species, tubers of 2 species, seeds of 7 species and fruits of 81 species (of which 15 species are consumed as vegetables) are consumed raw and/or cooked. Species such as *Cinnamomum tamala, Hypericum cernuum, Murraya koenigii, Solanum pseudocapsicum, Spirarea canescens, Spondias mangifera,* and *Zanthoxylum armatum* are used for flavouring (Table IV), and *Osyris arborea* and *Taxus baccata* as substitute for tea leaves (Table V). Notably, use of leaves and bark of *Taxus baccata* as substitute for tea leaves has not been recorded earlier and forms the first report of its type. It is estimated that in India about 3900 species are consumed as food plants, chiefly by the tribal inhabitants, out of which 300 species occur in the north-eastern region (Anon., 1994; Singh & Arora, 1978). It is worthy of mention that the present work in addition to the existing knowledge reveals 41 new species of plants as famine foods hitherto unknown to the science of ethnobotany (Arora & Pandey, 1996), thereby extending the base of our food security system and enhances our chances of survival in the event of any crisis. These new records are: *Cassia laevigata, Phoenix sylvestris, Thevetia neriifolia* (Buds and flowers); *Dioscorea bellophylla* (Tubers); *Euphorbia royleana, Leea aspera, Osbeckia stellata, Plumbago zeylanica, Smilax aspera, Zanthoxylum armatum* (Twigs and Tender Stem); *Ficus hispida, Ipomoea quamoclit, Moringa pterygosperma, Smilax aspera* (Leaves); *Achyranthes aspera, Hypericum cernuum, Murraya koenigii, Pinus roxburghii, Sechium edule, Spirarea canescens, Vigna unguiculata* (Seeds); *Bombax ceiba, Citrus jambhiri, Cordia dichotoma, Embellia tsjeriam-cottam, Ficus clavata, F. foveolata, F. roxburghii, Hedera helix, Polygonum chinense, Prinsepia utilis, Randia dumetorum, Rhamnus triquetra, Rubus lasiocarpus, Solanum pseudocapsicum, S. torvum, Taxus baccata, Terminalia bellerica, Thevetia neriifolia, Vitis himalayana, and V. rependa* (Fruits). For brewing local wine, 'sur', as many as

14 plant species are added to molasses obtained from *Saccharum officinarum* to impart a typical aroma so unique to the region. For doing so, necessary license has to be obtained from the Government. Species used are: *Acacia concinna, Achyranthes bidentata, Albizzia stipulata, Callicarpa macrophylla, Colebrookia oppositifolia, Eugenia jambolana, Hedera helix, Melothria heterophylla, Rhododendron arboreum, Senecio rufinervis, Skimmia laureola, Terminalia chebula, Trachelospermum lucidum* and *Woodfordia floribunda*. As is true for other indigenous people, the mystery for sound health of these people is also due to dependence on the wild edibles as an important source of indispensable nutrients. Besides, an interesting observation worth mentioning is that several of these food plants are also known for their therapeutic properties (Table III). So, these plants not only contribute to the diet of the people but also safeguard their health against a number of ailments. It will be interesting to analyse these plants chemically for their nutraceutical properties.

Table III. Edible Plants of Jogindernagar Tehsil

Plant Species	Edible Part/s
Achyranthes aspera$^{\oplus}$	Seeds
Actinidia chinense var. *deliocosa*	Fruits
Aegle marmelos$^{\oplus}$	Fruits
Aesculus indica$^{\oplus}$	Fruits
Bauhinia variegata$^{\oplus}$	Buds, Flowers
Berberis aristata$^{\oplus}$	Fruits
Berberis lycium$^{\oplus}$	Fruits
Bombax ceiba$^{\oplus}$	Fruits
Camellia sinensis$^{\oplus}$	Leaves (beverage)
Carica papaya$^{\oplus}$	Fruits
Carissa congesta$^{\oplus}$	Fruits
Carya illenoesis$^{\oplus}$	Fruits
Cinnamomum tamala$^{\oplus}$	Leaves (flavouring)
Citrus aurantium$^{\oplus}$	Fruits
Citrus grandis$^{\oplus}$	Fruits
Citrus jambhiri$^{\oplus}$	Fruits
Citrus limon$^{\oplus}$	Fruits
Cordia dichotoma$^{\oplus}$	Fruits
Coriaria nepalensis	Fruits

Cucumis sativus⊕	Fruits
Cucurbita hispida	Fruits
Cucurbita melo	Fruits
Cycas circinalis	Fruits
Cyphomandra betacea	Fruits
Dioscorea bellophylla⊕	Tubers
Dioscorea bulbifera	Tubers
Dioscorea deltoidea⊕	Tubers
Diospyros kaki⊕	Fruit
Ehretia acuminata	Fruit
Elaeagnus umbellata⊕	Fruits
Embellia tsjeriam-cottam⊕	Fruits
Emblica officinalis⊕	Fruits
Eriobotyrica japonica⊕	Fruits
Eugenia jambolana⊕	Fruits
Euphorbia royleana⊕	Stem (inner portion) cooked
Ficus bengalensis⊕	Fruits
Ficus clavata	Fruits
Ficus foveolata	Fruits
Ficus hispida⊕	Cooked fruits, Leaves
Ficus palmata⊕	Fruits
Flacourtia ramontchii⊕	Fruits
Gmelina arborea⊕	Fruits
Grewia oppositifolia⊕	Fruits
Hedera helix⊕	Berries
Hypericum cernuum	Seeds
Ipomoea eriocarpa	Seeds
Ipomoea quamoclit⊕	Leaves cooked
Juglans regia⊕	Fruits
Lantana camara⊕	Fruits
Leea aspera	Cooked fruits
Lettsomia thomsonii	Cooked
Litchi chinensis	Fruits

Luffa cylindrica	Cooked fruits
Mangifera indica⊕	Fruits
Momordica charantia⊕	Fruits
Moringa pterygosperma⊕	Leaves, Fruits
Morus alba⊕	Fruits
Myrica esculenta⊕	Fruits
Osbeckia stellata	Fruits
Osyris arborea⊕	Leaves (Beverage)
Phoenix sylvesteris⊕	Tender buds, Fruits
Pinus roxburghii⊕	Seeds
Plumbago zeylanica⊕	Young stems cooked
Polygonum chinense	Leaves, Ripe Fruits
Prinsepia utilis⊕	Seed oil
Prunus amygdalus⊕	Fruits
Prunus armeniaca⊕	Fruits
Prunus cerasoides⊕	Fruits
Prunus persica⊕	Fruits
Psidium guajava⊕	Fruits
Punica granatum⊕	Fruits
Pyrus communis⊕	Fruits
Pyrus malus⊕	Fruits
Pyrus pashia⊕	Fruits
Randia dumetorum⊕	Fruits
Rhamnus triquetra⊕	Fruits
Rhododendron arboreum⊕	Flowers
Rhus semialata⊕	Fruits
Rubus ellepticus⊕	Fruits
Rubus lasiocarpus⊕	Fruits
R. niveus⊕	Fruits
Saccharum officinarum⊕	Fruits
Sechium edule⊕	Cooked Fruits
Smilax aspera⊕	Cooked Leaves, Stem
Solanum melongena⊕	Fruits

Solanum pseudocapsicum	Fruits (spice)
Solanum torvum⊕	Cooked Fruits
Spirarea canescens	Seeds (flavouring curries)
Spondias mangifera	Fruits
Sterculia villosa⊕	Seeds
Taxus baccata⊕	Fruits, Leaves, Bark (substitute for tea leaves)
Terminalia bellerica⊕	Fruits
Thevetia neriifolia⊕	Fruits, Flowers cooked
Trichosanthes tricuspidata⊕	Fruits
Viburnum cotonifloium⊕	Fruits
Viburnum stellulatnm⊕	Fruits
Vigna unguiculata	Seeds
Vitis himalayana	Fruits
Vitis rependa⊕	Fruits
Zanthoxylum armatum⊕	Young Twigs, Fruits Flavouring curries
Zizyphus jujuba⊕	Fruits

⊕ denotes medicinal value

Table IV. Flavouring Plants of Jogindernagar Tehsil

Plant Species	Family	Parts Used
Cinnamomum tamala	Lauraceae	Leaves
Hypericum cernuum	Hypericaceae	Seeds
Murraya koenigii	Rutaceae	Leaves
Solanum pseudocapsicum	Solanaceae	Fruit
Spirarea canescens	Rosaceae	Seeds
Spondias mangifera	Anacardiaceae	Flowers
Zanthoxylum armatum	Rutaceae	Fruits

Table V. Beverage Plants of Jogindernagar Tehsil

Plant Species	Family	Parts Used
Camellia sinensis	Theaceae	Leaves
Osyris arborea	Santalaceae	Leaves
Taxus baccata	Taxaceae	Leaves and Bark

Altogether 27 plant species possess the curative properties for improving brain function, restoring stamina and vigour and vitality of the body (Table VI). For the desired effect, these are given to the patient for a period ranging from several days to months, and also these are easily available nearby. The commonly used ones are: *Aegle marmelos, Prunus armeniaca, Terminalia bellerica, Viburnum stellulatnm* (Tonic); *Carica papaya, Terminalia arjuna, Thevetia neriifolia* (Cardiac Tonic); *Carissa congesta* (Anaemia); *Cucumis sativus, Dendrocalamus hamiltonii, Moringa pterygosperma, Osyris arborea* (Post-delivery stress); *Ficus bengalensis, Prunus amygdalus, Rhamnus triquetra, Rubia cordifolia, Sida cordifolia, Terminalia bellerica, Vitex negundo, Xylosma longifolium* (Body tone); *Ficus bengalensis* (Sexual vigour); *Ficus roxburghii, Gardenia angusta, Mucuna pruriens, Prunus amygdalus* (Nervine tonic & Brain tonic); *Phoenix sylvestris* (Cooling tonic); *Prunus amygdalus, Sida cordifolia, Terminalia bellerica, Vitex negundo, Xylosma longifolium* (Vitality), *Daphne cannabina*, (Bitter tonic); *Mangifera indica* (Restorative tonic); *Ricinus communis* (Hair tonic); and *Saraca asoca* (Uterine tonic). All these plant species hold considerable scope for popularization and worth recommending for future use.

Table VI. Tonic Plants of Jogindernagar Tehsil

Plant Species	Curative Property
Aegle marmelos	Tonic
Carica papaya	Cardiac-tonic
Carissa congetsa	Anaemia
Cucumis sativus	Post-delivery weakness
Daphne cannabina	Bitter tonic
Dendrocalamus hamiltonii	Post-delivery weakness
Ficus bengalensis	Body tone, Sexual vigour
Ficus roxburghii	Brain tonic
Gardenia angusta	Nervine tonic
Mangifera indica	Restorative tonic

Moringa pterygosperma	Post-labour stress
Mucuna pruriens	Nervine tonic
Osyris arborea	Post-delivery stress
Phoenix sylvestris	Cooling tonic
Prunus amygdalus	Body tone, Nervine tonic, Vitality
Prunus armeniaca	Tonic
Rhamnus triquetra	Body tone
Ricinus communis	Hair tonic
Rubia cordifolia	Body tone
Saraca asoca	Uterine tonic
Sida cordifolia	Body tone, Vitality
Terminalia arjuna	Cardiac tonic
Terminalia bellerica	Tonic, Body tone, Vitality
Thevetia neriifolia	Cardiac tonic
Viburnum stellulatnm	Tonic
Vitex negundo	Body tone, vitality
Xylosma longifolium	Body tone, vitality

It is evident from the Table VII that as many as 67 species belonging to 55 genera under 38 families are considered nutritious and valued by the rural populace to supplement the fodder requirements. The diversity in forages predominantly belongs to Fabaceae (9 spp), followed by Moraceae (5 spp), Salicaceae (4 spp), Bambusaceae (3 spp), Convolvulaceae (3 spp), Caprifoliaceae (3 spp), Rhamnaceae (3 spp), Aquifoliaceae (2 spp), Betulaceae (2 spp), Cordiaceae (2 spp), Flacourtiaceae (2 spp), Urticaceae (2 spp), Vitaceae (2 spp) and the rest of the families viz., Amaranthaceae, Anacardiaceae, Apocynaceae, Araliaceae, Asteraceae, Buxaceae, Celastraceae, Combretaceae, Cornaceae, Corylaceae, Euphorbiaceae, Fagaceae, Hippocastanaceae, Leeaceae, Malvaceae, Meliaceae, Menispermaceae, Rosaceae, Rubiaceae, Sapindaceae, Saxifragaceae, Saurariaceae, Sterculiaceae, Symplocaceae, Tiliaceae and Ulmaceae are represented by one species each. Some of frequently used genera of the forages from the region are those of *Ficus* (5 spp), *Albizzia* (2 spp), *Alnus* (2 spp), *Ehretia* (2 spp), *Flemingia* (2 spp), *Ilex* (2 spp), *Ipomoea* (2 spp), *Populus* (2 spp), *Salix* (2 spp), *Vitis* (2 spp), etc. which illustrates the richness of genetic diversity in forages held by the people in this region of north-western Himalaya some of which are not even widely known or unknown in the country (Anon., 1994).

Table VII. Fodder Plants of Jogindernagar Tehsil

Family	Plant Species
Fabaceae	*Albizzia julibrissin, A. stipulata, Bauhinia variegata, Desmodium multiflorum, Flemingia strobilifera, F. macrophylla, Leaucaena glauca, Milletia extensa, Robinia pseudoacacia*
Moraceae	*Ficus clavata, F. foveolata, F. roxburghii, F. semicordata*
Salicaceae	*Populus ciliata, P. nigra, Salix babylonica, S. karelinii*
Bambusaceae	*Arunidinaria falcata, Bambusa nutans, Dendrcalamus hamiltonii*
Convolvulaceae	*Ipomoea cairica, I. eriocarpa, Lettsomia thomsonii*
Caprifoliaceae	*Leycesetria formosa, Lonicera quinquelocularis, Viburnum stellulatnm*
Rhamnaceae	*Rhamnus persicus, Sageretia oppositifolia, Zizyphus jujuba*
Aquifoliaceae	*Ilex dipyrena, I. excelsa*
Betulaceae	*Alnus nepalensis, A. nitida*
Cordiaceae	*Ehretia acuminata, E. laevis*
Flacourtiaceae	*Flacourtia ramontchi, Xylosma lonifolium*
Urticaceae	*Broussonetia papyrifera, Debregeasia hypoleuca*
Vitaceae	*Vitis himalayana, V. rependa*
Amaranthaceae	*Cyathula tomentosa*
Anacardiaceae	*Spondias mangifera*
Apocyanaceae	*Trachelospermum lucidum*
Araliaceae	*Hedera helix*
Asteraceae	*Senecio rufinervis*
Buxaceae	*Buxus wallichiana*
Celastraceae	*Euonymus echinatus*
Combretaceae	*Terminalia arjuna*
Cornaceae	*Cornus macrophylla*
Corylaceae	*Carpinus viminea*
Euphorbiaceae	*Sapium sebiferum*
Fagaceae	*Quercus galuca*
Hippocastanaceae	*Aesculus indica*
Leeaceae	*Leea aspera*
Malvaceae	*Kydia calycina*
Meliaceae	*Cedrella serrata*

Menispermaceae	*Stephania elegans*
Rosaceae	*Cotoneaster bacillaris*
Rubiaceae	*Wendlandia heynei*
Sapindaceae	*Dodonea viscosa*
Saxifragaceae	*Deutzia corymbosa*
Saurariaceae	*Sauraria nepalensis*
Sterculiaceae	*Sterculia villosa*
Symplocaceae	*Symplocos spicata*
Tiliaceae	*Grewia oppositifolia*
Ulmaceae	*Celtis australis*

With regard to the traditional veterinary herbal remedies, the rural populace of Jogindernagar Himalaya (present study) is well acquainted with the therapeutic value of the plant species (Table VIII). Various species used for curing the veterinary ailments are, namely, Anthrax (*Abrus precatorius*), mouth sores (*Acacia catechu*), easy shedding of placenta (*Bauhinia variegata, Cuscuta reflexa*), repelling worms (*Cordia dichotoma*), stomach disorders (*Euphorbia royleana, Oroxylum indicum*), spleen affections (*Oroxylum indicum, Vitex negundo*), gastro-intestinal (*Quercus leucotrichophora, Stephania glabra*), enhancing lactation (*Tinospora cordifolia*), cough and inflammation of lungs (*Trichosanthes tricuspidata*), paralysis (*Thuja orientalis)* and indigestion (*Vitex negundo*) thereby indicating consciousness for the health of their livestock. One of the reasons for inclination towards the use of these aforesaid plants for various aliments is due to their proved therapeutic efficiency evolved out of rich experience of the past and also due to availability in vicinity.

Table VIII. Ethnobotanically Important Plants Used for Veterinary Ailments

Plant Species	Ailment/s
Abrus precatorius	Anthrax
Acacia catechu	Mouth sores
Bauhinia variegata	Easy shedding of placenta
Cordia dichotoma	Repelling worms
Cuscuta reflexa	Easy shedding of placenta
Euphorbia royleana	Stomach disorders
Oroxylum indicum	Stomach discomforts, spleen affections
Quercus leucotrichophora	Gastro-intestinal disorders
Stephania glabra	Gastro-intestinal disorders

Thuja orientalis	Paralysis
Tinospora cordifolia	Enhancing lactation
Trichosanthes tricuspidata	Inflammation of lungs, cough
Vitex negundo	Indigestion, enlarged spleen

It is apparent from the table IX that the cordage requirements are largely met from 13 plant species belonging to 10 families. Some of the frequently used families in this regard are those of Bambusaceae (*Arundinaria falcata, Bambusa nutans, Dendrocalamus hamiltonii*), and Ranunculaceae (*Clematis buchananiana, Clematis grata*). Other families, namely Agavaceae, Convolvulaceae, Fabaceae, Moraceae, Rosaceae, Sterculiaceae, Tiliaceae and Urticaceae are represented by one species each. For their fuelwood requirements, the study highlights the use of as many as 31 species belonging to 19 families (Table X). Of these, the families viz., Fabaceae (4 spp: *Bauhinia vahlii, Dalbergia sissoo, Desmodium multiflorum, Indigofera gerardiana, Robinia pseudoacacia*), Rosaceae (3 spp: *Prunus domestica, Rubus ellipticus, Sorbaria tomentosa*), Saliaceae (3 spp: *Populus nigra, Salix karelinii, S. tetrasperma*), Caprifoliaceae (2 spp: *Lonicera quinquelocularis, Viburnum cotonifolium*), Fagaceae (2 spp: *Quercus glauca, Q. semicarpifolia*), Flacourtiaceae (2 spp: *Flacourtia ramontchi, Xylosma longifolium*) and Myrsinaceae (2 spp: *Embellia tsjeriam-cottam, Myrsine africana*) predominate in usage. Living close to nature, these people never extract them in unsustainable manner. It is felt that the rich knowledge of rural inhabitants of Jogindernagar Himalaya may well be utilized for setting several cottage and rural industries to improve their economic and social life.

Table IX. Fibre Yielding Plants of Jogindernagar Tehsil

Family	Plant Species
Agavaceae	*Agava americana*
Bambusaceae	*Arundinaria falcata, Bambusa nutans,Dendrocalamus hamiltonii*
Convolvulaceae	*Ipomoea eriocarpa*
Fabaceae	*Bauhinia vahlii*
Moraceae	*Ficus roxburghii*
Ranunculaceae	*Clematis buchananiana, Clematis grata*
Rosaceae	*Sorbaria tomentosa*
Sterculiaceae	*Sterculia villosa*
Tiliaceae	*Grewia oppostifolia*
Urticaceae	*Debreagesia hypoleuca*

Table X. Fuelwood Plants of Jogindernagar Tehsil

Family	Plant Species
Acanthaceae	*Adhatoda vasica*
Berberidaceae	*Berberis lycium*
Caprifoliaceae	*Lonicera quinquelocularis, Viburnum cotonifolium*
Cordiaceae	*Ehretia laevis*
Cornaceae	*Cornus macrophylla*
Elaeagnaceae	*Elaeagnus umbellata*
Euphorbiaceae	*Sapium sebiferum*
Fabaceae	*Bauhinia vahlii, Dalbergia sissoo, Desmodium multiflorum, Indigofera gerardiana, Robinia pseudoacacia*
Fagaceae	*Quercus glauca, Q. semicarpifolia*
Flacourtiaceae	*Flacourtia ramontchii, Xylosma longifolium*
Lamiaceae	*Colebrookia oppositifolia.*
Myrsinaceae	*Embellia tsjeriam-cottam, Myrsine africana*
Rhamnacae	*Sageretia oppositifolia*
Rosaceae	*Prunus domestica, Rubus ellipticus Sorbaria tomentosa*
Salicaceae	*Populus nigra, Salix karelinii, S. tetresperma*
Sapindaceae	*Dodonea viscosa*
Sterculiaceae	*Sterculia villosa*
Urticaceae	*Debreagesia hypoleuca*
Verbenaceae	*Caryopteris wallichiana*

From analysis of the data on total importance values (Table XI), it is evident that the plant species, namely *Emblica officinalis* (Euphorbiaceae), *Bombax ceiba* (Bombaceae), *Berberis arisata, B. lycium* (Berberidaceae), *Woodfordia floribunda* (Lythraceae), *Rubus ellepticus, Prinsepia utilis* (Rosaceae), *Zanthoxylum armatum* (Rutaceae), *Zizhphus jujuba* (Rhamnaceae), *Eugenia jambolana* (Myrtaceae), *Juglans regia* (Juglandaceae), *Prunus persica* (Rosaceae), *Quercus glauca* (Fagaceae), *Aesculus indica* (Hippocastanaceae), *Albizzia stipulata* (Fabaceae), *Bambusa nutans* (Bambusaceae), *Bauhinia variegata* (Fabaceae), *Dendrocalamus hamiltonii* (Bambusaceae), *Ficus bengalensis* (Moraceae), *Terminalia chebula* (Combretaceae), *Mangifera indica* (Anacardiaceae), *Carissa congesta* (Apocyanaceae), *Carya illenoesis* (Juglandaceae), *Glochiodion velutinum*

(Euphorbiaceae), *Pinus roxburghii* (Pinaceae), *Rosa moschata* (Rosaceae), *Salix tetrasperma* (Salicaceae), *Taxus baccata* (Taxaceae), *Tinospora cordifolia* (Menispermaceae), *Ficus hispida* (Moraceae), *Momordica charantia* (Cucurbitaceae), *Phoenix sylvestris* (Aracaceae), etc. have high economic potential. Species such as *Abrus precatorious, Achyranthes aspera, A. bidentata, Bauhinia variegata, Berberis aristata, B. lycium, Calotropis procera, Carica papaya, Cuscuta reflexa, Datura metel, Emblica officinalis, Ficus religiosa, Melia azedarach, Momordica charantia, Pogostemon benghalensis, Prunus amygdalus, Terminalia bellerica., T. chebula, Tinospora cordifolia, Vitex negundo, Woodfordia floribunda, Zanthoxylum armatum*, etc. (64 plants) rank high in therapeutic value; *Bauhinia variegata, Carica papaya, Carya illenoesis, Citrus aurantium, C. medica, Cucumis sativus, Emblica officinalis, Juglans regia, Luffa cylindrica, Mangifera indica, Momordica charantia, Prunus amygdalus, Prunus armeniaca, Psidium guajava, Punica granatum, Pyrus communis, P. malus, Rubus ellipticus, R. lasiocarpus, R. niveus, Saccharaum officinarum, Sechium edule, Solanum melongena, Vigna unguiculata, Zanthoxylum armatum, Zizhyphus jujuba*, etc. (58 plants) in food; *Bambusa nutans, Cedrella toona, Cedrus deodara, Dalbergia sissoo, Dendrocalamus hamiltonii, Abies pindrow, Albizzia stipulata, Broussonetia papyrifera, Eucalyptus umbellata, Eugenia jambolana, Pinus wallichiana, Pyrus pashia, Quercus glauca, Q. leucotrichophora, Q. semicarpifolia, Salix tetrasperma, Sorbaria tomentosa, Symplocos racemosa, Taxus baccata, Terminalia tomentosa, Ulmus laevigata*, etc. in timber; *Albizzia julibrissin, A. stipulata, Grewia oppostifolia, Bauhinia variegata, Bombax ceiba, Broussonetia papyrifera, Carissa congesta, Dendrocalamus hamiltonii, Ficus semicordata, Flemingia macrophylla, Ilex excelsa, Melia azeadarach, Rosa moschata, Rubus ellipticus, Terminalia tomentosa, Trichosanthes tricuspidata, Zizyphus jujuba*, etc. in fodder; *Carpinus viminea, Dodonea viscosa, Caryopteris wallichiana, Embellia tsjeriam-cottam, Indigofera gerardiana, Lonicera quinquelocularis, Rubus ellepticus, Symplocos recemosa, Viburnum cotonifolium, Berberis aristata, B. lycium, Colebrookia oppositifolia, Debregeasia hypoleuca, Desmodium multiflorum, Elaeagnus umbellata, Glochidion velutinum, Populus nigra, Rhamnus triquetra, Salix tetrasperma, Sapium sebiferum, Sorbaria tomentosa, Stephania glabra, Ulmus laevigata, Woodfordia floribunda, Xylosma longifloium*, etc. in fuel; *Grewia oppositifolia, Alnus nitida, Bombax ceiba, Bombax nutans, Bauhinia vahlii, Clematis buchananiana, C. grata, Ipomoea eriocarpa* in fibre; *Berberis aristata, B. lycium, Camellia sinensis, Carya illenoesis, Glochidion velutinum, Juglans regia, Woodfordia floribunda*, in dyes; and *Arundinaria falcata, Bambusa nutans, Bauhinia vahlii, Bougainvillea spectabilis, Butea monosperma, Callistemon lanceolatus, Camellia sinensis, Emblica officinalis, Gossypium arboreum, Hedera helix, Mangifera indica, Murraya koenigii, Phoenix sylvestris, Prunus amygdalus, Quercus glauca, Rosa moschata, Tinospora cordifolia, Vinca rosea, Woodfordia floribunda, Zanthoxylum armatum, Zizyphus jujuba*, etc. (58 plants) in others.

Therefore, the above details are of great interest particularly for those who are engaged in analysing the constituents of the plants to establish new active principles against various human ailments.

Table XI. Total Importance Values (TIV) of Ethnobotanically Important Plants of Jogindernagar Tehsil of Sino-Himalaya

S. No	Botanical Name	Dye	Fibre	Fodder	Fuel	Medicine	Oil	Tannin /Gum	Timber	Others	Edible	TIV
1.	*Abies pindrow*	-	-	-	-	6	4		8	-	-	18
2.	*Abrus precatorious*	-	-	-	-	10	-	-	-	4	-	14
3.	*Acacia catechu*	-	-	-	-	8	-	-	-	4	-	12
4.	*Acacia concinna*	-	-	-	-	6	-	-	-	8	-	14
5.	*Achyranthes aspera*	-	-		-	10	-	-		6	-	16
6.	*Achyranthes bidentata*	-	-	-	-	10	-	-	-	6	-	16
7.	*Actinidia chinense* var. *deliocosa*	-	-	-	-	-	-	-	-	-	6	6
8.	*Adhatoda vasica*	-	-	-	-	8	-	-	-	8	-	16
9.	*Aegle marmelos*	-	-	-	-	8	-	-	-	4	5	17
10.	*Aesculus indica*	-	-	4	-	8	-	-	6	5	6	29
11.	*Agave americana*	-	6	-	-	4	-	-	-	4	4	18
12.	*Albizzia julibrissin*	-	-	10	-	6	-	-	-	-	-	16
13.	*Albizzia lebbeck*	-	-	7	-	8	-	-	7	-	-	22
14.	*Albizzia stipulata*	-	-	10	-	4	-	-	8	6	-	28
15.	*Alnus nepalensis*	-	-	6	-	-	-	-	-	8	-	14
16.	*Alnus nitida*	6	-	6	-	-	-	-	-	-	-	12
17.	*Aloe vera*	-	-	-	-	8					5	13
18.	*Andrachne cordifolia*	-	-	7	-	4	-	-	-	5	-	16
19.	*Araucaria columnaris*	-	-	-	-	-		-	-	4	-	4
20.	*Arundinaria falcata*	-	-	-	-	-	-	-	-	10	-	10
21.	*Asclepias curassavica*	-	-	-	-	-		-	-	8	-	8
22.	*Asclepias physocarpa*	-	-	-	-	-		-	-	6	-	6
23.	*Atyloshia platycarpa*	-	-	-	-	7		-	-	-	6	13
24.	*Bambusa nutans*	-	6	-	-	8	-	-	10	10	-	28
25.	*Bauhinia vahlii*	-	6	-	-	8	-	-	-	10	-	24
26.	*Bauhinia variegata*	-	-	8	-	10	-	-	-	-	10	28
27.	*Berberis aristata*	8	-	6	6	10	-	-	-	-	8	38
28.	*Berberis lycium*	8	-	6	6	10	-	-	-	-	8	38
29.	*Boehmeria platyphylla*	-	-	-	-	5	-	-	-	-	-	5
30.	*Bombax ceiba*	-	-	8	-	8	-	8	-	8	8	40
31.	*Bougainvillea spectabilis*	-	-	-	-	-	-	-	-	10	-	10
32.	*Broussonetia papyrifera*	-	-	8	-	-	-	-	8	6	-	22
33.	*Buddleia crispa*	-	-	-	-	-	-	-	-	6	-	6
34.	*Buddleia paniculata*	-	-	6	-	2	-	-	-	8	-	16
35.	*Butea monosperma*	-	-	-	-	8	-	6	-	10	-	24
36.	*Buxus wallichiana*	-	-	6	-	-		-	4	-	-	10
37.	*Caesalpinia decapetala*	-	-	-	-	8	-	-	-	-	-	8
38.	*Callicarpa macrophylla*	-	-	-	-	8	-	-	-	8	-	16
39.	*Callistemon lanceolatus*	-	-	-	-	-	-	-	-	10	-	10
40.	*Calotropis procera*	-	-	-	-	10	-	-	-	-	-	10

41.	*Camellia sinensis*	8	-	-	-	4	-	-	-	10	-	22
42.	*Canna indica*	-	-	-	-	6	-	-	-	6	-	12
43.	*Carica papaya*	-	-	-	-	10	-	4	-	-	10	24
44.	*Carissa congesta*	-	-	8	-	4	-	-	-	6	8	26
45.	*Carpinus viminea*	-	-	5	8	-	-	-	7	4	-	24
46.	*Carya illenoesis*	8	-	-	-	8	-	-	-	-	10	26
47.	*Caryopteris wallichiana*	-	-	-	7	-	-	-	-	8	-	15
48.	*Cassia fistula*	-	-	-	-	8	-	-	6	-	-	14
49.	*Cassia laevigata*	-	-	-	-	7	-	-	-	-	8	15
50.	*Cedrella serrata*	-	-	5	-	-	-	-	7	-	-	12
51.	*Cedrella toona*	-	-	6	-	6	-	-	10	-	-	22
52.	*Cedrus deodara*	-	-	-	-	7	-	-	10	4	-	21
53.	*Celtis australis*	-	-	4	-	-	-	-	7	-	5	16
54.	*Cestrum nocturnum*	-	-	-	-	4	-	-	-	5	-	9
55.	*Cinnamomum camphora*	-	-	-	-	6	-	-	-	8	-	14
56.	*Cinnamomum tamala*	-	-	-	-	8	-	-	-	7	-	15
57.	*Cissampelos pareira*	-	-	-	-	8	-	-	-	-	-	8
58.	*Citrus aurantium*	-	-	-	-	6	-	-	-	-	10	16
59.	*Citrus grandis*	-	-	-	-	6	-	-	-	-	8	14
60.	*Citrus jambhiri*	-	-	-	-	6	-	-	-	7	6	19
61.	*Citrus limon*	-	-	-	-	4	-	-	-	8	5	17
62.	*Citrus medica*	-	-	-	-	4	-	-	-	7	10	21
63.	*Clematis buchananiana*	-	6	-	-	4	-	-	-	-	-	10
64.	*Clematis grata*	-	6	-	-	4	-	-	-	-	-	10
65.	*Clerodendron phillippinum*	-	-	-	-	4	-	-	-	6	-	10
66.	*Colebrookia oppositifolia*	-	-	-	6	4	-	-	-	4	-	14
67.	*Cordia dichotoma*	-	-	-	-	8	-	-	-	-	8	16
68.	*Coriaria nepalensis*	6	-	-	-	-	-	-	-	-	5	11
69.	*Cornus macrophylla*	-	-	6	-	-	-	-	7	-	-	13
70.	*Costus speciosus*	-	-	-	-	4	-	-	-	6	8	18
71.	*Cotoneaster bacillaris*	-	-	6	-	-	-	-	-	4	-	10
72.	*Cotoneaster macrophylla*	-	-	-	-	4	-	-	-	4	-	8
73.	*Cryptolepis buchanani*	-	-	-	-	6	-	-	-	-	6	12
74.	*Cucumis sativus*	-	-	-	-	4	-	-	-	6	10	20
75.	*Cucurbita hispida*	-	-	-	-	-	-	-	-	8	8	16
76.	*Cucurbita melo*	-	-	-	-	-	-	-	-	-	8	8
77.	*Cucurbita pepo*	-	-	-	-	-	-	-	-	6	8	14
78.	*Cuscuta reflexa*	-	-	-	-	10	-	-	-	-	-	10
79.	*Cyathula tomentosa*	-	-	6	-	-	-	-	-	6	-	12
80.	*Cycas circinalis*	-	-	-	-	-	-	-	-	5	-	5
81.	*Cyphomandra betacea*	-	-	-	-	-	-	-	-	4	6	10
82.	*Dalbergia sissoo*	-	-	-	-	6	4	-	10	-	-	20
83.	*Daphne cannabina*	-	-	-	-	6	-	-	-	-	-	6
84.	*Datura metel*	-	-	-	-	10	-	-	-	4	-	14
85.	*Datura stramonium*	-	-	-	-	8	-	-	-	8	-	16
86.	*Datura suaveolens*	-	-	-	-	6	-	-	-	8	-	14
87.	*Debregeasia hypoleuca*	-	-	6	6	6	-	-	-	-	-	18
88.	*Dendrocalmus hamiltonii*	-	-	8	-	4	-	-	10	6	-	28
89.	*Desmodium multiflorum*	-	-	6	6	-	-	-	-	6	-	18
90.	*Deutzia corymbosa*	-	-	6	-	-	-	-	-	6	-	12
91.	*Dioscorea bellophylla*	-	-	-	-	8	-	-	-	6	-	14
92.	*Dioscorea bulbifera*	-	-	-	-	-	-	-	-	8	-	8
93.	*Dioscorea deltoidea*	-	-	-	-	8	-	-	-	8	-	16
94.	*Diospyros kaki*	-	-	-	-	6	-	-	-	-	8	14
95.	*Dodonea viscosa*	-	-	6	8	8	-	-	-	-	-	22
96.	*Duranta repens*	-	-	-	-	6	-	-	-	6	-	12

97.	*Ehretia acuminata*	-	-	6	-	-	-	-	5	-	6	17
98.	*Ehretia laevis*	-	-	6	4	-	-	-	-	-	-	10
99.	*Elaeagnus umbellata*	-	-	-	6	6	-	-	-	-	8	20
100.	*Embellia tsjeriam-cottam*	-	-	-	7	-	-	-	-	-	6	13
101.	*Emblica officinalis*	6	-	-	5	10	-	-	-	10	10	41
102.	*Eriobotyrica japonica*	-	-	-	-	4	-	-	-	-	8	12
103.	*Erythrina suberosa*	-	-	-	-	7	-	-	-	8	-	15
104.	*Eucalyptus umbellata*	-	-	-	-	5	-	-	8	-	-	13
105.	*Eugenia jambolana*	-	-	-	-	8	-	-	8	6	8	30
106.	*Euonymus echinatus*	-	-	6	-	-	-	-	-	-	-	6
107.	*Euphorbia pulcherrima*	-	-	-	-	4	-	-	-	6	-	10
108.	*Euphorbia royleana*	-	-	-	-	4	-	5	-	4	6	19
109.	*Ficus bengalensis*	-	-	-	-	8	-	4	-	8	8	28
110.	*Ficus clavata*	-	-	4	-	-	-	-	-	-	4	8
111.	*Ficus foveolata*	-	-	4	-	-	-	-	-	-	-	4
112.	*Ficus glomerata*	-	-	6	-	7	-	-	-	8	-	21
113.	*Ficus hispida*	-	-	-	-	7	-	4	-	6	8	25
114.	*Ficus palmata*	-	-	-	-	8	-	5	-	-	8	21
115.	*Ficus religiosa*	-	-	-	-	10	-	-	-	8	-	18
116.	*Ficus roxburghii*	-	-	7	-	4	-	6	-	-	7	24
117.	*Ficus rumphii*	-	-	-	-	-	-	-	-	8	-	8
118.	*Ficus semicordata*	-	-	8	-	4	-	-	-	4	-	16
119.	*Flacourtia ramontchi*	-	-	6	4	5	-	-	-	-	7	22
120.	*Flemingia macrophylla*	-	-	8	-	6	-	-	-	-	-	14
121.	*Flemingia strobilifera*	-	-	-	-	5	-	-	-	-	-	5
122.	*Fuchsia splendens*	-	-	-	-	-	-	-	-	5	-	5
123.	*Gardenia angusta*	-	-	-	-	-	-	-	-	5	-	5
124.	*Glochidion velutinum*	8	-	-	6	4	-	-	5	3	-	26
125.	*Gmelina arborea*	-	-	-	-	6	-	-	6	-	8	20
126.	*Gossypium arboreum*	-	-	-	-	-	-	-	-	10	-	10
127.	*Grevillea robusta*	-	-	-	-	-	-	-	6	5	-	11
128.	*Grewia oppositifolia*	-	8	9	-	6	-	-	-	-	-	23
129.	*Hedera helix*	-	-	6	-	4	-	-	-	10	-	20
130.	*Hibiscus rosa-sinensis*	-	-	-	-	5	-	-	-	6	-	11
131.	*Holarrhena antidysentrica*	-	-	-	-	6	-	-	-	-	-	6
132.	*Hydrangea grandiflora*	-	-	-	-	-	-	-	-	5	-	5
133.	*Hypericum cernuum*	-	-	-	-	-	-	-	-	6	-	6
134.	*Hypericum patulum*	-	-	-	-	4	-	-	-	-	-	4
135.	*Ilex dipyrena*	-	-	7	-	4	-	-	4	-	-	15
136.	*Ilex excelsa*	-	-	8	-	-	-	-	-	-	-	8
137.	*Indigofera gerardiana*	-	-	-	7	6	-	-	-	6	-	19
138.	*Indigofera pulchella*	-	-	-	-	7	-	-	-	-	-	7
139.	**Inula cappa**	-	-	-	-	-	-	-	-	6	-	6
140.	*Ipomoea cairica*	-	-	6	-	-	-	-	-	-	-	6
141.	*Ipomoea carnea*	-	-	-	-	4	-	-	-	8	-	12
142.	*Ipomoea eriocarpa*	-	6	6	-	-	-	-	-	-	-	6
143.	*Ipomoea muricata*	-	-	-	-	4	-	-	-	7	-	11
144.	*Ipomoea quamoclit*	-	-	-	-	5	-	-	-	-	-	5
145.	*Jacaranda acutifolia*	-	-	-	-	5	-	-	6	8	-	19
146.	*Jasminum dispersum*	-	-	-	-	7	-	-	-	6	-	13
147.	*Jasminum grandiflorum*	-	-	-	-	8	-	-	-	6	-	14
148.	*Jasminum humile*	-	-	-	-	8	4	-	-	6	-	18
149.	*Jasminum multiflorum*	-	-	-	-	5	4	-	-	6	-	15
150.	*Juglans regia*	8	-	-	-	5	-	-	-	7	10	30
151.	*Kydia calycina*	-	-	7	-	-	-	-	6	-	-	13
152.	*Lagerstroemia indica*	-	-	-	-	4	-	-	-	6	-	10
153.	*Lantana camara*	-	-	-	-	6	-	-	-	5	4	15
154.	*Leea aspera*	-	-	6	-	-	-	-	-	-	5	11
155.	*Leptodermis lanceolata*	-	-	3	-	4	-	-	-	-	-	7
156.	*Lettsomia thomsonii*	-	-	6	-	-	-	-	-	-	7	13

157.	*Leucaena glauca*	-	-	7	-	-	-	-	-	7	-	14
158.	*Leycesteria formosa*	-	-	6	-	-	-	-	-	4	-	10
159.	*Litchi chinensis*	-	-	-	-	2	-	-	-	-	8	10
160.	*Litsea polyantha*	-	-	-	-	5	-	-	-	-	-	5
161.	*Litsea sebifera*	-	-	6	-	5	-	-	-	-	-	11
162.	*Litsea umberosa*	-	-	-	-	-	-	-	6	4	-	10
163.	*Lonicera japonica*	-	-	-	-	-	-	-	-	5	-	5
164.	*Lonicera quinquelocularis*	-	-	6	7	-	-	-	4	-	-	17
165.	*Loranthus ligustrinus*	-	-	4	-	5	-	-	-	-	-	9
166.	*Luffa cylindrica*	-	-	-	-	4	-	-	-	4	10	18
167.	*Lyonia ovalifolia*	-	-	-	-	6	-	-	-	-	-	6
168.	*Machilus odoratissima*	-	-	-	-	-	-	-	6	-	-	6
169.	*Mallotus phillipinensis*	-	-	7	-	6	-	-	-	8	-	21
170.	*Malvaviscus conzatti*	-	-	-	-	3	-	-	-	7	6	16
171.	*Mangifera indca*	-	-	-	-	7	-	-	-	10	10	27
172.	*Melia azedarach*	-	-	8	-	10	-	-	5	-	-	23
173.	*Melothria heterophylla*	-	-	-	-	5	-	-	-	7	6	18
174.	*Michelia champaca*	-	-	-	-	4	-	-	-	-	4	8
175.	*Milletia extensa*	-	-	5	-	-	-	-	-	6	-	11
176.	*Momordica charantia*	-	-	-	-	10	-	-	-	5	10	25
177.	*Moringa pterygosperma*	-	-	-	-	8	-	7	-	-	8	23
178.	*Morus alba*	-	-	-	-	8	-	-	-	8	8	24
179.	*Mucuna pruriens*	-	-	-	-	8	-	-	-	-	-	8
180.	*Murraya koenigii*	-	-	-	-	9	-	-	-	10	-	19
181.	*Myrica esculenta*	-	-	-	-	8	-	-	-	8	8	24
182.	*Myrsine africana*	-	-	6	-	8	-	4	-	-	-	18
183.	*Nerium indicum*	-	-	-	-	8	-	-	-	-	-	8
184.	*Opuntia dillenii*	-	-	-	-	6	-	-	-	8	4	18
185.	*Oroxylum indicum*	-	-	5	-	7	-	-	-	-	-	12
186.	*Osbeckia stellata*	-	-	-	-	-	-	-	-	6	4	10
187.	*Osyris arborea*	-	-	-	-	4	-	-	-	6	-	10
188.	*Pavetta indica*	-	-	-	-	3	-	-	-	6	-	9
189.	*Phoenix sylvestris*	-	-	5	-	4	-	-	-	10	6	25
190.	*Phyllanthus parviflolius*	-	-	-	-	5	-	-	-	4	-	9
191.	*Picea smithiana*	-	-	-	-	5	-	-	-	5	-	10
192.	*Pinus roxburghii*	-	-	-	-	4	8	-	-	8	6	26
193.	*Pinus wallichiana*	-	-	4	-	-	-	-	8	-	-	12
194.	*Pistacia integrimma*	-	-	-	-	8	-	-	-	6	-	14
195.	*Plumbago zeylanica*	-	-	-	-	7	-	-	-	-	6	13
196.	*Plumeria rubra*	-	-	-	-	-	-	-	-	6	-	6
197.	*Pogostemon plectranthoides*	-	-	-	-	10	-	-	-	-	-	10
198.	*Polygonum chinense*	-	-	-	-	-	-	-	-	-	6	6
199.	*Populus ciliata*	-	-	6	-	2	-	-	4	-	-	12
200.	*Populus nigra*	-	-	-	6	2	-	-	6	-	-	14
201.	*Premna latifolia*	-	-	-	-	8	-	-	-	-	-	8
202.	*Prinsepia utilis*	-	-	-	-	4	8	7	-	10	7	36
203.	*Prunus amygdalus*	-	-	-	-	10	-	-	-	-	10	20
204.	*Prunus armeniaca*	-	-	-	-	6	-	8	-	-	10	24
205.	*Prunus cerasoides*	-	-	6	-	4	-	-	-	8	4	22
206.	*Prunus domestica*	-	-	-	4	3	-	-	-	-	8	15
207.	*Prunus persica*	-	-	-	-	4	8	8	-	2	8	30
208.	*Psidum guajava*	-	-	-	-	6	-	-	-	-	10	16
209.	*Punica granatum*	-	-	-	-	8	-	-	-	6	10	24
210.	*Pyrus communis*	-	-	-	-	4	-	-	-	4	10	18
211.	*Pyrus malus*	-	-	-	-	-	-	-	-	4	10	14
212.	*Pyrus pashia*	-	-	-	-	4	-	-	8	-	8	20
213.	*Quercus glauca*	-	-	6	-	6	-	-	8	10	-	30
214.	*Quercus leucotrichophora*	-	-	-	-	6	-	-	8	9	-	23
215.	*Quercus semicarpifolia*	-	-	-	-	-	-	-	8	4	-	12
216.	*Randia dumetorum*	-	-	-	-	8	-	-	-	6	-	14
217.	*Reinwardtia indica*	-	-	-	-	2	-	-	-	6	-	8

218.	*Rhamnus persicus*	-	-	4	-	-	-	-	-	6	-	10
219.	*Rhamnus triquetra*	-	-	-	6	3	-	-	-	-	6	15
220.	*Rhododendron arboreum*	-	-	-	-	8	-	-	-	6	8	22
221.	*Rhus cotinus*	-	-	-	-	-	-	-	-	4	-	4
222.	*Rhus semialata*	-	-	-	-	3	-	-	-	4	-	7
223.	*Rhus succedanea*	-	-	-	-	2	-	-	-	4	-	6
224.	*Ricinus communis*	-	-	-	-	7	-	-	-	8	-	15
225.	*Robinia pseudoacacia*	-	-	4	4	2	-	-	-	4	-	14
226.	*Rosa alba*	-	-	-	-	8	-	-	-	8	-	16
227.	*Rosa moschata*	-	-	8	-	8	-	-	-	10	-	26
228.	*Roylea cinerea*	-	-	-	-	6	-	-	-	4	-	10
229.	*Rubia cordifolia*	-	-	-	-	5	-	-	-	-	-	5
230.	*Rubus ellipticus*	-	-	8	7	6	-	-	-	6	10	37
231.	*Rubus lasiocarpus*	-	-	-	-	5	-	-	-	-	10	15
232.	*Rubus niveus*	-	-	4	-	-	-	-	-	-	10	14
233.	*Saccharum officinarum*	-	-	-	-	8	-	-	-	4	10	22
234.	*Sageretia oppositifolia*	-	-	4	-	-	-	-	-	-	3	7
235.	*Salix babylonica*	-	-	-	-	2	-	-	-	4	-	6
236.	*Salix karelinii*	-	-	2	2	-	-	-	-	-	2	6
237.	*Salix tetrasperma*	-	-	6	6	2	-	-	8	4	-	26
238.	*Sapindus mukorossi*	-	-	-	-	6	-	-	-	8	-	14
239.	*Sapium insigne*	-	-	-	-	6	-	4	-	2	-	12
240.	*Sapium sebiferum*	-	-	7	6	4	-	-	-	-	-	17
241.	*Saraca asoca*	-	-	-	-	5	-	-	-	8	-	13
242.	*Sarcococca saligna*	-	-	-	-	4	-	-	-	-	2	6
243.	*Saurauia nepaulensis*	-	-	4	-	-	-	-	-	-	2	6
244.	*Sechium edule*	-	-	-	-	4	-	-	-	3	10	17
245.	*Senecio rufinervis*	-	-	4	-	-	-	-	-	5	-	9
246.	*Side cordifolia*	-	-	-	-	6	-	-	-	-	-	6
247.	*Skimmia laureola*	-	-	-	-	-	-	-	6	6	-	12
248.	*Smilax aspera*	-	-	-	-	3	-	-	-	5	4	12
249.	*Solanum erianthum*	-	-	-	-	2	-	-	-	-	-	2
250.	*Solanum melongena*	-	-	-	-	2	-	-	-	-	10	12
251.	*Solanum pseudocapsicum*	-	-	-	-	-	-	-	-	2	6	8
252.	*Solanum torvum*	-	-	-	-	-	-	-	-	-	6	6
253.	*Solidago canadensis*	-	-	-	-	-	-	-	-	5	-	5
254.	*Sorbaria tomentosa*	-	-	-	6	3	-	-	8	2	-	19
255.	*Spirarea canescens*	-	-	-	-	-	-	-	5	5	-	10
256.	*Spondias mangifera*	-	-	6	-	-	-	-	-	4	8	18
257.	*Stephania elegans*	-	-	6	-	-	-	-	-	2	-	8
258.	*Stephania glabra*	-	-	-	6	-	-	-	-	-	-	6
259.	*Sterculia villosa*	-	-	4	-	-	-	-	-	-	-	4
260.	*Strobilanthes glutinosis*	-	-	4	-	-	-	-	-	-	-	4
261.	*Symplocos racemosa*	-	-	-	7	4	-	-	8	-	-	19
262.	*Symplocos spicata*	7	-	4	-	-	-	-	-	-	-	11
263.	*Tabernaemontana divaricata*	-	-	-	-	4	-	-	-	6	-	10
264.	*Taxus baccata*	7	-	-	-	7	-	-	8	4	-	26
265.	*Tecoma grandiflora*	-	-	-	-	-	-	-	-	6	-	6
266.	*Terminalia arjuna*	-	-	-	-	8	-	-	-	7	-	15
267.	*Terminalia bellerica*	6	-	-	-	10	-	-	-	-	6	22
268.	*Terminalia chebula*	6	-	-	-	10	-	-	-	8	4	28
269.	*Terminalia tomentosa*	-	-	8	-	-	-	-	8	-	-	16
270.	*Thevetia neriifolia*	-	-	-	-	6	-	-	-	-	4	10
271.	*Thuja orientalis*	-	-	-	-	5	-	-	-	8	-	13
272.	*Thunbergia coccinea*	-	-	-	-	-	-	-	-	6	-	6
273.	*Tinospora cordifolia*	-	-	6	-	10	-	-	-	10	-	26
274.	*Trachelospermum lucidum*	-	-	7	-	-	-		-	5	-	12
275.	*Trichosanthes tricuspidata*	-	-	8	-	8	-	-	-	8	-	24
276.	*Tropaeolum mazus*										6	6

277.	*Ulmus laevigata*	-	-	-	6	-	-	-	8	-	-	14
278.	*Viburnum cotonifolium*	-	-	-	7	-	-	-	-	-	8	15
279.	*Viburnum stellulatnm*	-	-	7	-	-	-	-	6	-	8	21
280.	*Vigna unguiculata*	-	-	-	-	-	-	-	-	-	10	10
281.	*Vinca rosea*	-	-	-	-	-	-	-	-	10	-	10
282.	*Vitex negundo*	-	-	-	-	10	-	-	-	8	-	18
283.	*Vitis himalayana*	-	-	6	-	-	-	-	-	-	4	10
284.	*Vitis jacuemontii*	-	-	-	-	6	-	-	-	-	8	14
285.	*Vitis rependa*	-	-	7	-	4	-	-	-	-	8	19
286.	*Wendlendia heynei*	-	-	-	-	-	-	-	6	3	-	9
287.	*Woodfordia floribunda*	8	-	-	6	10	-	-	-	10	4	38
288.	*Xylosma longifolium*	-	-	6	6	2	-	-	-	6	-	20
289.	*Zanthoxylum armatum*	-	-	-	-	10	-	-	6	10	10	36
290.	*Ziziphus jujuba*	-	-	8	-	4	-	-	-	10	10	32

Thus, it is evident that the rural people are the inheritors and developers of rich traditions of plants and ecosystem management and the whole gamut of ethnobotany rests in their hands. Further, we are still unaware of the potential and economic worth of many plant species and attempts should be made by one and all to safeguard the natural genetic variability and traditional uses of plants with an eye to potential future use.

❑❑❑

Chapter **4**

Literature Cited

Adegoke, E.A., Akisanya, A. & Naqvi, S.H.Z. 1968. Studies of Nigerian medicinal plants-I. A preliminary survey of plant alkaloids. *J. West. Afr. Sci. Assoc.,* 13: 13-33.

Ahluwalia, K.S. 1952. Medicinal plants of Kangra valley. *Indian Forester* 78(4): 181-194.

Ahuja, B.S. 1965. *Medicinal Plants of Saharanpur.* Survey of Medicinal Plants, Central Council of Ayurvedic Research Gurukul Kangri Vishwavidalya, Hardwar.

Aitchison, J.E.T. 1868. Lahul, its flora and vegetable products etc. *J. Linn. Soc. Bot.* 10: 69-101.

Alcorn, J.B. 1995. The scope and aims in ethnobotany in developing world, 23-29. *In*: Schultes, R.E. & Reis, S.V. (ed.) *Ethnobotany, Evolution of a Discipline.* Dioscorides Press, Porolana.

Alcorn, J.B. 1996. Is biodiversity conserved by indigenous people? : 233-238. *In*: Jain, S.K. (ed.) *Ethnobiology in Human Welfare.* Deep Publ., New Delhi.

Ambasta, S.P. (ed.)1986. *The Useful Plants of India.* C.S.I.R., New Delhi.

Aminuddin & Girach, R.D. 1991. Ethnobotanical studies on Bondo tribe of district Karaput (Orissa), India. *Ethnobotany* 3: 15-19.

Aminuddin & Girach, R.D. 1993. Observations on ethnobotany of the Bhujia-a tribe of Sonabera plateau. *Ethnobotany* 5: 83-86.

Aminuddin & Girach, R.D. 1996. Native phytotheraphy among the Paudi Bhuinya of Bonai Hills. *Ethnobotany* 8: 66-70.

Anonymous, 1994. *Ethnobiology in India- A Status Report.* Ministry of Environment and Forest, Govt. of India.

Anonymous, (1948-1976). *The Wealth of India, Raw Materials* Vol.–I-XI Publications & Information Directorate, C.S.I.R., New Delhi

Anonymous, 1948. *Report of the Committee on Indigenous Systems of Medicine* Vol.–I. Ministry of Health, Govt. of India, New Delhi.

Anonymous, 1990. *Phytochemical Investigation of Certain Medicinal Plants Used in Ayurveda.* Central Council for Research in Ayurveda & Sidha, New Delhi.

Anonymous, 2002-2003. *Statistical Outline of Himachal Pradesh.* Directorate of Economics and Statistics, Himachal Pradesh, Shimla-1.

Ansari, A.A. 1991. Ethnobotanical notes on some plants of Khirsu, Pauri Garhwal, U.P. *Ethnobotany* 3: 105-106.

Arizpe, L. 1996. Culture and environment. *Nature and Resources* 32: 1-30.

Arora, R.K. & Pandey, A. 1996. *Wild Edible Plants of India- Diversity, Conservation and Use.* NBPGR, New Delhi.

Arora, R.K., Maheswari, M.L., Chandel, K.P.S. and Gupta, R. 1980. Mano (*Inula racemosa*): Little known aromatic plant of Lahaul valley, India. *Eco. Bot.* 34(2): 174-180.

Asolkar, L.V., Kakkar, K.K. & Chakre, O.J. 1992. *Second Supplement to Glossary of Indian Medicinal Plants with Active Principles,* Part–I (A-K). CSIR, New Delhi-2.

Aswal, B.S. & Mehrotra, B.N. 1994. *Flora of Lahaul Spiti (A Cold Desert in North West Himalaya).* B.S.M.P.S., Dehradun.

Aswal, B.S. 1996. Conservation of ethno-medicinal plants diversity of Garhwal Himalaya in India, 133-135. *In*: Jain, S.K. (ed.) *Ethnobiology in Human Welfare.* Deep Publ., New Delhi.

Atkinson, E.T. 1882. *Economic Botany of the Himalayan Region.* Cosmo Publ., New Delhi.

Badola, H.K. 2001. Medicinal plant diversity in Himachal Pradesh, 407-430. *In*: Samant, S.S., Dhar, U. and Palani, L.M.S. (eds.) *Himalayan Medicinal Plants: Potential and Prospects.* Himvikas Occasional Publication, No. 14, Gyanodya Prakashan.

Badola, H.K. and Pal, M. 2002. Endangered medicinal plant species in Himachal Pradesh. *Current Science* 83: 797-798.

Badola, H.K. and Pal, M. 2003. Threatened medicinal plants and their conservation in Himachal Himalayas. *Indian Forester* 129(1):55-68.

Bajpayee, Kaptain Kishor & Dixit, Gopal 1996. Ethnobotanical Studies on Food stuffs of Tribals of Tarai-Region, Uttar Pradesh. *J. Econ. Taxon. Bot.* Additional Series 12: 128-132.

Balasubramaniam, P. & Prasad, Narendra 1996. Medicinal plants among the Irulars of Attapady and Bolumampatti forests in the Nilgiri Biosphere Reserve. *J. Econ. Taxon. Bot.* Additional Series 12: 253-259.

Balick, M.K. & Cox, P.A. 1976. *Plants, People and Culture: The Science of Ethnobotany*. Scientific American Library, New York.

Banerjee, Archana 1996. Medicinal uses of some flowers by Santhals of West Bengal. *J. Econ. Taxon. Bot.* Additional Series 12: 314-317.

Banerjee, D.K. & Pal, D.C. 1996. Plant used by the tribals of North Indian plains for hair and scalp protection, 187-188. *In*: Jain, S.K. (ed.) *Ethnobiology in Human Welfare*. Deep Publ., New Delhi.

Barford, A.S. & Kvist, L.P. 1996. *Comparative Ethnobotanical Studies of the Amerindian Groups in Coastal Equador*. The Royal Danish Academy of Sciences and Letters. Copenhegen, 166 pp.

Baruah, P. & Sarma, G.C. 1984. Studies on the medicinal uses of plants by the Boro tribals of Assam II. *J. Econ. Tax. Bot.* 5: 599-604.

Baruah, P. & Sarma, G.C. 1987. Studies on the medicinal uses of plants by the Northeast tribes-III. *J. Econ. Tax. Bot.* 11: 71-76.

Belal, Ahmed Esmat & Springuel, Irina 1996. Economic value of plant diversity in arid environments. *Nature & Resources* 32 (1): 33-39.

Bellany, R. 1995. *Ethnobiology- Expedition Field Techniques*. Expedition Advisory Center, Royle Geographic Societym London.

Bennet, S.S.R. 1986. *Name Changes in Flowering Plants of India and Adjacent Regions*. Triseas Publishers, 94A, Indra Colony Dehra Dun-248001, India.

Bentley, Robert & Trimen, Henry 1980. *Medicinal Plants* Vol. I-IV. J. and A. Churchill, New Bulington Street, London.

Bhakta, Tejendra, Mukerjee, Pulok K., Saha, K., Pal, M. & Saha, B.P. 1997. Hypoglycaemic Activity of *Cassia fistula* Linn. (Leguminosae) Leaf (Methanolic extract) in alloxan-induced Diabetic Rats. *Ethnobotany* 9: 35-38.

Bhandary, M.J., Chandrasekhar, K.R. & Kaveriappa, K.M. 1996. Ethnobotany of Gowlis of Uttar Kannada District, Karnataka. *J. Econ. Taxon. Bot.* Additional Series 12: 244-249.

Bhattacharyya, Goutam 1996. Medico-ethno-botanical value of Saurashtra weeds. *J. Econ. Taxon. Bot. Additional Series* 12: 166-168.

Binoj Kumar, M.S. & Balakrishnan, N.P. 1996. Ethnobotanical studies of one genus Euphorbial (Euphorbiaceae). *J. Econ. Taxon. Bol.* Additional Series 12: 46-49.

Bisset, Norman Grainger & Vichtl, Max (ed.) 2001. *Herbal Drugs and Pharmaceuticals*. Medpharm Scientific Publi., Stuttgart.

Boef, de S.W. (ed.) 1992. Local Knowledge and Agriculture Research Report on WAU/ ENDA-Zimbabwe/CGN/GRAIN seminar (28 Sept-2 Oct. 1992) at Nyanga, Zimbabwe. Dept. of Sociology and rural Development, Wageningen Agricultural Univ. & CPRO-DLO centre for Genetic Resources, wageningen, The Netherlands.

Borthakur, S.K. & Sarma, U.K. 1996. Ethnoveterinary medicine with special reference to cattle prevalent among the Nepalis of Assam, India, 197-199. *In*: Jain, S.K. (ed.) *Ethnobiology in Human Welfare*. Deep Publ., New Delhi.

Borthakur, S.K. 1993. Native phytotherapy for child and women diseases from Assam. *Ethnobotany* 5(87): 87-91.

Borthakur, S.K. 1996a. Postnatal care of women in traditional system in Assam. *Ethnobotany* 8: 51-55.

Borthakur, S.K. 1996b. Wild edible plants in markets of Assam, India-an ethnobotanical investigation, 31-34. *In*: Jain, S.K. (ed.) *Ethnobiology in Human Welfare*. Deep. Publ., New Delhi.

Borthakur, S.K., & Gogoi, P. 1994. Indigenous technology of making writing materials among the Tai Khamtis. *Ethnobotany* 6: 5-8.

Brahmam, M. & Saxena, H.O. 1990 Ethnobotany of Gardamardan Hills–some noteworthy folk-medicinal uses. *Ethnobotany* 2: 71-79.

Brahmam, M., Dhal, N.K. & Saxena, H.O. 1996. Ethnobotanical studies among the Tanla of Malayagiri Hills in Dhenkanal District, Orissa, India, 393-396. *In*: Jain, S.K. (ed.) *Ethnobiology in Human Welfare*. Deep Publ., New Delhi.

Brussell, D.E. 1998. *Potions, Poisons and Panaceas: An Ethnobotanical Study of Montserrat*. Southern Illinois University Press, Carbondale.

Chakravarthy, B.K., Gupta, S. & Godke, K.D. 1983. Preliminary report on pharma-cological & toxicological studies of *Acacia catechu* Willd. *Indian Drugs* 20: 397.

Chandra, S. 1990. *Foundations of Ethnobotany (Pre-1900 Ethnobotany)*. A Review and Bibliography. Deep Publ., New Delhi.

Chandrasekhar, K. & Srivastava, S.K. 2003. Ethnomedicinal studies in Pin valley national park, Lahaul-spiti, Himachal Pradesh. *Ethnobotany* 15: 44-47.

Chaudhuri, Rai H.N. & Pal, D.C. 1981. Plants in folk religion and mythlogy, 59-68. *In*: Jain, S.K. (ed.) *Glimpses of Indian Ethnobotany*. Oxford & IBH Publ. Co. New Delhi.

Chauhan, N.S. 1974. Medicinal plants of Una forest division, Una district, H.P. lying in Shivalik range. *Nagarjun* 27: 31-39.

Chauhan, N.S. 1984. *Medicinal Wealth of Pabbar Valley in Himachal Pradesh*, M.Sc. Thesis, HPKVV, Palampur.

Chauhan, N.S. 1988. Ethnobotanical study of medicinal plants of Himachal Pradesh. 187-198. *In*: Kaushik, P. (ed.) *Indigenous Medicinal Plants*. Today and Tomorrow Printers & Publ., New Delhi.

Chauhan, N.S. 1989a. Herbal wealth of cold desert area of Himachal Pradesh. *Proceedings, Natinal Seminar-cum-workshop on cold desert area in India.* 65-68, Govt. Press, H.P. Shimla.

Chauhan, N.S. 1989b. Potential of aromatic plants flora in Himachal Pradesh. *Indian Perfumer* 33(2): 118-122.

Chauhan, N.S. 1990. Medicinal orchids of Himachal Pradesh. Orchid Society of India (Special Number). *J. Orchid Soc. India* 4 (1,2): 99-105.

Chauhan, N.S. 1996. Ethno-medicobotany of Pabbar valley in District Shimla, Himachal Pradesh (India). *In: UHF-IUFRO International workshop in Prospects of Medicinal Plants.* 4-9 Nov. p.30.

Chauhan, N.S. 1999. *Medicinal and Aromatic Plants of Himachal Pradesh.* Indus Publ. Co., New Delhi.

Chauhan, N.S. 2003. Important Medicinal and Aromatic Plants of Himachal Pradesh. *Indian Forester* 129(8): 979-998.

Chhetri, R.B. 1994. Further observations on ethnomedicobotany of Khasi Hills in Meghalaya, India. *Ethnobotany* 6: 33-36.

Chopra, I.C. & Handa, K.L. 1961. *Review of Research on Indian Medicinal and Allied Plants.* ICAR, New Delhi.

Chopra, R.N., Badhwar, R.L. & Gosh, S. 1965. *Poisonous Plants of India.* 2 Vols., ICAR, New Delhi.

Chopra, R.N., Chopra, I.C. & Verma, B.S. 1969. *Supplement to Glossary of Indian Medicinal Plants.* C.S.I.R., New Delhi.

Chopra, R.N., Nayar, S.L. & Chopra, I.C. 1956. *Glosary of Indian Medicinal Plants.* CSIR, New Delhi.

Chowdhery, H.J. 1996. Arunachal Pradesh, its people, ethnobotanical diversity and conservation, 323-328. *In:* Jain, S.K. (ed.) *Ethnobiology in Human Welfare.* Deep. Publ., New Delhi.

Chowdhury, H.J & Wadhwa, B.M. 1984. *Flora of Himachal Pradesh.* Vol. 1-3. Bot. Sur. India, Calcutta.

Chowdhury, Jasim Uddin, Alam, M.K., & Hassan, M.A. 1996. Some traditional folk formularies against dysentery and diarrhoea in Bangladesh. *J. Econ. Taxon. Bot.* Additional Series 12: 20-23.

Cohen, J.I., Alcorn, J.B. & Potter, C.S. 1991. Utilization and conservation of genetic resources. International Projects for Sustainable agriculture. *Econ. Bot.* 45: 190-198.

Collett, H. 1902. *Flora Simlensis.* Thacker Spink and Co. Calcutta and Shimla, Reprinted 1971. B.S.M.P.S. Dehradun.

Craford, G.W. 1983. *Palaeoethnobotany of the Kameda Peninsula.* Jamon Univ., Michigan, Ann. Abor.

Cyong, J.C. & Takahashi, M. 1982. Guanosine-3¢5¢-monophosphate activity in fruits of *Zizyphus jujuba*. *Chem. Pharma. Bull.* 30: 1081.

Dagar, H.S. & Dagar, J.C. 1996. Ethnobotanical studies of the Nicobarese of Chowra island of Nicobar group of islands. *J. Econ. Tax. Bot.* Addl. Series 12: 381-388.

Dagar, H.S. 2003. Ethnobotanical evaluation of some endemic plants in Andaman and Nicobar Islands. 134-147. *In*: *Vistas in Ethnobotany*, Vol.-I, *Indian Journal of Pure and Applied Biology*.

Dagar, J.C. & Dagar, H.S. 1999. *Ethnobotany of Aborigines of Andaman-Nicobar Islands*. Surv. of India, Port Blair, Andaman & Nicobar Islands, India.

Dastur, J.F. 1970. *Medicinal Plants of India and Pakistan*. D.B. Taraporevala Sons & Co., Private Ltd. 2nd ed. Bombay, p. 212.

Datt, Bhaskar 1996. Ethnobotanical resources of Chhatarpur district (Madhya Pradesh), *In*: Jain, S.K. (ed.) *Ethnobiology in Human Welfare*, Deep Publ., New Delhi.

Dey, A.C. 1980. *Indian Medicinal Plants Used in Ayurvedic Preparations*. Bishen Singh Mahender Pal Singh, Dehradun.

Dhaliwal, D.S. & Sharma, M. 1999. *Flora of Kullu District (Himachal Pradesh)*. Bishen Singh Mahender Pal Singh, Dehradun.

Dhiman, D.R. 1976. *Himachal Pradesh Ki Vanoshdhiya Sampada*. Imperial Printing Press. Dharamsala, H.P.

Dixit, R.D., & Pal, D.C. 1996. Role of household remedies in primary health care, 108-109. *In*: Jain, S.K. (ed.) *Ethnobiology in Human Welfare*, Deep Publ., New Delhi.

Dobremez, J.F. 1976. Exploration and prospects of medicinal plants in eastern Nepal. *In*: *Mountain Environment and Development*. Swiss Association for Technical Assistance in Nepal, Kathmandu.

Dobriyal, R.M., Singh, G.S., Rao, K.S. & Saxena, K.G. 1997. Medicinal plant resources in Chhakinal watershed in north-western Himalaya. *J. Herbs Spices & Medicinal Plants* 5: 15-27.

Duhoon, S.S., Kopper, M.N., and Chandra, U. 1996. Seabuckthorn (*Hippophae spp*)- A less known wonder plant of ethno-medico-botanical importance in cold desert of India. *J. Econ. Tax. Bot.* Addl. Series 12:43-45.

Duke, J.A. 1968. *Darine Ethnobotanical Dictionary*. (3rd Edition) Sci. Publ., Jodhpur.

Duke, J.A. 1986. *An Isthmian Ethnobotanical Dictionary*. Sci. Publ., Jodhpur.

Dwarakan, P. & Ansari, A.A. 1996. Less known uses of plants of Kollimalai (Salem Dt., Tamil Nadu) in South India. *J. Econ. Taxon. Bot.* Additional Series 12: 284-286.

Faulks, P.J. 1958. *An Introduction to Ethnobotany*. Moredale, London.

Felger, R.S. & Moser, M. B. 1985. *People of the Desert and Sea: Ethnobotany of the Serj Indians*. Univ. Arizon Press, Tucson.

Ford, R. 1980. Ethnobotany: historical diversity and synthesis, 33-49. *In*: Ford, R. (ed.) *The Nature and Status of Ethnobotany*. Univ. Michgan, USA.

Ford, R.I. (ed.) 1978. *The Nature and Scope of Ethnobotany*. Anthropological Paper no. 67. Museum of Anthro. Univ. of Michigan, Ann. Arbor.

Gaur, R.D. & Singh, P.B. 1993. Ethno-medicinal plants of Mandi district, Himachal Pradesh. *Bull. Medico Ethno. Bot. Res.* 14: 1-14.

Geetha, S., Lakshmi, G. & Ranjithakani, P. 1996a. Ethnobotanical review: wild fibre yielding plants of Kolli Hills, Tamil Nadu. *J. Eco. Taxon. Bot.* Additional Series. 12 : 250-252.

Geetha, S., Lakshmi, G. & Ranjithkani, P. 1996b. Ethnoveterinary medicinal plants of Kolli Hills, Tamil Nadu. *J. Econ. Taxon. Bot.* Additional Series. 12: 289-191.

Gefu, Jerome O., Abdu, Paul A. & Alawa, Clement B. (eds.) 2000. *Ethnoveterinary Practices, Research and Development*. Publ. by National Animal Production Research Institute, Shika Ahmadu Bello University, PMB 1096, Zaria, Nigeria.

Ghosh, Ashis, Maity, Sarathi & Maity, Malaty 1996. Ethnomedicine in Bankura district, West Bengal. *J. Econ. Taxon. Bot.* Additional Series 12: 318-320.

Gill, L.S. & Nyawuame, H.G.K. 1994. Leguminosae in ethnomedicinal practices of Nigeria. *Ethnobotany* 6: 51-64.

Gill, L.S. 1990. A survey of the plants of southern Nigeria for tannins. *Nig. J. Appl. Sci.* 8: 5-8.

Gill, L.S., Idu, M. & Ogbor, D.N. 1997. Folk medicinal plants; practices and beliefs of the Bini People in Nigeria. *Ethnobotany* 9: 1-5.

Gill, L.S., Nyawuame, H.G.K., Esezobor, E.I., & Osagie, I.S. 1993. Nigerian Folk Medicine : practices and beliefs of Esan people. *Ethnobotany* 5: 129-142.

Girach, R.D. & Aminuddin 1995. Ethnomedicinal uses of plants among the tribals of Singhbhum district, Bihar, India. *Ethobotany* 7: 103-107.

Girach, R.D. 1992. Medicinal plants used by Kondh tribe of district Phulbani (Orissa) in Eastern India. *Ethnobotany* 4: 53-66.

Girach, R.D., Aminuddin, Ahmed, M., Brahmam, M. & Misra, M.K. 1996. Native phytotherapy among rural population of district Bhadrak, Orissa, 162-164. *In*: Jain, S.K. (ed.) *Ethnobiology in Human Welfare*. Deep Publ., New Delhi.

Girach, R.D., Aminudhin, Brahmam, M. & Mishra, M.K. 1997. Observations on ethnomedicinal plants of Bhadrak district, Orissa, India. *Ethnobotany* 9: 44-46.

Goel, A.K., Sahoo, A.K. & Mudgal, V. 1984. A contribution to the ethnobotany of Santal Paragana. *Bot. Surv. India*, Howrah.

Gogoi, J.C., Sharma, D.K. & Borthakur, S.K. 2003. Folk-lore lepatoprotective herbal recipes for Assam in Northeastern India. 1-21. *In*: *Vistas in Ethnobotany* Vol.-1. *Indian Journal of Applied and Pure Biology*.

Gogoi, P. & Boissya, C.L. 1984. Information about a few herbal medicines used by the people of Assam (India) against Jaundice. *Himal. Res. Dev.* 2: 41-44.

Gogoi, P. & Borthakur, S.K. 1991. Plants in religio-cultural beliefs of Tai Khamtis of Assam (India). *Ethnobotany* 3: 89-95.

Goud, P. Saiprasad & Pullaiah, T. 1996a. Ethnobotany of Kurnool district– some wild plants used as a food. *J. Econ. Taxon. Bot.* Additional Series. 12: 224-227.

Goud, P. Saiprasad & Pullaiah, T. 1996b. Folk veterinary of Kurnool district, Andhra Pradesh. *Ethnobotany* 8: 71-74.

Grotenhermen, Franjo & Russo, Ethan (ed.) 2002. *Cannabis and Cannabinoids Pharmacology, Toxicology and Therapeutic Potential.* The Haworth Press, Inc., alice Street, Binghamton, New York.

Gunther, Erna 1945. *Ethnoboatny of Western Washington.* Univ. of Washington Press, London.

Gupta, M.P. 1988. Present status and future prospects for the cultivation and collection of Medicinal plants of Himachal Pradesh. *Indian Forester* 114(1): 19-25.

Gupta, R. 1966. *Orchis latifolia* Linn. A little known economic plant of north-western Himalaya. *Indian Forester* 92(11): 701-705.

Gupta, R. 1971. Medicinal and aromatic plants of Bhandal ranges, Churah forest division, Chamba district, Himachal Pradseh. *J. Bomb. Nat. His. Soc.* 68: 791-803.

Gupta, Rajendra 1961. Flora of Lam Dal. *Ind. Forester* 87(5): 316-324.

Gupta, Rajendra 1981. Plants in folk medicine of the Himalaya, 83-90. In Jain, S.K. (ed.) *Glimpses of Indian Ethnobotany.* Oxford & IBH Publ. Co., New Delhi.

Hartwell, J.L. 1967-71. Plants used against cancer: A survey. *Lloydia* 30: 379-436; 31: 71-170; 32: 79-107, 153-205, 247-296; 33: 97-104, 280-392; 34: 103-106.

Heiser, C.B. 1995. Edgar Anderson botanist and curator of useful plants. *Ann. Miss. Bot. Gard.* 82(1): 54-60.

Hembrom, P.P. 1991. Tribal medicine in Chotanagpur and Santhap Parganas of Bihar, India. *Ethnobotany* 3: 97-99.

Hembrom, Peter Paul 1996. Contact therapy practised by Mundas of Chotanagpur (Bihar). *Ethnobotany* 8: 36-39.

Henry, A.N., Hosagoudar, V.B. & Ravi Kumar, K. 1996. Ethno-medico-botany of the Southern Western Ghats of India, 173-180. *In*: Jain, S.K. (ed.) *Ethnobiology in Human Welfare*, Deep Publ., New Delhi.

Hooker, J.D. 1872-1897. *The Flora of British India,* Vol. I-VIII. Lalit Mohan Basu, Allahabad.

Hosagoudar, V.B. & Henry, A.N. 1996a. Ethnobotany of Soligas in Biligiri Rangana Betta, Karnataka, Southern India. *J. Econ. Taxon. Bot.* Additional Series 12: 228-243.

Hosagoudar, V.B. & Henry, A.N. 1996b. Ethnobotany of Kadars, Malasars and Kuthuvans of the Anamalais in Coimbatore District, Tamil Nadu, India. *J. Econ. Taxon. Bot.* Additional Series 12: 260-267.

Hosagoudar, V.B. & Henry, A.N. 1996c. Ethnobotany of Tribes Irular, Kurumban and Paniyan of Nilgiris in Tamil Nadu, Southern India. *J.Econ. Taxon. Bot.* Additional Series 12: 364-366.

Huidrom Singh, B.K. 1996. Plants used in medico-sexual purposes by Meitein community in Manipur state, India. *J. Econ. Taxon. Bot.* Addln. Series 12: 364-366.

Islam, M. 1996. Ethnobotany of certain under ground parts of plants of north-eastern region, India. *J. Econ. Taxon. Bot.* Additional Series 12: 338-343.

Jain A.K. 1992. Clans of Sor tribals of Madhya Pradesh and their role in conservation. *Ethnobotany* 4: 67-69.

Jain, A.K. & Sharma, H.O. 1996. Ethnobiological studies of Shariya tribe of Central India, 397-399. *In*: Jain, S.K. (ed.) *Ethnobiology in Human Welfare*. Deep Publ., New Delhi.

Jain, S.K. & Rao, R.R. (eds.) 1977. *A Handbook of Field and Herbarium Methods*. Today's & Tommorrow's Printers & Publ., New Delhi.

Jain, S.K. & Saklani, Arvind 1992. *Cross-Cultural Ethnobotanical Studies in North East India. Ethnobotany* 4: 25-38.

Jain, S.K. & Srivastava, S. 1999. *Dictionary of Ethnoveterinary Plants of India*. Deep Publications, New Delhi.

Jain, S.K. 1967. Ethnobotany- its scope and study. *Ind. Mus. Bull.* 2: 39-43.

Jain, S.K. 1968. *Medicinal Plants*. National Book Trust, New Delhi.

Jain, S.K. 1981. Observations on ethnobotany of the tribals of Central India, 193-198. *In*: Jain, S.K. (ed.) *Glimpses of Indian Ethnobotany*. Oxford & IBH, Publi. Co., New Delhi.

Jain, S.K. 1986. Ethnobotany. *Interdiscipl. Sci. Rev.* II: 285-292.

Jain, S.K. 1987a. *A Manual of Ethnobotany*. Sci. Publ., Jodhpur.

Jain, S.K. 1987b. Ethnobotany its concepts and relevance. pp. 1-12 *Pres. Add. X Bot. Conf.* Patna.

Jain, S.K. 1991. *Dictionary of Indian Folk Medicine and Ethnobotany*. Deep Publications, New Delhi.

Jain, S.K. 1995. Ethnobotanical diversity in zingibers of India. *Ethnobotany* 1: 83-88.

Jain, S.K., Mudgal, V., Banerjee, D.K., Guha, A., Pal, D.C. & Das. D. 1984. *Bibliography of Ethnobotany*. B.S.I., Howrah.

Jain, S.K. Sinha, B.K. & Gupta, R.C. 1991. *Notable Plants in Ethnomedicine of India*. Deep Publications, New Delhi.

Jain, S.K., Ranjan, Vinay, Sikarwar, R.K.S., & Sakīani, A. 1994. Botanical distribution of psychoactive plants of India. *Ethnobotany* 6: 65-75.

Jain, S.K. & Sastry, A.R.K. 1980. *Threatened Plants of India. A-State-of-the-Art-Report*. BSI & MAB, New Delhi. P.18.

Jain, S.P. 1984. Ethnobotany of Morni & Kalesar (Ambala, Harayana). *J. Econ. Tax. Bot.* 5: 809-813.

Jain, S.P. 1996. Ethno-medico-botanical survey of Chaibasa, Singh Bhum District, Bihar. *J. Econ. Taxon. Bot.* Additional Series 12: 403-407.

Jain, S.P. & Puri, H.S. 1994. An ethno-medico-botanical survey of Parvati valley in Himachal Pradesh (India). *J. Econ. Tax. Bot.* 10: 231-222.

Jamir, N.S. & Rao, R.R. 1990. Fifty new or interesting medicinal plants used by the Zelings of Nagaland (India). *Ethnobotany* 2: 11-18.

Janaki Ammal, E.K. 1955. *An Introduction to the Subsistence Economy of India. Background paper No. 10.* Winner Gref Foundation International Symposium on 'Man's Role in Changing the Face of the Earth'. Princeton Inn. Princeton, N.J. June 16-22.

Jha, R.R. & Varma, S.K. 1996. Ethnobotany of Sauria Paharias of Santhal pargana, Bihar: I medicinal plants. *Ethnobotany* 8: 31-35.

Jha, Vidyanath & Basak, P.K. 1994. Ethnobotanical aspects of Mithila Painting. *Ethnobotany* 6: 9-18.

Jha, Vidyanath, Choudhary, U.N. & Saraswati, K.C. 1991. Botanical Aspects of an Ethnoveterinary Prescription in Mithila, North Bihar (India). *Ethnobotany* 3: 101-104.

Jha, Vidyanath, Mishra, Sunita, Kargupta, A.N. & Jha, Abha 1996. Leaves and flowers utilised as supplementary vegetables in Darbhanga (North Bihar) and their ethnobotanical significace. *J. Econ. Taxon. Bot.* Additional series 12: 395-402.

Joseph, J. & Kharkonger, P. 1981. A preliminary ethnobotanical survey in the Khasi and Jaintia hills, Meghalaya, 115-123. *In*: S.K. Jain (ed.): *Glimpses of Indian Ethnobotany*. Oxford & IBH Publ. Co., New Delhi.

Joshi, M.C., Patel, M.B. & Mehta, P.J. 1980. Some Folk Medicines of rural Khasi and Jaintia tribes in Meghalaya. 124-136. *In:* Jain, S.K. (ed.) *Glimpses of Indian Ethnobotany*. Oxford and IBH Publ. Co., New Delhi.

Joshi, P. 1995. *Ethnobotany of the Primitive Tribes in Rajasthan*. Rupa Books Pvt. Ltd., Jaipur.

Kapoor, Promila 1978. Exploration of plant resources of Himachal Pradesh as food source, 153-163. *In: Proceeding of National Seminar on Resources Development and Environment in the Himalayan Region*. D.S.T. Govt. of India.

Kapur, S.K. 1993. Ethnomedico plants of Kangra valley (Himachal Pradesh). *J. Econ. Tax. Bot.* 17(2): 395-408.

Kapur, S.K. 1996. Traditionally important medicinal Plants of Bhaderwah Hills - Jammu Province–III. *J. Econ. Taxon. Bot.* Additional Series 12: 62-74.

Kapur, S.K. & Nanda, S. 1996. Traditionally important medicinal plants of Bhaderwah Hills (Jammu Province) Part–II. *J. Econ. Taxon. Bot.* Additional Series 12: 56-61.

Kapur, S.K. & Singh, Paviter, 1996. Traditionally important medicinal plants of Udhampur district (Jammu Province) Part–I. *J. Econ. Taxon. Bot.* Additional Series 12: 75-81.

Kapur, S.K. & Srivastava, T.N. 1996. Traditionally important medicinal plants of Udampur District (Jammu Province) Part–II. *J. Econ. Taxon. Bot.* Additional Series 12: 82-88.

Kapur, S.K., Nanda, S. & Srivastava, T.N. 1996. Ethnobotanical uses of PRL-Herbarium - III. *J. Econ. Taxon. Bot.* Additional Series 12: 50-55.

Katewa, S.S. & Arora, Asha 1997. Some plants in folk medicine of Udapur district (Rajasthan). *Ethnobotany* 9: 48-51.

Kaushik, P. 1988. *Indigenous Medicinal Plants (Including Microbes and Fungi).* Today's & Tomorrow's Printer's & Publishers, New Delhi.

Khanna, K.K., Shukla, G. & Mudgal, V. 1996a. New traditional medicinal uses of plants from Jalaun district, Uttar Pradesh. *J. Econ. Taxon. Bot.* Additional Series 12: 108-111.

Khanna, K.K., Srivastava, P.K. & Mudgal, V. 1996b. Noteworthy medicinal plants uses from rural folklore of Raebareli district, Uttar Pradesh. *J. Econ. Taxon. Bot.* Additional Series 12: 118-122.

Kharkongor, P. & Joseph, J. 1981. Folklore medicobotany of rural Khasi and Jaintia tribes in Meghalaya, 124-136. *In*: Jain, S.K. (ed.) *Glimpses of Indian Ethnobotany.* Oxford & IBH, Publi. Co., New Delhi.

Khoshoo, T.N. 1995. Census of India's biodiversity: tasks ahead. *Curr. Sci.* 69: 14-17.

Khory, Rustomjee Naserwamjee & Katrak, Nanabhai Nawroji (Reprint, 1984). *Materia Medica of India and Their Therapeutics.* Neeraj Publishing House, Delhi-52.

Kindscher, K. 1987. *Edible plants of the Prame: An Ethnobotanical Guide.* Univ. Press Kanas, Lawrence.

Kirtikar, K.R. & Basu, B.D. 1935. *Indian Medicinal Plants.* Vol. I-IV. Lalit Mohan Basu, 49, Leader Road, Allahabad, India.

Koelz, W.N. 1979. Notes on the Ethnobotany of Lahaul, a province of the Punjab. *Quar. J. Crude Drug. Res.* 17: 1-56.

Kosambi, D.D. 1962. *Myth and Reality.* Popular Press, Bombay.

Kothari, M.J. & Moorthy, S. 1996. Ethnobotany in human welfare of Raigad district–in Maharastra State, India, 403-407. *In* : Jain, S.K. (ed.) *Ethnobiology in Human Welfare.* Deep Publ., New Delhi.

Kulkarni, D.K. & Kumbhojkar, M.S. 1996. Pest control in tribal areas of western Maharastra - an ethnobotanical approach. *Ethnobotany* 8: 56-59.

Kulkarni, D.K. & Kumbhojkar, M.S. 1993. Kitchen garden plants of Mahadeokali tribe in Maharastra. *Ethnobotany* 5: 119-127.

Kumar, Anand. 1996. Some ethnomedicinal plants of the Murias of the Indravati Tiger reserve, Bastas (Madhya Pradesh). *J. Econ. Taxon. Bot.* Additional Series 12: 201-205.

Kumar, R. Vijaya & Pullaiah, T. 2003. A survey of wild edible plants of Chenchu tribes of Prakasam District (Andhra Pradesh), 22-31. *In*: *Vistas in Ethnobotany* Vol.–1. *Indian Journal of Applied and Pure Biology*.

Kumar, Y., Fancy, S. & Rao, R.R. 1987. Further contribution to the ethnobotany of Meghalaya: Plants used by War Jaintia of Jaintia Hill district. *J. Econ. Tax. Bot.* 11: 65-70.

Kurian, J.C. 1995. *Plants That Heal*. Oriental Watchman Publishing House, Pune, India.

Lal, Brij, Vats, S.K. Singh, R.D. & Gupta, Akshey K. 1996. Plants used as Ethnomedicine and Supplementary food by Gaddis of Himachal Pradesh, India. 384-387. *In* : Jain, S.K. (ed.) *Ethnobiology in Human Welfare*. Deep Publications, New Delhi.

Lalramnghinglova, J.H. 1996. Ethnobotany of Mizoram- A preliminary survey. *J. Econ. Taxon. Bot.* Additional Series 12: 439-459.

Lata, Sneh 1996. *Grewia Optiva* Drumm. 'Bhimal' : A boon for rural people in Kumaun Hills, India, 471-471. *In*: Jain, S.K. (ed.) *Ethnobiology in Human Welfare*, Deep Publications, New Delhi.

Lile, Jiang, J., Chen, H. & Chen, Z. 1960. Antio-influenza virus effect of Chinese traditional medicines. *Act Microbiologica Sinica* 8: 164-69.

Mahato, Ashwini Kumar & Mahato, Pushapa 1996. Ethnobotanical wealth of Chhotanagpur Plateau–IV. Some medicinal plants used against intestinal warms. *J. Econ. Taxon. Bot.* Additional Series 12: 389-391.

Maheshwari, J. K. & Singh, J. P. 1984. Contribution to the ethnobotany of Bhoxa tribe of Hijnor and Pauri Garhwal districts, U.P. *J. Econ. Tax. Bot.* 5: 251-259.

Maheshwari, J.K., Painuli, R.M. & Dwivedi, R.P. 1990. Notes on ethnobotany of the Oraon and Korwa tribes of M.P., 75-90. *In*: Jain, S.K. (ed.): *Contribution to Indian Ethnobotany*. Sci. Publ., P. Box. 91, Jodhpur, India 342001.

Maheshwari, J.K., Singh, K.K. & Saha, S. 1981. *The Ethnobotany of the Tharus of Kheri District, Uttar Pradesh*. Nation. Bot. Res. Inst. Lucknow.

Maheshwari, J.K., Singh, K.K. & Saha, S. 1986. *Ethnobotany of Tribals of Mirzapur District, U.P.* Nation. Bot. Res. Inst., Lucknow.

Maheswari, J.K. & Singh, J.P. 1987. Contribution to the ethnobotany of Bhoxa tribe of Bijnor and Pauri Gahwal District U.P. *J.Econ Tax. Bot.* 5(2): 251-259.

Maheswari, J.K. (ed.) 1992. *Ethnobotany in India*. Scientific Publ., Jodhpur.

Malhotra, S.K. & Moorthy, S. 1973. Some useful medicinal plants of Chandrapur district (Maharastra State). *Bull. Bot. Surv. India* 15: 13-21.

Malik, Seema 1996. Belief and health care : A study among the Saoras of Koraput distt. Orisa, 151-152. *In*: Jain S.K. (ed.) *Ethnobiology in Human Welfare*, Deep. Publ., New Delhi.

Manandhar, N.P. 1980a. *Medicinal Plants of Nepal Himalaya*. Ratna Pustak Bhandar, Kathmandu, Nepal.

Manandhar, N.P. 1980b. Some less known medicinal plants of Rasuwa district (Nepal). *Quart. J. Crude Drug Res.* 18(3): 147-151.

Manandhar, N.P. 1985. Ethnobotanical notes on certain medicinal plants used by Tharus of Dang-deokhuri district, Nepal. *Int. J. Crude Drug Res.* 25(4): 153-159

Manandhar, N.P. 1987a. Traditional medicinal plants used by tribals of Lamjung district, Nepal. *Int. J. Crude Drug Res.* 25(4): 236-240

Manandhar, N.P. 1987b. An ethhnobotanical profile of Manang valley, Nepal. *J. Econ. Taxon. Bot.* 10(1): 207-213

Manandhar, N.P. 1989a. Medicinal plants used by Chepang tribes of Makawanpur district, Nepal. *Fitotreapia* 60(1): 61-68

Manandhar, N.P. 1989b. *Useful Wild Plants of Nepal*. Franz Steiner Verlag Wiesbaden GMBH, Stuttgart, W. Germany.

Manandhar, N.P. 1990. Folk-lore medicine of Chitwan district, Nepal. *Ethnobotany* 2: 31-38.

Manandhar, N. P. 1996. Traditional practice for oral health care in Nepal. *J. Econ. Taxon. Bot.* Additional Series 12: 408-413.

Mandal, S.K. & Basu, S.K. 1996. Ethnobotanical studies among some tribals of Niligiri district, Tamil Nadu. *J. Econ. Taxon. Bot.* Additional series 12: 268-271.

Mao, A.A. 1993. A preliminary report on the folklore botany of Mao Nagas of Manipur (India). *Ethnobotany* 5: 143-147.

Martin, M. 1972. *Introduction a 1' Ethnobotanique du combodge*. Paris.

Martin, Marina, Mathias, Evelyn & McCorkle, Constance (ed.) 2001. *Ethnoveterinary Medicine-An Annotated Bibliography of Community Animal Healthcare*. Pub. ITDG Publishing, London.

Masih, S.K. 2003, Diversity of ethnomedicinal wealth in Amar Kantak plateau region (India), 204-214. *In*: *Vistas in Ethnobotany*, Vol.-I, *Indian Journal of Applied and Pure Biology*.

Megoneitso & Rao, R.R. 1983. Ethnobotanical studies in Nagaland-4. Sixty two medicinal plants used by the Angami Nagar. *J. Econ. Tax. Bot.* 4(1-3): 167-172.

Mehrotra, Shanta, Rawat, A.K.S., Singh, H.K. & Shome, Usha 1995. Standardization of popular Ayurvedic adaptogenic preparation 'Chayavanprash' and ethnobotany of its ingredients. *Ethnobotany* 7: 1-15.

Minnis, Paul. E. (ed.) 2000. *Ethnobotany- A Reader*. University of Okhahoma, Norman, U.S.A.

Mishra, D. K., Samanta, Godhuli & Mishra, T. K. 1996a. Ethnomedicine of the tribe Kharia of Midnapore District, West Bengal. *J. Econ. Taxon. Bot.* Additional Series 12: 329-331.

Mishra, Sunita, Jha, Vidyanath, & Jha, Shaktinath 1996b. Plants in Ethnoveteinary Practices in Darbhanga (North Bihar). *In*: Jain, S.K. (ed.) *Ethnobiology in Human Welfare*, Deep Publ., New Delhi.

Mishra, T.K., Bhunia, A., Sahoo, T.K. & Roy, S.B. 1997. Ethnobotany vis-à-vis joint forest management movement of Midnapore (West Bengal), India. *Ethnobotany* 9: 52-55.

Mohanty, R.B. & Padhy, S.N. 1996. Traditional phytotheraphy for diarrhoeal diseases in Ganjam and Phulbani districts of South Orissa, India. *Ethnobotany* 8: 60-65.

Molla, M.A. & Roy, B. 1985. Traditional uses of some medicinal plants by the Robha tribals in Jalpaiguri district, West Bengal. *J. Econ. Tax. Bot.* 7(3): 578-580.

Molla, H.A. & Roy, B. 1996. Some ethnobotanical claims from the Jalpaiguri district of West Bengal. *J. Econ. Taxon. Bot.* Additional Series. 12: 322-324.

Mudgal, V. & Pal, D.C. 1980. Medicinal plants used by tribals of Mayurbhanj (Orissa). *Bull Bot. Surv. India* 22: 59-62.

Mudgal, V. 1987. Literature on ethnobotany- Indian and foreign, 45-57. *In*: Jain, S.K. (ed.) *A Manual of Ethnobotany*. Scientific Publishers, Jodhpur.

Mudgal, V., Pal, D.C., Kayal, R.N. & Saha, S. 1999. *Ethnobotany of Totopora*. Bishen Singh Mahender Pal Singh, Dehradun.

Mukherjee, A., Mukherjee, S. & Mukherjee, M. 1986. Less known medicinal plants of Rubiaceae in Darjeeling Hills, West Bengal (India). *Indian J. Appl. & Pure Bot.* 1: 86-91.

Mukherjee, Ambarish & Namhata, Debashis 1990. Medicinal plant of tribals of Sundargarh district, Orissa. *Ethnobotany* 2: 57-60.

Nair, N.C. 1964. On a botanical tour to lahaul and Spiti (Punjab Himalaya). *Bull. Bot. Surv. India*. 6: 219-235.

Nair, N.C. 1977. *Flora of Bashahr Himalaya*. Int. Bioscience Publi., Hisar.

Nazarudeen, A., Seeni, S., Koshy, K.C. & Pushangadan, P. 1996. Folk plants of food, medicine, adornment and repellent used by the Paniyar community in North Kerela. *J. Econ. Taxon. Bot.* Additional Series 12: 299-305.

Negi, K.S. & Gaur, R.D. 1991. A contribution to the edible wild fruits of Uttar Pradesh hills. *Bull. Bot. Surv. India* 43: 233-266.

Negi, K.S. & Pant, K.C. 1990. Notes on ethnobotany of Gangwal- A tribe of Garhwal Himalaya. *Ethnobotany* 2: 81-85.

Negi, K.S., Tiwari, J.K. & Gaur, R.D. 1985. Economic importance of some common trees in Garhwal Himalaya: An ethnobotanical study. *Indian J. Forestry* 8: 276-289.

Negi, K.S., Tiwari, J.K., Gaur, R.D. & Pant, K.C. 1993. Notes on ethnobotany of five districts of Garhwal Himalaya, Uttar Pradesh, India. *Ethnobotany* 5: 73-81.

Negi, P.S. & Subramani, S.P. 2002. Ethnobotanical study in the village Chhitkul of Sangla valley, district Kinnaur, Himachal Pradesh. *J. Non-Timber Forest Prod.* 9(3-4): 113-120.

Negi, Y.S. & Bhalla, P. 2002. Collection and marketing of important medicinal and aromatic plants in tribal areas of Himachal Pradesh. *Indian Forester* 128(6): 641-649.

Neuwinger, H.D. 1996. *African Ethnobotany- Poisons and Drug*. Chapman & Hall.

Neuwinger, H.D. 2000. *African Traditional Medicine: A Dictionary of Plant Uses and Applications.* Medphrm Scientific Publishers, Stuttgart, Germany, x + 589 pp + 46 pp Supplement, ISBN 3-88763-086-6.

Newall, C.A., Anderson, L.A. & Phillipson, J.D. 1996. *Herbal Medicine: A Guide for Health Care professionals.* The Pharmaceutical Press, London.

Painuli, R.M. & Maheswari, J.K. 1996. Some interesting ethnomedicinal plants used by Sahariya tribe of Madhya Pradesh. *J. Econ. Taxon. Bot.* Additional Series. 12: 179-185.

Panda, S. 1996. Plant use–A recent perspective from Eastern-Himalayas. *J. Econ. Taxon. Bot.* Additional Series 12: 332-337.

Pandey, Ashok K., Bora, H.R. & Deka, S.C. 1996. An ethno-medico botanical study of Golghat district, Assam : native plant remedies for jaundice. *J. Econ. Taxon. Bot.* Additional Series 12: 344-349.

Pandey, D.N. 1998. *Ethnoforestry (local knowledge for sustainable forestry and livelihood security).* Himanshu Publ., Udaipur, New Delhi.

Pandey, Prakash H. & Verma, B.K. 2003. Magico-therapeutic plants of Uttar Pradesh, India, 178-187. *In: Vistas in Ethnobotany,*Vol.-I, *Indian Journal of Pure and Applied Biology.*

Parmar, C. & Kaushal, M. K. 1982. *Wild Fruits of the Sub-Himalayan Region.* Kalyani Publishers New Delhi- Ludhiana.

Payee, Gabriell DeBear (ed.) 2000. *Cultural Uses of Plants: A Guide to Learning About Ethnobotany.* The New York Botanic Garden Press, New York.

Pieroni, A. 1999. Gathered wild food plants in the upper valley of the Serchio River (Garfagnana), Central Italy. *Econ. Bot.* 53(3): 327-341.

Polunin, O. & Stainton, A. 1984. *Flowers of the Himalaya.* Oxford University Press, Delhi.

Prasad, P. Nagendra, Ranjit Singh, A.J.A. Narayanan, L.M. & Natarajan, C.R. 1996. Ethnobotany of the Kanikkars of South Tamil Nadu-1. *J. Econ. Taxon. Bot.* Additional Series 12: 292-298.

Pullaiah, T. & Kumar, T. Dharma Chandra 1996. Herbal plants in Mannanur forest Mahaboobnagar district, Andra Pradesh. *J. Econ. Taxon. Bot.* Additional Series 12: 218-220.

Radhakrishnan, K., Pandurangan, A. G., Pushpangandan P. 1996. Ethnobotany of Wild Edible plants of Kerala, India, 48-51. *In* : Jain, S.K. (ed.) *Ethnobotany in Human Welfare,* Deep Publ., New Delhi.

Ragupathy, S. & Mahadevan, A. 1991. Ethnobotany of Kodiakkarai reserve forest, Tamil Nadu, South India. *Ethnobotany* 3: 79-82.

Rajendran, A., Rama Rao, N. & Henry, A.N. 1996. Hepatic stimulant plants of Andhra Pradesh, India. *J. Econ. Taxon. Bot.* Additional Seires 12: 221-223.

Rama Rao, N. & Henry, A. N. (ed.) 1996. *Ethnobotany of Eastern Ghats in Andra Pradesh, India.* Bot. Surv. of India, Calcutta.

Ramachandran, V.S. & Nair, V.J. 1981. Ethnobotanical observation on 'Irulars' of Tamil Nadu (India). *J. Econ. Tax. Bot.* 2: 183-190.

Rana, T.S., Dutt, Bhaskar & Rao, R.R. 1996. Strategies of sustainable utilization of plant resources by the tribals of the Tons valley, Western Himalaya. *Ethnobotany* 8: 96-104.

Ranjan, Vinay 1996. Some ethnomedicinal plants of Lalitpur district, Uttar Pradesh, India 149-150. *In*: Jain S.K. (ed.). *Ethnobiology in Human Welfare,* Deep Publ., New Delhi.

Rao, N.S. Balaji, Rajasekhar, D., Raju, D. Chengal & Nagaraju, J. 1996. Ethnomedicinal notes on some plants of Tirumala Hills for dental disorders. *Ethnobotany* 8: 88-91.

Rao, R. R. 1996. Traditional knowledge and sustainable development: key role of ethnobiologists. *Ethnobotany* 8: 14-24.

Rao, R.R. & Jamir, N.S. 1982. Ethnobotanical Studies in Nagaland-II; 54 Medicinal Plants used by Nagas. *J. Econ. Tax. Bot.* 3: 11-17.

Rao, R.R. 1981a. Ethnobotany of Meghalaya-medicinal plants used by Khasi and Garo tribes. *Econ. Bot.* 35: 4-9.

Rao, R.R. 1981b. Ethnobotanical studies on the flora of Meghalaya-Some interesting reports of herbal medicines, 137-148. *In*: Jain, S.K. (ed.) *Glimpses of Indian Ethnobotany.* Oxford & IBH, Publi. Co., New Delhi.

Rastogi, M.A. 1960. Medicines from the wild. A case study of the Great Himalayan Park. *The Indian Magazine of her people and culture.* pp. 74-75.

Rau, M.A. 1961. A recent survey of medicinal plant resources of Lahaul valley (Panjab India). *Pakistan J. Soc. & Ind. Res. Spl. Nr. International Symp.* 4: 201-205.

Ravishankar, T. 1996. Role of indigenous people in the conservation of plant genetic resources, 310-314. *In*: Jain, S.K. (ed.) *Ethnobiology in Human Welfare.* Deep Publ., New Delhi.

Ravishankar, T. & Henry, A.N. 1992. Ethnobotany of Aditlabad District, Andra Pradesh, India. *Ethnobtany* 4: 45-52.

Ray, S. 2003. Ethbomedicinal plants of Dewas district (M.P.) India. 58-68. *In*: *Vistas in Ethnobotany,* Vol.-1, *Indian Journal of Pure and Applied Biology.*

Reddy, H. Hemambara, Vijayalakshmi, K. & Venkata Raju, R.R. 1996. Native phytotherapy for snake bite in Nallamalais Eastern Ghats, India. *J. Econ. Taxon. Bot.* Additional Series. 12: 214-217.

Redford, K.H. & Mansur, J.A. 1997. *Traditional Peoples and Biodiversity Conservation in Large Tropical Landscapes.* America Verde Publications, Arlington, Virginia, USA.

Renfrew, J.M. 1973. *Palaeoethnobotany- The Prehistoric Food Plants of the Near East and Europe.* Columbia Univ., New York.

Sagwal, S.S. & Gupta, N.K. 1988. Role of forestery extension in exploitation on medicinal and aromatic plants of Himachal Pradesh, 50-54. *In:* Khosla, P.K. & Sehgal, R.N. (eds.) *Trends in Tree Science.* ISTS Publ., Solan.

Sahoo, A.K. & Mudgal, V. 1995. Less known ethnobotanical uses of plants of Phulbani district, Orissa, India. *Ethbobotany* 7: 63-67.

Sahu, T.R. 1996. Life support promising food plants among aboriginals of Bastar (M.P.), India, 26-30. *In*: Jain, S.K. (ed.) *Ethnobiology in Human Welfare* Deep Publ. New Delhi.

Saini, D.C. 1996a. Ethnobotany of Tharus of Basti District, Uttar Pradesh. *J. Econ. Taxon. Bot.* Additional Series 12: 138-153.

Saini, V.K. 1996b. Plants in the Welfare of Tribal Women and Children in Certain Areas of Central India 140-144. *In:* Jain, S.K. (ed.) *Ethnobiology in Human Welfare,* Deep. Publ., New Delhi.

Saklani, S. & Jain,. S.K. (ed.) 1994. *Cross-Cultural Ethnobotany of North-East India.* Deep Publ., New Delhi.

Samwatsar, S. & Diwanji, V.B. 1996a. Plants used for skin diseases, cuts, wounds, and Bruishes by the Tribals of Western M.P. *J. Econ. Taxon. Bot.* Additional Series. 12: 192-195.

Samwatsar, S. & Diwanji, V.B. 1996b. Plants used in snake, scorpion and insect bites/ stings by Adibasis of Jhabua (M.P.) *India. J.Econ Taxon. Bot.* Additional Series 12: 199-200.

Sarin, Y.K. & Chopra, C.L. 1985. Medicinal and aromatic plant resources of Himachal Pradesh, present status and future prospects. *Workshop on Sci. and Tech., Shimla,* Oct. 29-30, 1985.

Sarin, Y.K. 1967. A survey of vegetable raw resources of Lahaul. *Indian Forester* 93(7): 489-499.

Satapathy, K.B. & Brahmam, M. 1996. Some medicinal plants used by tribals of Sundargarh district, Orissa, India, 153-155. *In*: Jain, S.K. (ed.) *Ethnobiology in Human Welfare,* Deep Publications, New Delhi.

Saxena, H.O., Brahmam, M. & Dutta, P.K. 1981. Ethnobotanical studies in Orissa, 232-244. *In*: Jain, S.K. (ed.) *Glimpses of Indian Ethnobotany*. Oxford & IBH, Publi. Co., New Delhi.

Schoen, Allen M.& Wynu, Susan G. (eds.) 1998. *Complementary and Alternative Veterinary Medicine: Principles and Practice*. Mosby, Inc. St. Louis, Miss, USA.

Schultes, R.E. & Raffauf, R.F. 1990. *The Healing Forest*. Dioscorides Press, Portland, Oregon.

Schultes, R.E. & Reis, S.V. (ed.) 1997. *Ethnobotany- Evolution of a Discipline*. Dioscorides Press. Portland, Oregon.

Schultes, R.E. 1960. Tapping our heritage of ethnobotanical lore. *Econ. Bot.* 8: 224-227.

Schultes, R.E. 1962. The role of ethnobotanists in the search for new medicinal plants. *Lloydia* 35: 257-266.

Schultes, R.E. 1967. The place of ethnobotany in the ethnopharmacologic search for psychotomimetic drugs, *In*: Efron, D. (ed.) *Ethnopharacologic Search for Psychoactive Drugs*. Public Health Serv. Publ. No. 1645: 35-57, Washington.

Schultes, R.E. 1979. Amazonia as a source of new economic plants. *Econ. Bot.* 32: 259-266.

Schultes, R.E. 1983. Richard Spruce: An early ethnobotanist and explorer of the northwest Amazon and northern Andes. *J. Ethnobiol.* 3: 139-147.

Schultes, R.E. 1986. The reason for ethnobotanical conservation. *Bull. Bot. Surv. India* 28: 203-324.

Seitel, Peter (ed.) 2001. *Safeguarding Traditional Cultures: A Global Assessment*. Centre for Folklife and Cultural Heritage, Smithsonian Institution, Washington, DC.

Sensarma, Priydarshan 1991. Herbal veterinary medicines in an ancient-Sanskrit work - The Garuda Purana. *Ethnobotany* 3: 83-87.

Shabnam, S.R. 1964. Medicinal Plants of Chamba. *Indian Forester* 90: 50-63.

Shah, G.L., Yadav, S.S. & Badri, N. 1983. Medicinal plants from Dahanu forest division in Maharashtra State. *J. Econ. Tax. Bot.* 4: 141-151.

Shah, N.C. & Joshi, M.C. 1971. An ethnobotanical study of Kumaon region of India. *Econ. Bot.* 25: 414-422.

Shah, N.C. & Singh, S.C. 1990. Hitherto unreported phytotherapeutical uses from tribal pockets of Madhya Pradesh (India). *Ethnobotany* 2: 91-95.

Sharma, B.D. & Rana, J.C. 1999. Traditional medicinal uses of plants of Himachal hills. *J. Econ. Tax. Bot.* 23(1): 173-176.

Sharma, K.R. & Sood, M. 1997. *Important Medicinal Plants of Himachal Pradesh.* Vol-I & II. Dr. Y.S. Parmar University of Horticulture and Forestry, Nauni, Solan, H.P.

Sharma, O.P. 1976. *Some Useful Plants of Himachal Pradesh.* Himachal Pradesh University.

Sharma, P.K., Dhyani, S.K. & Shanker, V. 1979. Some useful and medicinal plants of the district Dehradun and Siwalik. *J. Sci. Res. Pl. Med.* 1: 17-43.

Sharma, S.C. 1996. A medicobotanical study in relation to veterinary medicines of Shahjahanpur district (Uttar Pradesh). *J. Econ. Taxon. Bot.* Additional Series 12: 123–127.

Sharma, T.R. & Borthakur, S.K. 2003. Inquire into Old Lead-II: Ethnoveterinary medicine for treatment of horse in Assam, 69-85. *In*: *Vistas in Ethnobotany*, Vol. I, *Indian Journal of Pure and Applied Biology*.

Shome, Usha, Rawat, A.K.S. & Mehrotra, Shanta, 1996. Time tested household remedies, 96-100. *In*: Jain, S.K. (ed.) *Ethnobiology in Human Welfare*, Deep Publ., New Delhi.

Shrestha, P. 1985. Research note: contribution to the ethnobotany of the Palpa area. *Contribution to the Nepalese Studies* 12(2): 63-74.

Shukla, Gyanesh & Verma, B.K. 1996. Roots–A vital plant part to cure body ailments among tribals/rural folklore of Western Bihar. *J. Econ. Taxon. Bot.* Additional Series 12: 392-394.

Sikarwar, R.L.S. & Kaushik, J.P. 1992. Some less known medicinal uses of trees among the Sahariyas of Morena district, M.P., India. *Ethnobotany* 4: 71-74.

Sikarwar, R.L.S. 1996. Ethnoveterinary herbal medicines in Morena district of Madhya Pradesh, India, 194-196. *In*: Jain, S.K. (ed.) *Ethnobiology in Human Welfare*, Deep Publ., New Delhi.

Singh, A.K., Singh, R.N. & Singh, S.K. 1987. Some ethnobotanical plants of Terai region of Gorakhpur district I. *J. Econ. Tax. Bot.* 23: 185-198.

Singh, Brahma, Chaurasia, O.P. & Jadhav, K.L. 1996a. An ethnobotanical study of Indus valley (Ladakh). *J. Econ. Taxon. Bot.* Additional Series 12: 92-101.

Singh, G.S., Rao, K.S. & Saxena, K.G. 1998. Environment, ecological and socio-economic impact of introduced crops in Western Himalaya: A case study of Kullu valley. *J. Human Ecol.* 9: 63-72.

Singh, G., Singh, H.B. & Mukherjee, T.K. (eds.) 2003a. Ethnomedicine of North-East India. Proceedings of National Seminar on *Traditional Knowledge Base on Herbal Medicines and Plant Resources of North-East India: Protection, Utilization and Conservation.* National Institute of Sciences Communication and Information Resources, New Delhi.

Singh, G.S. 1990. Ethnobotanical studies of useful plants of the Kullu district in North-western Himalaya, India. *J. Econ. Tax. Bot.* 23: 185-188.

Singh, H. 1988. Ethnobiological treatment of piles by Bhoxas of Uttar Pradesh. *Anc. Sci. Life* 8: 167-170.

Singh, H. Kumar, Bir, Hyniewta, T.M. & Bora, P.J. 1997. Ethno-medico-botanical studies in Tripura, India. *Ethnobotany* 9: 56-58.

Singh, H.B., Singh, R.S. & Sandhu, J.S. (eds.) 2003b. *Herbal Medicine of Manipur: A Colour Encyclopedia.* Daya Publ. House, New Delhi.

Singh, J., Bhuyan, T.C. & Ahmed, A. 1996b. Ethnobotanical studies on the Mishing tribes of Assam with special reference to food and medicinal plants–I. *J. Econ. Taxon. Bot.* Additional Series 12: 350-356.

Singh, K.K. & Kumar, K. 2000a. *Ethnobotanical Wisdom of Gaddi Tribe of Western Himalayas.* Bishen Singh Mahender Pal Singh, Dehradun.

Singh, K.K. & Kumar, K. 2000b. Observations in ethnoveterinary medicine among the Gaddi tribe of Kangra valley, Himachal Pradesh. *Ethnobotany* 12: 42-44.

Singh, K.K. & Maheshwari, J. K. 1983. Traditional phytotherapy amongst the tribals of Varanasi district, U. P. *J. Econ. Tax. Bot.* 4: 829-838.

Singh, K.K. & Maheshwari, J. K. 1992. Some less known medicinal uses of trees among the Sahariyas of Morena district, M.P., India. *Ethnobotany*: 71-74.

Singh, K.K. & Prakash, Anand 1996. Observations on ethnobotany of the Kol tribe of Varanasi District, Uttar Pradesh. *J. Econ. Taxon. Bot.* Additional Series 12: 133-137.

Singh, K.K. & Prakash, Anand 1994. Indigenous phytotherapy among the Gond tribe of Uttar Pradesh, India. *Ethnobotany* 6: 37-41.

Singh, K.K. & Rao, R.R. 2003. Studies on some potential ethnomedicine among the tribals of Uttar Pradesh, India, 42-57. *In : Vistas in Ethnobotany* Vol.-I, *Indian Journal of Applied & Pure Biology.*

Singh, K.K. 1996a. Ethnobotanical observations on Sonbhadra district of Southern Uttar Pradesh, India : utilization and conservation, 145-148. *In:* Jain, S.K. (ed.) *Ethnobiology in Human Welfare,* Deep Publ., New Delhi.

Singh, K.K., Palvi, S.K. & Singh, H.B. 1980. Survey of some medicinal plants of Dharchula block-I Pithoragarh district of U.P. *Bull. Medico-ethnobot. Res.* 1: 1-7.

Singh, P.B. 1991. *Illustrated Field Guide to Commercially Important Medicinal and Aromatic Plants of Himachal Pradesh.* Society for Herbal Medicine and Himalayan Biodiversity, 65/12 Ram Nagar, Mandi, Himachal Pradesh.

Singh, P.B. 1993. Medicinal plants of ayurvedic importance from Mandi district of Himachal Pradesh. *B.M.E.B.R.* 14:126-136.

Singh, P.B. 1996b. Wild edible plants of Mandi district in North-West Himalaya, 22-25. *In:* Jain, S.K (ed.) *Ethnobiology in Human welfare.* Deep Publ., New Delhi.

Singh, P.B. 1999. *Illustrated Field Guide to Commercially Important Medicinal and Aromatic Plants of Himachal Pradesh* (with special reference to Mandi district). Published by Society for herbal medicine and Himalayan biodiversity, 65/12, Ramnagar, Mandi, Himachal Pradesh.

Singh, P.B. & Aswal, B.S. 1992. Medicinal plants of Himachal Pradesh used in Indian pharmaceutical industry. *B.M.E.B.R.* 13: 172-208.

Singh, V. & Pandey, R.P. 1996. Ethno-medicinal plants used for venereal and gynaecological diseases in Rajasthan (India). *J. Econ. Taxon. Bot.* Additional Series 12: 154-165.

Singh, V. & Pandey, R.P. 1998. *Ethnobotany of Rajasthan, India.* Scientific Publ., Jodhpur.

Singh, V.K. & Zaheer Anwar Ali 1996. Folk medicinal plants used for family planning in India, 184-186. *In* : Jain, S.K. (ed.). *Ethhnobiology in Human Welfare,* Deep Publ., New Delhi.

Sinha, B.K., Main, Vinod & Padhye, P.M. 1996. Ethno-medicinal plants of Bay Island for skin care. *J. Econ. Taxon. Bot.* Additional Series 12: 375-380.

Sinha, S.C. 1986. Ethnobotanical study of Manipur, Ph.D. Thesis, Manipur University, Imphal, *In* : Jain, S.K. (ed.) *Ethnobiology in Human Welfare,* Deep Publ., New Delhi.

Sinha, S.C. 1987. Ethnobotany-medicinal Plants. *Frontier Bot.* 1 : 123-152.

Sinha, S.C. 1996. Wild edible plants of Manipur, India, 42-47. *In* : Jain, S.K. (ed.) *Ethnobiology in Human Welfare.* Deep Publ., New Delhi.

Siwakoti, M. & Varma, S.K. 1996. Medicinal plants of the Terai of Eastern Nepal. *J. Econ. Taxon. Bot.* Additional Series 12: 423-438.

Sood, S.K. & Thakur, S. 2004. *Ethnobotany of Rewalsar Himalaya (Dist. Mandi, Himachal Pradesh, India).* Deep Publi., New Delhi.

Sood, S.K., Nath, R., & Kalia, D.C. 2001. *Ethnobotany of Cold Desert Tribes of Lahaul-Spiti (N.W. Himalaya).* Deep Publi., New Delhi.

Stainton, A. 1988. *Flowers of Himalaya, A Supplement.* Oxford University Press, Delhi.

Sundaravelu, S., Selvaraju, M. & Sundaramoorthy, P. 2003. Medicinal weed flora of South Arcot, Cuddalore district in Tamil Nadu, 87-94. *In* : *Vistas in Ethnobotany* Vol.-I, *Indian Journal of Pure & Applied Biology.*

Swami, Ajai & Gupta, B.K. 1996. A note on some commonly occurring medicinal weeds of Udhampur district (J&K). *J. Econ. Taxon. Bot.* Additional Series 12: 89-91.

Tarafder, C.R. 1986. Ethnobotany of Chhotanagpur (Bihar). *Folklore* 27: 119-124.

Taylor, P.M. 1990. *The Folk Biology of the Tobelo People- A Study in Folk Classification.* Smiths Inst. Press, Washington.

Thakur, R.S., Puri, H.S, & Husani, Akhtar, 1989. *Major Medicinal Plants of India.* Central Instutute of Medicinal & Aromatic Plants, Lucknow, India.

Tirkey, Amia, Khan, Fatima & Khan, S.S. 2003a. Medicinal plants used in the treatment of hair in Raijarh District of Madhya Pradesh (India), 104-110. *In: Vistas in Ethnobotany,* Vol.-I, *Indian Journal of Pure and Applied Biology.*

Tirkey, Amia, Khan, Fatima & Khan, S.S. 2003b. Medicinal plants used by the tribals of Raijarh Disteict (M.P.) for Jaundice, 163-170. *In: Vistas in Ethnobotany,* Vol. I, *Indian Journal of Pure and Applied Biology.*

Tiwari, K.C. & Tiwari, V.P. 1996. Some important medicinal plants of the tropical, sub tropical and temperate region of Siang, Subansiri and Tirap district of Arunchal Pradesh. *J. Econ. Taxon. Bot.* Additional Series 12: 359-363.

Tosh, Jayananda 1996. Ethnobotanical study of Western Maharastra. *J. Econ. Taxon. Bot.* Additional Series 12: 169-174.

Towle, M.A. 1961. *The Ethnobotany of Pre-Columbian Peru.* Aldine Publ., Chicago.

Turner, N.J., Thompson, L.C., Thompson, M.T. & York, A.Z. 1990. *Thompson Ethnobotany.* Royal Bristish Columbia Museum, Canada.

Uniyal, M.R. & Chauhan, N.S. 1971. Medicinal plants of Uhal valley in Kangra Forest Division. H.P. *Jour. Res. Ind. Med.* 6(3): 287-299.

Uniyal, M.R. & Chauhan, N.S. 1972. Commercially important medicinal plants of Kullu forest division. H.P. *Nagarjun* 4(2): 28-32.

Uniyal, M.R. & Chauhan, N.S. 1973. Traditionally important medicinal plants of Kullu forest Division, H.P. *Jour. Res. Ind. Med.* 7(1): 76-85.

Uniyal, M.R. & N.S. Chauhan, 1973. Traditionally important medicinal plants of Kangra Valley in Dharamsala Forest Circle, Himachal Pradesh. *J. Res. Indian Med.* 8(1): 76-85.

Uniyal, M.R. 1968. Medicinal plants of Bhagirathi valley lying in Uttarkashi forest division. *Indian Forester* 94: 407-420.

Uniyal, M.R., Bhat, A.V. & Chaturvedi, 1973. Preliminary observations on medicinal plants of Lahaul Spiti division in Himachal Pradesh. *Bull. Medico-Ethnobot. Res.* 3(1): 1-26.

Upadhye, Anuradha & Kumbhojkar, M.S. 1996. Ethnobotany of genus *Sterculia* L. from India. *J. Econ. Taxon. Bot.* Additional Series 12 : 40-42.

Upadhye, Anuradha S., Vartak, V.D. & Kumbhojkar, M.S. 1994. Ethno-medicobotanical studies in Western Maharastra, India. *Ethnobotany* 6: 25-31.

Uphof, J.C. Th. 1968. *Dictionary of Economic Plants*. H.R. Eugelmann (J. Cramer), Weinheim.

Varma, S.K. 1997. Comparitive studies on folk drugs of tribals of Chotanagpur and Santhal Pargana of Bihar, India. *Ethnobotany* 9: 70-76.

Ved Prakash, & Mehrotra, B.N. 1987. Ethnobotanical studies on the flora of Khandala, Maharastra state. *J. Econ. Tax. Bot.* 9(1): 205-208.

Verma, P., Khan, A.A. & Singh, K.K. 1995. Traditional phyotherapy among the Baiga tribe of Shahdol district of Madhya Pradesh, India. *Ethnobotany* 7: 69-73.

Vihari, Vishwa 1995. Ethnobotany of Cosmetics of Indo-Nepal border. *Ethnobotany* 7: 89-94.

Virjee, G., Dar, H., Kachroo, P. & Bhat, G.M. 1984. Taxo-ethonobotanical studies of the rural areas in district Rajouri (Jammu). *J. Econ. Tax. Bot.* 5(4): 831-838.

Vishwanathan, M.V. & Singh, H.B. 1996. Plants used as household remedies In India. 105-107. *In*: Jain, S.K. (ed.) *Ethnobiology in Human Welfare*. Deep Publ., New Delhi.

Vishwanthan, M.B. 1997. Ethnobotany of the Malyalis in North Arcot District, Tamil Nadu, India. *Ethnobotany* 9: 77-79.

Wambebe, C.O.N. 1990. Natural products in developing economy. *In*: Igbocchi, A.C. & Osisigu, I.U.W. (eds.) *National Workshop on Natural Products*. Univ. of Benin Press, Nigeria.

Watt, G. 1972. *A Dictionary of Economic Products of India* Vol.-1-VI. Periodical Experts, Delhi (India).

Wielgorskaya, Tatiana 1995. *Dictionary of Generic Names of Seed Plants*. B.S.M.P.S. Dehra Dun-248001 (India). Reprint, 2003.

Yang, M.H. & Blunden, G. 1993. Compounds with Antiviral Activity from Chines Medicinal Plants. *Ethnobotany* 5: 67-72.

Yanzone, G.S., Yonzone, Aruna & Bhujel, R.B. 1996. Contribution to the ethnobotany of Darjeeling District, India, 388-389. *In*: Jain, S.K. (ed.) *Ethnobiology in Human Welfare*, Deep Publ., New Delhi.

□□□

Chapter **5**

Appendices

Appendix-I

Index to Local Names

Appendix-II

Index to Uses

ABORTIFACIENT — *Abrus precatorius, Cissampelos pareira, Cuscuta reflexa, Plumbago zeylanica, Ricinus Communis.*

ACIDITY — (see Gastro-intesinal Disorders).

ADHESIVE — *Aegle marmelos, Bombax ceiba, Vitis jacquemontii.*

ADULTERANT — *Carica papaya.*

ANAEMIA — *Carissa congesta, Psidium guajava, Vitis jacquemontii.*

ANTHELMINTIC — *Albizzia lebbeck, Bauhinia vahlii, Butea monosperma, Carica papaya, Cassia fistula, Citrus grandis, Citrus jambhiri, Cordia dichotoma, Cuscuta reflexa, Flemingia strobilifera, Mallotus phillipinensis, Mangifera indica, Melia azedarach, Momordica charantia, Myrsine africana, Pinus roxburghii, Pistacia integrimma, Prunus armeniaca, Ricinus communis, Tabernaemontana divaricata, Tinospora cordifolia, Thevetia neriifolia.*

ANTIALLERGIC	*Albizzia lebbeck.*
ANTIEMETIC	*Citrus grandis, Citrus medica, Ficus bengalensis, Rosa moschata.*
ANTIDOTE	*Achyranthes aspera, Achyranthes bidentata, Cassia laevigata, Eugenia jambolana, Flemingia strobilifera, Ipomoea quamoclit, Pinus roxburghii, Stephania elegans Vitex negundo.*
ANTIFERTILITY	*Atyloshia platycarpa, Moringa pterygosperma, Solanum torvum, Dioscorea deltoidea.*
ANTISEPTIC	*Agave americana.*
ANTISPASMODIC	*Taxus baccata.*
APHRODISIAC	*Abrus precatorius, Achyranthes aspera, Bauhinia variegata, Bombax ceiba, Costus speciosus, Litsea sebifera, Malvaviscus conzatii, Moringa pterygosperma, Mucuna pruriens, Pavetta indica, Populus nigra, Prunus amygdalus, Randia dumetorum, Sida cordifolia, Tinospora condifolia, Vitex negundo, Woodfordia floribunda.*
APPETISER (Increasing appetite, Loss of appetite, Reducing Appetite)	*Berberis lycium, Cassia fistula, Emblica officinalis, Eriobotyrica japonica, Ficus palmata, Michelia Champaca, Pyrus communis, Viburnum cotonifolium.*
ARTHRITIS	(see Joint Diseases).
ASTHMA	*Acacia concinna, Achyranthes aspera, Adhatoda vasica, Cinnamomum tamala, Datura stramonium, Ficus reliogosa, Michelia champaca, Opuntia dillenii, Pistacia integrimma, Quercus glauca. Trichosanthes tricuspidata.*
ASTRINGENT	*Cotoneaster microphylla, Eugenia jambolana.*
BALDNESS	(see Hair Care).
BED WETTING	*Juglans regia.*
BEVERAGE	***Camellia sinensis, Osyris arborea, Taxus baccata.***
BITES	*Vinca rosea.*
BLOOD PURIFIER	*Agave americana, Berberis aristata, Berberis lycium, Carissa congesta, Duranta repens, Melia azedarach, Momordica cbarantia, Myrsine africana, Murraya koenigii, Nerium indicum, Picea smithiana, Populus ciliata, Prunus cerasoides, Pyrus pashia, Rhamnus triquetra, Terminalia arjuna, Thevetia neriifolia.*
BLOOD SUGAR	*Abrus precatorius, Butea monosperma, Carica papaya, Eugenia jambolana, Ficus palmata, Mangifera indica, Momordica charantia, Murraya koenigii, Premna latifolia, Rubus lasiocarpus, Tinospora cordifolia.*

BODY ACHE	(see Pain).
BOILS	(see Skin Diseases).
BONE FRACTURE	*Grewia oppositifolia Litsea polyantha, Litsea sebifera, Oroxylum indicum, Prunus cerasoides.*
BRONCHITIS (Respiratory Tract Infection)	*Cinnamomum tamala, Taxus baccata, Terminalia bellirica.*
BRUSH	*Agave americana.*
BURNS	*Caesalpinia decapetala.*
CANCER	*Taxus baccata* (Breast cancer), *Vinca rosea* (cancer).
CARDIAC COMPLAINTS	*Camellia sinensis, Carica papaya, Elaeagnus umbellata, Opuntia dillenii, Rhododendron arboreum, Terminalia arjuna, Thevetia neriifolia.*
CARMINATIVE	(see Gastro-intesinal Disorders).
CHEST INFECTION	*Saraca asoca.*
COLD	*Adhatoda vasica, Cinnamomum tamala, Eucalyptus umbellata, Taxus baccata, Vitis jaequemontii, Woodfordia floribunda.*
COLIC	(see Gastro-intesinal Disorders).
COOLANT	(see Refrigerant).
CONCEPTION	(see Fertility).
CONSTIPATION	*Berberis aristata, Cassia fistula, Embellia tsjeriam-cottam, Ficus palmata, Prunus armeniaca, Rubus ellepticus, Rubus niveus, Saurauia nepaulensis, Terminalia arjuna, Terminalia bellerica, Thevetia neriifolia.*
COUGH	(see Lung Diseases).
DEHYDRATION	*Ficus semicordata.*
DEMULCENT	*Cordia dichotoma.*
DEPRESSANT	(see Nervous Disorders).
DENTAL PROBLEMS	*Achyranthes aspera, Albizzia lebbeck, Calotropis procera, Clematis grata, Duranta repens, Euphorbia royleana, Ficus hispida, Indiofera gerardiana, Ipomoea carnea, Jasminum grandiflorum, Jasminum humile, Juglans regia, Psidium guajava, Thuja orientalis, Zanthoxylum armatum.*
DETERGENT & SHAMPOO	*Acacia concinna, Aesculus indica, Discorea deltoidea, Grewia oppositifolia, Randia dumetorum, Sapindus mukorossi.*

DIARRHOEA (see Gastro-intesinal Disorders).

DIGESTIVE DISEASES (see Gastro-intesinal Disorders)

DIURETIC *Achyranthes aspera, Andrachne cordifolia, Bauhinia variegata, Cassia fistula, Emblica officinalis, Gmelina arborea, Murraya koenigii, Rhododendron arboreum, Rosa alba, Rubus niveus, Saccharum officinarum, Saraca asoca, Vitis rependa.*

DYE *Alnus nitida, Berberis aristata, Camellia sinensis, Carya illenoesis, Coriaria nepalensis, Glochidion velutinum, Juglans regia, Osbeckia stellata, Prinsepia utilis, Rubia cordifolia, Symplocos spicata, Taxus baccata, Woodfordia floribunda.*

DYSENTERY (see Gastro-intesinal Disorders).

ENLARGED SPLEEN (see Spleen Complaints).

EPILEPSY *Bauhinia variegata, Cordia dichotoma, Debregeasia hypoleuca, Emblica officinalis, Loptodermis lanceolata, Terminalia bellerica.*

EYE DISEASES *Berberis lycium, Jasminum grandiflorum, Opuntia dillenii, Rosa alba, Rosa moschata, Teminalia chebula, Zizyphus jujuba.*

EYE LINER *Euphorbia royleana.*

FACIAL COSMETICS *Cucumis sativus.*

FEBRIFUGE *Bombax ceiba, Cedrella toona, Cedrus deodara, Cinnamomum camphara, Datura stramonium, Dioscorea deltoidea, Oroxylum indicum, Smilax aspera, Tabernaemontana divaricata, Tinospora cordifolia.*

FERTILITY *Ficus religiosa, Randia dumetorum.*

FEVER (see Febrifuge).

FIBRE *Agave americana, Arundinaria falcata, Bambusa nutans, Bauhinia vahlii, Clematis buchaniana, C. grata, Debreagesia hypoleuca, Dendrocalamus hamiltonii, Ficus roxburghii, Grewia oppositifolia, Ipomoea eriocarpa, Sorbaria tomentosa, Sterculia villosa.*

FISH POISON *Euphorbia royleana.*

FLATULENCE (see Gastro-intesinal Disorders).

FUEL ***Adhatoda vasica,*** *Bauhinia vahlii, Berberis lycium, Caryopteris wallichiana, Colebrookia oppositifolia, Cornus macrophylla, Dalbergia sissoo, Debregeasia hypoleuca, Desmodium multiflorum, Dodonea viscosa, Ehretia laevis,*

	Elaeagnus umbellata, Embelia tsjeriam- cottam, Flacourtia ramontchii, Indigofera gerardiana, Lonicera quinquelocularis, Myrsine africana, Populus nigra, Prunus domestica, Quercus glauca, Q. semicarpifolia, Robinia pseudoacacia, Rubus ellipticus, Sageretia oppositifolia, Salix karelinii, S. tetrasperma, Sapium sebiferum, Sorbaria tomentosa, Sterculia villosa, Viburnum cotonifolium, Xylosma longifolium.
GALL BLADDER DISORDER	*Citrus jambhiri.*
GASTRO-INTESTINAL DISORDERS	*Aegle marmelos, Aloe vera, Berberis lycium, Calotropis procera, Cassia fistula, Cassia laevigata, Cinnamomum camphora, Cinnamomum tamala, Cissampelos pareira, Citrus medica, Cordia dichotoma, Coteneastar microphylla, Cycas circinalis, Dalbergia sissoo, Dendrocalamus hamiltonii, Embellia tsjeriam-cottam, Emblica officinalis, Eugenia jambolana, Ficus glomerata, Ficus hispida, Ficus palmata, Ficus reliogosa, Ficus roxburghii, Hedera helix, Holarrhena antidysenterica, Indigofera pulchella, Ipomoea cairica, Jasminum dispersum, Litsea polyantha, Litsea sebifera, Mangifera indica, Moringa pterygosperma, Morus alba, Myrica esculenta, Opuntia dillenii, Oroxylum indicum, Prunus amygdalus, Prunus domestica, Psidium guajava, Quercus leucotrichophora, Rhus semialata, Ricinus communis, Rosa alba, Rosa moschata, Rubus lasiocarpus, Sapium sebiferum, Sechium edule, Sida cordifolia, Smilax aspera, Stephania glabra, Sterculia villosa, Symplocos racemosa, Terminalia bellerica., Terminalia bellerica, Terminalia chebula, Terminalia tomentosa, Tinospora cordifolia, Trachelospermum lucidum, Trichosanthes tricuspidata, Vitex negundo, Vitis jacquemontii, Woodfordia floribunda, Zanthoxylum armatum, Zizyphus jujuba..*
GIDDINESS	*Punica granatum.*
GLANDULAR ADENYTIS	*Bauhinia variegata.*
GOITRE	*Datura metel.*
HAIR CARE (Baldness, Hair fall)	*Ipomoea muricata, Opuntia dillenii.*
HERNIA	*Buddleia paniculata, Debregeasia hypoleuca, Murraya koenigii, Salix tetrasperma.*
HYDROCELE	*Caesalpinia decapetala, Myrica esculenta.*
ILLUMINANT	*Prinsepia utilis.*
INCENSE	*Skimmia laureola, Tabernaemontana divaricata, Taxus baccata.*

INJURIES *Albizzia julbrissin, Aloe vera, Boehmeria platyphylla, Callicarpa macrophylla, Cissampelos pareira, Colebrookia oppositifolia, Ficus bengalensis, F. glomerata, Jasminum dispersum, Leea aspera, Reinwardtia indica, Solanum erianthum, Sorbaria tomentosa, Thuja orientalis, Tinospora cordifolia, Zizyphus jujuba.*

INK *Quercus leucotrichophora, Rhus semialata.*

INTESTINAL DISEASES (see Gastro-intesinal Disorders).

INTOXICANTS *Datura stramonium, Datura suaveolens.*

INTERNAL SWELLINGS *Canna indica, Caesalpinia decapetala, Rubia cordifolia, Trahelospermum lucidum, Vinca rosea.*

INTERNAL INJURIES *Grewia oppositifolia, Indigofera gerardiana, Vitex negundo.*

JAUNDICE (see Liver Complaints).

JOINT DISEASES (Arthritis, Rheumatism, Knee Pain) *Abies pindrow, Adhatoda vasica, Aesculus indica, Calotropis procera, Carissa congesta, Cuscuta reflexa, Dioscorea deltoidea, Dodonea viscosa, Duranta repens, Eugenia jambolana, Grewia oppositifolia, Hedera helix, Litsea sebifera, Loranthus ligustrinus, Melia azedarach, Moringa pterygospema, Morus alba, Mucuna pruriens, Pogostemon benghalensis, Prinsepia utilis, Prunus amygdalus, Prunus armeniaca, Pogostemom benghalensis, Quercus leucotrichophora, Ricinus communis, Rosa moschata, Sapium sebiferum, Sarcococca saligna, Sida cordifolia, Smilax aspera, Thevetia neriifolia, Trachelospermum lucidum, Vitex negundo.*

KIDNEY DISORDERS *Aloe vera, Berberis lycium, Citrus medica, Ficus reliogosa, Mallotus phillipinesis, Saccharum officinarum, Tinospora cordifolia.*

LAXATIVE (see Constipation).

LEUCORRHOEA *Bauhinia variegata, Dioscorea bellophylla, Plumbago zeylanica, Saraca asoca, Terminalia arjuna, Tinospora cordifolia, Woodfordia floribunda.*

LIVER COMPLAINTS *Albizzia stipulata, Aloe vera, Berberis lycium, Carica papaya, Citrus jambhiri, Citrus medica, Emblica officinalis, Ficus glomerata, Ficus reliogosa, Gardenia angusta, Indigofera pulchella, Mallotus phillipinesis, Punica granatum, Saccharum officinarum, Terminalia bellerica, Tinospora cordifolia.*

LUNGS DISEASES *Acacia concinna, Adhatoda rasica, Cinnamomum tamala, Citus aurantium, Clerodendrum philippinum, Datura metel,*

	Eucalyptus umbellata, Euphorbia pulcherrima, Ficus palmata, Indigofera pulchella, Pinus roxburghii, Pistacia intregrimma, Punica granatum, Rhus succedanea, Taxus baccata, Terminalia bellerica, Terminalia chebula, Trichosanthes tricuspidata, Woodfordia floribunda.
MAGICO-RELIGIOUS	*Abrus precatorius, Aegle marmelos, Asclepias curassavica, Butea monosperma, Canna indica, Caryopteris wallichiana, Cestrum nocturnum, Cinnamomum camphora, Clerodendron phillipinum, Costus specious, Cucurbita hispida, Datura stramonium, D. suaveolens, Dioscorea bellophylla, D. bulbifera, Emblica officinalis, Erythrina suberosa, Ficus bengalensis, F. glomerata, F. religiosa, F. rumphii, Gossypium arboreum, Grewia oppositifolia, Hibiscus rosa-sinensis, Ipomoea muricata, Jasminum dispersum, J. grandiflorum, J. humile, J. multiflorum, Malvavsicus conzatii, Mangifera indica, Pistacia integrimma, Prinsepia utilis, Prunus cerasoides, Reinwardtia indica, Ricinus communis, Rosa alba, Rosa moschata, Saraca asoca, Tecoma grandiflora, Tinospora cordifolia, Zanthoxylum armatum.*
MALARIA	*Melothria heterophylla.*
MASTICATORY	*Acacia catechu.*
MENSTRUATION COMPLAINTS	*Aesculus indica, Cedrella toona, Cinnamonum tamala, Cassia fistula, Cuscuta reflexa, Dodonea viscosa, Ficus reliogosa, Ipomoea quamoclit, Myrsine africana, Randia dumetorum, Rubia cordifolia, Sida cordifolia.*
MENTAL DISEASES	(see Nervous Diseases).
MIGRAINE	*Calotropis procera.*
MOUTH SORES/ULCERS	*Acacia catechu, Callicarpa macrophyla, Colebrookia oppositifolia, Cryptolepis buchaniana, Diospyros kaki, Ficus reliogosa, Ipomoea carnea, Leptodermis lanceolata, Myrica esculenta, Phyllanthus parvifolious, Pyrus pashia, Roylea cinera, Solanum melongena, Vitex negundo, Zanthoxylum armatum.*
NAIL DISEASES	*Ilex dipyrena.*
NASAL PROBLEMS	*Cestrum nocturnum, Holarrhena antidysenterica, Indigofera pulchella, Mangifera indica, Myrica esculenta, Rhododendron arboreum.*
NERVOUS DISORDERS	*Datura metel, Ficus religiosa.*
NOSE BLEEDING	(see Nasal Problems).

PAINS AND ACHES	*Citrus limon, Clematis buchananiana, Datura metel, Eucalyptus umbellata, Flemingia macrophylla, Erythrina suberosa, Myrica esculenta, Nerium indicum, Pistacia integrimma, Prisepia utilis, Pyrus pashia, Sorbaria tomentosa, Vitis jacquemontii, Zanthoxylum aramtum, Zizyphus jujuba.*
PARALYSIS	*Nerium indicum.*
PERFUMERY	
PHLEGM	*Cissampelos pareira, Datura metel.*
PHTHSIS	*Rhus succedeana.*
PILES	*Berberis lycium, Bombax ceiba, Calotropis procera, Embica officinalis, Ficus glomerata, Hibiscus rosa-sinensis, Jasminum grandiflorum, Lagerstroemia indica, Mangifera indica, Melia azedarach, Morus alba, Murraya koenigii, Pavetta indica, Phoenix sylvestris, Prunus domestca, Psidium guajava, Quercus glauca, Sapindus mukorossi, Symplocos racemosa, Terminalia bellerica, Tinospora cordifolia.*
REFRIGERANT	*Aegle marmelos, Aloe vera, Camellia sinensis, Clerodendron philippinum, Dodonea viscosa, Emblica officinalis, Litchi chinensis, Litsea sebifera, Mallotus phillipinensis, Mangifera indica, Morus alba, Myrsine africana, Pyrus communis, Rhododendron arboreum, Rosa alba, Rosa moschata, Rubus lasiocarpus, Saccaharum officinarum, Tabernaemontana divaricta, Terminalia bellerica, Tinospora cordifolia, Vitis rependa, Zizyphus jujuba.*
REPELLENT	*Cedrus deodara, Deutzia corymbosa, Eucalyptus umbellata, Ficus rumphii, Lantana camara, Milletia extensa, Quercus glauca.*
RUBEFACIENT	*Prinsepia utilis.*
SCAITICA	*Lagerstroemia indica, Plumbago zeylanica, Terminalia arjuna.*
SHAMPOO/ DETERGENT	*Acacia concinna, Acsculus indica, Dioscorea deltoidea, Grewia oppsitifolia, Randia dumetorum, Sapindus mukorossi.*
SINUS	(see Nasal Problems).
SKIN DISEASES	*Acacia catechu, Adhatoda vasica, Albizzia julibrissin, Aloe vera, Butea monosperma, Calotropis procera, Carica papaya, Carya illenoesis, Cedrus deodara, Cissampelos pareira, Clematis grata, Cryptolepis buchaniana, Cuscuta reflexa, Dalbergia sissoo, Daphne cannabina, Dodonea viscosa, Emblica officinalis, Erythrina suberosa, Ficus bengalensis, Ficus palmata, Ficus*

	religiosa, Ficus roxburghii, Flemingia microphylla, Flocourtia ramontchi, Hypericum patulum, Jasminum grandiflorum, Jasminum humile, Jasminum multiflorum, Juglans regia, Lyonia oralifolia, Mallotus phillipensis, Melia azedarach, Melothria heterophyla, Momordica charantia, Plumbago zeylanica, Premna latifolia, Prunus amygdalus, Pyrus pashia, Ricinus communis, Rosa alba, Roylea cinerea, Rubia cordifolia, Salix babylonica, Sapium insigne, Sarcococca saligna, Solanum torvum, Stephania glabra, Terminalia chebula,. Thevetia neriifolia, Trichosanthes tricuspidata, Vinca rosea, Woodfordia floribunda, Zanthoxylum armatum, Zizyphus jujuba..
SPERMATORRHOEA	*Cissampelos pareira, Cordia dichotoma, Emblica officinalis, Eugenia jambolana, Ficus hispida, Ficus roxburghii, Rosa alba, Rubus lasiocarpus, Sapium sebiferum, Tinospora cordifolia.*
SPITTING	*Morus alba.*
SPLEEN COMPLAINTS	*Butea monosperma, Terminalia bellerica, Trichsanthes tricuspidata.*
STOMACHACHE	(see Gastro-intesinal Disorders)
STIMULANT	*Camellia sinensis.*
SUPPRESSING HUNGER	*Achyranthes aspera.*
SUR	***Acacia concinna,*** *Achyranthes bidentata, Albizzia stipulata, Callicarpa macrophylla, Colebrookia oppositifolia, Eugenia jambolana, Hedera helix, Melothria heterophylla, Rhododendron arboreum, Saccharum officinarum, Senecio rufinervis, Skimmia laureola, Terminalia chebula, Trachelospermum lucidum, Woodfordia floribunda.*
SWELLINGS	*Caesalpinia decapetala, Canna indica, Flemingia macrophylla, Lantana camara, Rubia cordifolia, Trachelospermum lucidum, Vinca rosea.*
THROAT COMPLAINTS	*Acacia catechu, Achyranthes aspera, Cassia laevigata, Cryptolepis buchaniana, Embellia tsjeriam-cottam, Leptodermis lanceolata, Myrica esculenta, Punica granatum, Rubus ellipticus.*
TONSILITIS	(see Throat Complaints).
TUBERCULOSIS	(see Lung Diseases).
TYPHOID	*Phyllanthus parvifolius.*
ULCER	*Clematis buchananiana.*

URIC ACID ACCUMULATION	*Achyranthes bidentata.*
URINARY COMPLAINTS	(see Diuretic).
URTICARIA	(see Skin Diseases).
VENEREAL DISEASES	*Ficus glomerata, Jacaranda ovalifolia, Quercus leucotrichrphora, Sarcococca saligna.*
WHOOPING COUGH	*Abrus precatorius, Pogostemon benghalensis.*
WOUNDS	(see Injuries).

Appendix- III

Index to Families

Name of the Family	Total Number of Genera and spp.	Total Number of Spp. in the Genera
Acanthaceae	*Adhatoda* (1), *Strobilanthes* (1), *Thunbergia* (1).	3/3
Actinidaceae	*Actinidia* (1).	1/1
Agavaceae	*Agave* (1).	1/1
Amaranthaceae	*Achyranthes* (2), *Cyathula* (1).	2/3
Anacardiaceae	*Mangifera* (1), *Rhus* (3), *Spondias* (1).	3/5
Apocyanaceae	*Carissa* (1), *Holarrhena* (1), *Nerium* (1), *Plumeria* (1), *Tabernaemontana* (1), *Thevetia* (1), *Vinca* (1).	8/8
Aquifoliaceae	*Ilex* (2).	1/2
Araliaceae	*Hedera* (1).	1/1
Araucariaceae	*Araucaria* (1).	1/1
Arecaceae	*Phoenix* (1).	1/1
Asclepiadaceae	*Asclepias* (2), *Calotropis* (1), *Cryptolepis* (1).	3/4
Asteraceae	*Inula* (1), *Senecio* (1), *Solidago* (1).	3/3
Bambusaceae	*Arundinaria* (1), *Bambusa* (1), *Dendrocalamus* (1).	3/3
Berberidaceae	*Berberis* (2).	1/2
Betulaceae	*Alnus* (2).	1/2

Bignoniaceae	*Jacaranda* (1), *Oroxylum* (1), *Tecoma* (1).	3/3
Bombacaceae	*Bombax* (1).	1/1
Buxaceae	*Buxus* (1), *Sarcococca* (1).	2/2
Cactaceae	*Opuntia* (1).	1/1
Cannaceae	*Canna* (1).	1/1
Caprifoliaceae	*Leycestria* (1), *Lonicera* (2), *Viburnum* (2).	3/5
Caricaceae	*Carica* (1).	1/1
Celastraceae	*Euonymus* (1).	1/1
Combretaceae	*Terminalia* (4).	1/4
Convolvulaceae	*Ipomoea* (5), *Lettsomia* (1).	2/6
Cordiaceae	*Cordia* (1), *Ehretia* (2).	2/3
Coriariaceae	*Coriaria* (1).	1/1
Cornaceae	*Cornus* (1).	1/1
Corylaceae	*Carpinus* (1).	1/1
Cucurbitaceae	*Cucumis* (1), *Cucurbita* (3), *Melothria* (1), *Momordica* (1), *Trichosanthes* (1), *Luffa* (1), *Sechium* (1).	7/9
Cupressaceae	*Thuja* (1).	1/1
Cuscutaceae	*Cuscuta* (1).	1/1
Cycadaceae	*Cycas* (1).	1/1
Dioscoreaceae	*Dioscorea* (3).	1/3
Ebenaceae	*Diospyros* (1).	1/1
Elaeagnaceae	*Elaeagnus* (1).	1/1
Ericaceae	*Lyonia* (1), *Rhododendron* (1).	2/2
Euphorbiaceae	*Andrachne* (1), *Emblica* (1), *Euphorbia* (2), *Glochiodion* (1), *Mallotus* (1), *Phyllanthus* (1), *Ricinus* (1), *Sapium* (2).	8/10
Fabaceae	*Abrus* (1), *Acacia* (2), *Albizzia* (3), *Atyloshia* (1), *Bauhinia* (2), *Butea* (1), *Caesalpinia* (1), *Cassia* (2), *Dalbergia* (1), *Desmodium* (1), *Erythrina* (1), *Flemingia* (2), *Indigofera* (2), *Leucaena* (1), *Milletia* (1), *Mucuna* (1), *Robinia* (1), *Vigna* (1), *Saraca* (1).	19/26
Fagaceae	*Quercus* (3).	1/3
Flacourtiaceae	*Flacourtia* (1), *Xylosma* (1).	2/2
Hippocastanaceae	*Aesculus* (1).	1/1

Hydrangaceae	*Hydrangea* (1).	1/1
Hypericaceae	*Hypericum* (2).	1/2
Juglandaceae	*Carya* (1), *Juglans* (1).	2/2
Lamiaceae	*Colebrookia* (1), *Pogostemon* (1), *Roylea* (1).	3/3
Lauraceae	*Cinnamomum* (2), *Litsea* (3), *Machilus* (1).	3/6
Leeaceae	*Leea* (1).	1/1
Liliaceae	*Aloe* (1), *Smilax* (1).	2/2
Linaceae	*Reinwardtia* (1).	1/1
Logainaceae	*Buddleia* (2).	1/2
Lorconthaceae	*Loranthus* (1).	1/1
Lythraceae	*Lagerstroemia* (1), *Woodfordia* (1).	2/2
Malvaceae	*Gossypium* (1), *Hibiscus* (1), *Malvaviscus* (1), *Sida* (1), *Kydia* (1).	5/5
Magnoliaceae	*Michelia* (1).	1/1
Melastomataceae	*Osbeckia* (1).	1/1
Meliaceae	*Cedrella* (2), *Melia* (1).	2/3
Menispermaceae	*Cissampelos* (1), *Stephania* (2), *Tinospora* (1), *Trachelospermum* (1).	4/5
Moraceae	*Ficus* (10), *Morus* (1).	2/11
Moringaceae	*Moringa* (1).	1/1
Myricaceae	*Myrica*.	1/1
Myrsinaceae	*Embellia* (1), *Myrsine* (1).	2/2
Myrtaceae	*Callistemon* (1), *Eucalyptus* (1), *Eugenia* (1), *Psidium* (1).	4/4
Nyctaginaceae	*Bougainvillea*. (1)	1/1
Oleaceae	*Jasminum* (4).	1/4
Onagraceae	*Fuchsia* (1).	1/1
Pinaceae	*Abies* (1), *Cedrus* (1), *Picea* (1), *Pinus* (2).	4/5
Pistaceae	*Pistacia* (1).	1/1
Plumbagonacene	*Plumbago* (1).	1/1
Poaceae	*Saccharum* (1).	1/1
Polygonaceae	*Polygonum* (1).	1/1
Protoeaceae	*Grevillea* (1).	1/1
Punicaceae	*Punica* (1).	1/1

Ranunculaceae	*Clematis* (2).	1/2
Rhamnaceae	*Rhamnus* (2), *Sageretia* (1), *Zizyphus* (1).	3/4
Rosaceae	*Cotoneaster* (2), *Eriobotyrica* (1), *Prinsepia* (1), *Prunus* (5), *Pyrus* (3), *Rosa* (2), *Rubus* (3), *Sorbaria* (1), *Spirarea* (1).	9/19
Rubiaceae	*Gardenia* (1), *Leptodermis* (1), *Pavetta* (1), *Randia* (1), *Rubia* (1), *Wendlendia* (1).	6/6
Rutaceae	*Aegle* (1), *Citrus* (5), *Murraya* (1), *Skimmia* (1), *Zanthoxylum* (1).	5/9
Saliaceae	*Populus* (2), *Salix* (3).	2/5
Santalaceae	*Osyris* (1).	1/1
Sapindaceae	*Dodonaea* (1), *Litchi* (1), *Sapindus* (1).	3/3
Saurauiaceae	*Saurauia* (1).	1/1
Saxifragaceae	*Deutzia* (1).	1/1
Solanaceae	*Cestrum* (1), *Cyphomandra* (1), *Datura* (3), *Solanum* (4).	4/9
Sterculiaceae	*Sterculia* (1).	1/1
Symplocaceae	*Symplocos* (2).	1/2
Taxaceae	*Taxus* (1).	1/1
Theaceae	*Camellia* (1).	1/1
Thymelaeaceae	*Daphne* (1).	1/1
Tiliaceae	*Grewia* (1).	1/1
Tropaeolaceae	*Tropaeoleum* (1).	1/1
Ulmaceae	*Celtis* (1), *Ulmus* (1).	2/2
Urticaceae	*Boehmeria* (1), *Broussonetia* (1), *Debregeasia* (1).	3/3
Verbenaceae	*Callicarpa* (1), *Caryopteris* (1), *Clerodendron* (1), *Duranta* (1), *Gmelina* (1), *Lantana* (1), *Premna* (1), *Vitex* (1).	8/8
Vitaceae	*Vitis* (3).	1/3
Zingiberaceae	*Costus* (1).	1/1

□□□